OPTIMIZATION METHODS IN METABOLIC NETWORKS

OPTIMIZATION METHODS IN METABOLIC NETWORKS

COSTAS D. MARANAS
Department of Chemical Engineering
The Pennsylvania State University

ALI R. ZOMORRODI
Bioinformatics Program
Boston University

Published by John Wiley & Sons, Inc., Hoboken, New Jersey
Published simultaneously in Canada

Library of Congress Cataloging-in-Publication Data:

Names: Maranas, Costas D., 1967–, author. | Zomorrodi, Ali R., author.
Title: Optimization methods in metabolic networks / Costas D. Maranas, Ali R. Zomorrodi.
Description: Hoboken, New Jersey : John Wiley & Sons Inc., [2016] | Includes bibliographical references and index. | Description based on print version record and CIP data provided by publisher; resource not viewed.
Identifiers: LCCN 2015040810 (print) | LCCN 2015040236 (ebook) | ISBN 9781119188919 (Adobe PDF) | ISBN 9781119189015 (ePub) | ISBN 9781119028499 (cloth)
Subjects: | MESH: Metabolic Networks and Pathways–physiology. | Computer Simulation. | Linear Models.
Classification: LCC QP171 (print) | LCC QP171 (ebook) | NLM QU 120 | DDC 612.3/9–dc23
LC record available at http://lccn.loc.gov/2015040810

Cover image courtesy of Costas D. Maranas and Ali R. Zomorrodi

Set in 10/12pt Times by SPi Global, Pondicherry, India

Printed in the United States of America

10 9 8 7 6 5 4 3 2 1

1 2016

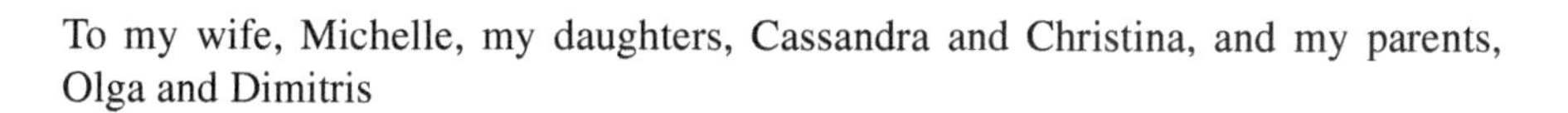

To my wife, Michelle, my daughters, Cassandra and Christina, and my parents, Olga and Dimitris

Costas D. Maranas

To my wife, Assieh, my daughter, Tara, and my parents, Fatemeh and Hossein

Ali R. Zomorrodi

CONTENTS

The ancillary material for the book including a set of computer codes for a number of examples in the book can be found at http://maranasgroup.com. The solutions to exercises are available to instructors upon request.

PREFACE

The objective of this book is to provide a tutorial on the computational tools that use mathematical optimization concepts and representations for the curation, analysis and redesign of metabolic networks. Emphasis is placed on explaining the optimization formulation fundamentals and relevant algorithmic solution techniques. This book does not aim to serve as a comprehensive optimization text as it provides only a condensed treatment of different classes of optimization problems. The main goal here is to use the language and tools of mathematical programming to describe and solve frequently occurring problems in the analysis and redesign of metabolic networks. The interested reader is thus encouraged to refer to dedicated optimization textbooks for a more thorough treatment of different classes of mathematical programming problems. Similarly, the description of metabolism in the book is limited, requiring frequent referral to relevant biochemistry and molecular biology resources. The reader is encouraged to consult the original journal publications for all described techniques. Nevertheless, the treatment presented here has benefited from many years of application and customization on a variety of projects. As a consequence, the day-to-day application, implementation and integration of these techniques may differ from their original exposition in the journal publications. Care is exercised to ensure that the metabolic network descriptions, nomenclature and assumptions are consistent throughout. Many important optimization-based techniques in metabolic networks are absent from this treatment due to space limitations, lack of working experience with them or difficulty in integrating them with the rest of the material. Course notes for a graduate elective on "Optimization in Biological Networks" taught at Pennsylvania State University – University Park formed the basis of the techniques described here.

This book can be used by itself or in combination with other books to support coursework on special topics in network analyses of biological and in particular metabolic networks. It can be used to introduce students with knowledge of metabolism to formal mathematical treatments of core computational tasks in metabolic networks or alternatively to expose students with a mathematical programming background to metabolism. The hope is that the book will serve as a starting point for the students for more in-depth investigations of relevant techniques and concepts found in the cited literature.

Chapter 1 uses a regulatory network inference problem to introduce modeling using mathematical programming formulations. Key concepts such as sets, parameters, variables, constraints and optimization formulations are introduced. This is followed by a formal introduction and definition of fundamental concepts in optimization that establish criteria for the existence of a local or global optimal solution. Chapter 2 introduces linear programming (LP), which underpins many analysis techniques in metabolic networks, and linear duality theory, which is used for solving bilevel optimization problems described in Chapter 8. Chapter 3 provides an introduction to metabolic network modeling and flux balance analysis (FBA) in particular. A number of core FBA analysis techniques that rely on LP are described here. Chapter 4 unveils the concept of discrete or binary variables alongside continuous variables for modeling a variety of tasks in metabolic networks. Both representations and solution techniques of mixed-integer linear programming (MILP) problems are highlighted. Chapter 5 describes how reaction free energy of change considerations can be used to restrict reaction directionality and eliminate thermodynamically infeasible cycles in metabolic networks. Chapter 6 addresses the task of resolving metabolic network gaps and growth prediction inconsistencies in metabolic networks using computational techniques relying on MILP problem formulations. Chapter 7 reviews optimization-based approaches for finding paths that link a source to a target metabolite. Chapter 8 addresses the use of bilevel programming for describing and solving metabolic network redesign problems to maximize the yield of a target product in microbial strain design. Chapter 9 introduces basic concepts, optimality criteria, and solution techniques for both unconstrained and constrained nonlinear programming problems (NLPs). Chapter 10 provides examples of how NLP underpins metabolic network analysis problems such as integration of kinetic expressions into genome-scale metabolic models and metabolic flux elucidation using carbon-labeled isotopes in metabolic flux analysis (MFA). Finally, Chapter 11 lays the foundation for the description and solution of mixed-integer NLP (MINLP) problems and highlights how the nonlinearities of kinetic expressions can be integrated in the strain design bilevel optimization formulations introduced in Chapter 8. A number of example problems are addressed in the text and suggested problems are provided at the end of every chapter. The source GAMS code for all examples in the book are available through a dedicated website for this book (www.maranasgroup.com). Alternative representations using MATLAB, LINGO, etc. can also be derived in a straightforward manner. In addition, Appendix A provides a short guide to GAMS.

The book is intended as a textbook for a graduate or advanced undergraduate elective in the use of optimization in metabolic networks. Topics covered in the book start with a formal treatment of the relevant optimization problem class followed by application in the context of metabolic network analysis. The class of optimization problems becomes progressively more complex starting with LP and MILP and concluding with NLP and MINLP problems. Many of the proofs of convergence and optimality theorems are geared toward the students with a more in-depth interest in optimization and can be skipped without loss of teaching material congruity. Hands-on implementation of all the discussed methods is encouraged to reinforce the introduced concepts.

We would like to acknowledge the people in the C. Maranas group for providing feedback and tirelessly helping with the preparation of the book. In particular, we would like to acknowledge the contribution of Ali Khodayari for assembling the initial set of lecture notes from the related graduate course at Penn State and for helping to create and improve figures; Anupam Chowdhury for performing simulations for examples of Chapters 3, 4, 7, 8, 10, and 11 and for preparing the related figures and computer codes; Chiam Yu Ng and Margaret Simons for helping to create figures and drafting the initial versions of the chapters; and Sarat Ram Gopalakrishnan for helpful feedback on section 10.3 of Chapter 10. Finally, we would like to express our gratitude to Margaret Simons for designing the cover image and for carefully editing the proofs and making numerous suggestions.

Costas D. Maranas
Ali R. Zomorrodi

1

MATHEMATICAL OPTIMIZATION FUNDAMENTALS

This chapter reviews the fundamentals of mathematical optimization and modeling. It starts with a biological network inference problem as a prototype example to highlight the basic steps of formulating an optimization problem. This is followed by a review of some basic mathematical concepts and definitions such as set and function properties and convexity analysis.

1.1 MATHEMATICAL OPTIMIZATION AND MODELING

Mathematical optimization (programming) systematically identifies the best solution out of a set of possible choices with respect to a pre-specified criterion. The general form of an optimization problem is as follows:

$$\begin{aligned}
&\text{minimize (or maximize) } f(\boldsymbol{x}) \\
&\text{subject to} \\
&h(\boldsymbol{x}) = 0 \\
&g(\boldsymbol{x}) \leq 0 \\
&\boldsymbol{x} \in S
\end{aligned}$$

where

- $\boldsymbol{x}$ is a N-dimensional vector referred to as, the vector of *variables*.
- S is the set from which the elements of $\boldsymbol{x}$ assume values. For example, S can be the set of real, nonnegative real or nonnegative integers. In general, variables in an optimization problem can be continuous, discrete (integer) or combinations

Optimization Methods in Metabolic Networks, First Edition. Costas D. Maranas and Ali R. Zomorrodi.

thereof. The former is used to capture the continuously varying properties of a system (e.g., concentrations), whereas the latter is used for discrete decision making (e.g., whether or not to eliminate a reaction).

- $f(\boldsymbol{x})$ is referred to as the *objective function* and serves as a mathematical description of the desired property of the system that should be optimized (i.e., maximized or minimized).
- $h(\boldsymbol{x}) = [h_1(\boldsymbol{x}), h_2(\boldsymbol{x}), \ldots, h_L(\boldsymbol{x})]^{\mathrm{T}}$ and $g(\boldsymbol{x}) = [g_1(\boldsymbol{x}), g_2(\boldsymbol{x}), \ldots, g_M(\boldsymbol{x})]^{\mathrm{T}}$ are constraints that must be satisfied as equalities or one-sided inequalities, respectively, and represent the feasible space of decision variables.

Any vector $\boldsymbol{x}$ that lies in S and satisfies $h(\boldsymbol{x})$ and $g(\boldsymbol{x})$ is called a feasible solution. In addition, if vector $\boldsymbol{x}$ minimizes (maximizes) the objective function, it is an optimal *solution point* to the optimization problem with an associated optimum *solution value* $f(\boldsymbol{x})$. There are different classes of optimization problems depending on the (non)linearity properties of the objective function and constraints as well as the presence or absence of discrete (i.e., binary or integer) and/or continuous variables. Standard classes of optimization problems that generally require different solution techniques are as follows:

(i) Linear programming (LP) problems involve a linear objective function $f(\boldsymbol{x})$ and constraints $h(\boldsymbol{x})$ and $g(\boldsymbol{x})$ as well as only continuous variables $\boldsymbol{x}$ (Chapter 2).

(ii) Mixed-integer LP (MILP) problems are LP problems with some of the variables assuming only discrete values (Chapter 4).

(iii) Nonlinear programming (NLP) problems involve a nonlinear objective and/or some nonlinear constraints while all variables are continuous (Chapter 9).

(iv) Mixed-integer nonlinear programming (MINLP) problems are NLPs with some variables assuming only discrete values (Chapter 11).

Mathematical optimization has been used extensively to model a wide variety of problems in science and engineering. The development of an optimization formulation modeling a real-life problem often needs to be traversed multiple times as new data, modified problem descriptions and re-interpretations due to unanticipated optimal solutions come to play. This book concentrates on mathematical optimization applications for the analysis and redesign of biological systems, with a special emphasis on metabolic networks. Example 1.1 describes the basic steps for formulating a biological network inference task as an optimization problem.

Example 1.1

Given a set of genes and time-course DNA microarray data, formulate an optimization problem to identify the regulatory interaction coefficients between genes best explaining the observed gene expression levels. The schematic representation of time-course DNA microarray data for a sample gene interaction network is given in Figure 1.1.

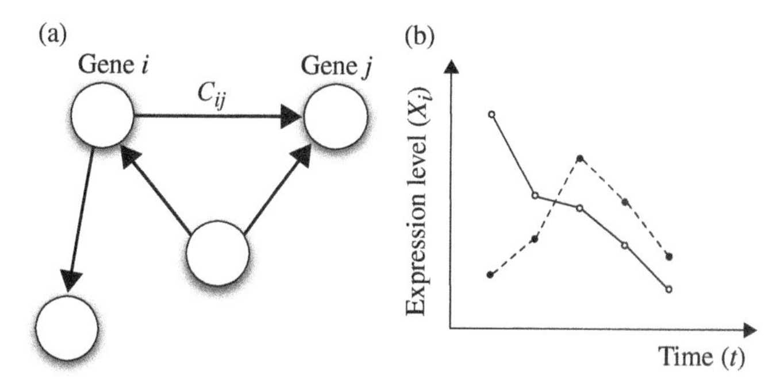

FIGURE 1.1 (a) An example of a simple gene regulatory network. Nodes and edges represent genes and interactions between genes, respectively. C_{ij} denotes the interaction coefficient of genes i and j (i.e., how gene j is affecting gene i). (b) A schematic representation of time-course DNA microarray expression data for two genes. These data are usually presented as the log ratio of the expression level of a gene at each time point with respect to a reference.

Solution: A stepwise description is provided that codifies the sequence of tasks carried out for constructing the optimization model whose solution answers the problem.

Sets:
The first task in deriving the optimization formulation of a problem is defining a number of sets indicating the essential elements of the problem over which the parameters, variables and/or constraints are defined. Two sets can be defined for this problem as follows:

$$\text{Set of genes}: \boldsymbol{I} = \{i \mid i = 1,2,\ldots,N\}$$
$$\text{Set of time points}: \boldsymbol{T} = \{t \mid t = 1,2,\ldots,T_\mathrm{f}\}$$

Here, N denotes the number of genes in the network.

Parameters:
Parameters (some of which are indexed over sets) encode the available data for the problem. The parameters that can be defined for this problem include the following:

X_{it}: Expression level of gene $i \in I$ at time point $t \in T$.

LB_{ij}: Lower bound on the interaction coefficient C_{ij} (see the next section for the definition of C_{ij}). Subscript j assumes values from set I.

UB_{ij}: Upper bound on the interaction coefficient C_{ij}.

Δt: Sampling interval assuming that it is constant throughout the experimental DNA microarray data. For convenience we set $\Delta t = 1$.

Variables:
In contrast to parameters that have known values, variables typically only have initial values and/or lower/upper bounds, and their optimal values are obtained upon solving the optimization problem. As was the case with parameters, the introduction of sets allows for the grouping of multiple unknowns under the same variable name. For the

problem in hand, we define two different categories of variables including a continuous and a discrete set of variables. The continuous variable set is defined as follows:

C_{ij}: Interaction coefficient between genes $j \in I$ and $i \in I$ (effect of gene j on gene i), where

$$\begin{cases} C_{ij} > 0 & \text{if gene } j \text{ activates gene } i \\ C_{ij} < 0 & \text{if gene } j \text{ represses gene } i \\ C_{ij} = 0 & \text{if gene } j \text{ has no effect on gene } i \end{cases}$$

The binary variables y_{ij} capture the presence or absence of an interaction between genes i and j as follows:

$$y_{ij} = \begin{cases} 1 & \text{if gene } j \text{ affects gene } i \\ 0 & \text{otherwise} \end{cases}$$

Constraints:
Constraints are defined to enforce the conditions that need to be satisfied for the problem. A constraint for this example is needed to impose the assumption that the rate of change in the expression level of each gene is a linear function of the contributions of all genes in the network (including itself):

$$\frac{dX_{it}}{dt} = \sum_{j \in I} C_{ij} X_{jt}, \quad \forall i \in I, \quad t \in T \tag{1.1}$$

Here, we approximate the derivative terms with algebraic linear constraints using a finite (forward) difference approximation:

$$\frac{dX_{it}}{dt} \approx \frac{X_{i,t+1} - X_{i,t}}{\Delta t}, \quad \forall i \in I, t \in \{1, \ldots, T_f - 1\} \tag{1.2}$$

This implies that the identification of C_{ij} requires solving the following under-determined set of linear equalities (note that the system of equations is under-determined because the number of pairwise interactions is much larger than the number of equations):

$$X_{i,t+1} - X_{i,t} = \Delta t \sum_{j \in I} C_{ij} X_{jt}, \quad \forall i \in I, t \in \{1, \ldots, T_f - 1\} \tag{1.3}$$

An additional constraint is introduced to model the presence or absence of an interaction for each pair of genes enforcing the definition of binary variables y_{ij}:

$$LB_{ij} y_{ij} \le C_{ij} \le UB_{ij} y_{ij}, \quad \forall i, j \in I \tag{1.4}$$

Observe that if y_{ij} is equal to zero, then C_{ij} is forced to assume a value of zero; whereas when y_{ij} is equal to one, then C_{ij} is free to assume any value between LB_{ij} and UB_{ij}.

Objective function:
Given that this is an under-determined system of equations, there can be infinite sets of C_{ij} all satisfying the given constraints. Optimization can be used to select one out

of the many feasible values for C_{ij} that satisfies an optimality criterion. Here, we invoke the parsimony assumption whereby we accept as the most relevant solution the one that minimizes the total number of regulatory interactions. The total number of regulatory coefficients can be obtained by summation over all binary variables.

Optimization model (formulation):
By collecting all the constraints described earlier, the optimization problem is stated as follows:

$$\text{minimize} \sum_{i \in I} \sum_{j \in I} y_{ij} \qquad [P1]$$

subject to

$$X_{i,t+1} - X_{i,t} = \Delta t \sum_{j \in I} C_{ij} X_{jt}, \quad \forall i \in I, \quad t \in \{1, \ldots, T_{\mathrm{f}} - 1\} \tag{1.3}$$

$$LB_{ij}\, y_{ij} \le C_{ij} \le UB_{ij} y_{ij}, \quad \forall i, j \in I \tag{1.4}$$

$$y_{ij} \in \{0,1\}, C_{ij} \in \mathbb{R}, \qquad \forall i, j \in I$$

The solution of this problem will provide the presence or absence of a regulatory interaction between each pair of genes in the network (captured by binary variable y_{ij}) and the magnitude and sign (i.e., activation vs. inhibition) of these interactions (captured by the continuous variables C_{ij}).

Exploring trade-offs between prediction error and model complexity:
It is important to emphasize that the solution of an optimization problem always needs to be scrutinized in terms of both mathematical accuracy and the relevance to the problem. For example, a key concern for this example is whether the obtained coefficients indeed capture biologically relevant interactions or are simply artifacts of the parameter fitting process. In addition, one might be interested to know whether the identified regulatory coefficients are unique or there exists alternate optimal sets. Optimization provides ways to address these types of questions by trading-off accuracy versus model complexity (i.e., parsimony in this case). This can be accomplished for this example by exploring how the total number of non-zero C_{ij}'s decrease upon allowing for some degree of violation in the equality constraints. Introducing slack variables S_{it}^{+} and S_{it}^{-} $\left(\text{with } S_{it}^{+}, S_{it}^{-} \ge 0\right)$ for Constraint 1.3 allows for both positive and negative departures from equality:

$$X_{i,t+1} - X_{i,t} - \Delta t \sum_{j \in I} C_{ij} X_{jt} = S_{it}^{+} - S_{it}^{-}, \quad \forall i \in I, t \in \{1, \ldots, T_{\mathrm{f}} - 1\} \tag{1.5}$$

In addition, since we would like to identify a regulatory network with fewer interactions (e.g., one less than the interactions identified for the original problem represented by $y^{\max}$), we can add the following constraint:

$$\sum_{i \in I} \sum_{j \in I} y_{ij} \le y^{\max} - 1 \tag{1.6}$$

The re-formulated optimization problem aims to identify a more compact regulatory interaction network while minimizing the departure from experimental data and is described as follows:

$$\text{minimize} \sum_{i \in I} \sum_{t \in T-\{T_f\}} \left(S_{it}^{+} + S_{it}^{-}\right) \qquad [P2]$$

subject to

$$X_{i,t+1} - X_{i,t} - \Delta t \sum_{j \in I} C_{ij} X_{jt} = S_{it}^{+} - S_{it}^{-}, \quad \forall i \in I, t \in \{1, \ldots, T_f - 1\} \tag{1.5}$$

$$\mathrm{LB}_{ij} y_{ij} \le C_{ij} \le \mathrm{UB}_{ij} y_{ij}, \quad \forall i, j \in I \tag{1.4}$$

$$\sum_{i \in I} \sum_{j \in I} y_{ij} \le y^{\max} - 1 \tag{1.6}$$

$$y_{ij} \in \{0, 1\}, C_{ij} \in \mathbb{R}, \; S_{it}^{+}, S_{it}^{-} \ge 0, \quad \forall i, j \in I$$

Note that in contrast to [$P1$], [$P2$] minimizes the total violation of the Constraint 1.3. By solving [$P2$] for different network sizes specified by the right-hand side of Constraint 1.6 and by subsequently plotting the total error in prediction (i.e., the objective function value of [$P2$]) against the number of nonzero regulatory interactions (i.e., sum of the binary variables), a monotonically decreasing curve is obtained, as shown in Figure 1.2. The error will be quite high for very sparse models, but will approach zero as the total number of regulatory interactions approaches $y^{\max}$. In general, there tends to be a *break point* in the curve beyond which additional nonzero regulatory interactions improve the error only slightly as shown in Figure 1.2. This implies that once this point (or a desired accuracy threshold) is reached, additional parameters are likely to "overfit" rather than capture information in the data.

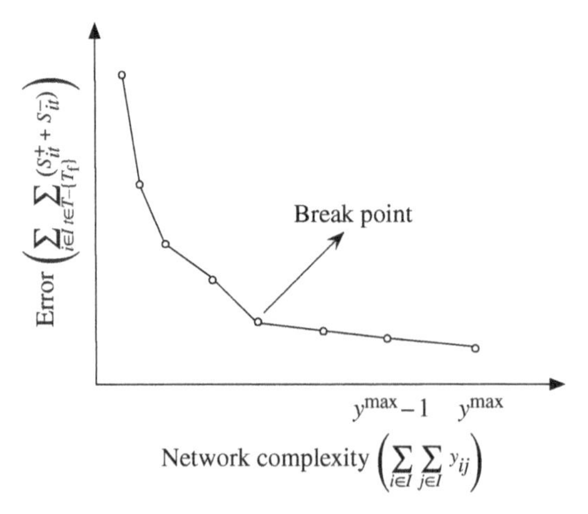

FIGURE 1.2 Schematic representation of the error as a function of the number of nonzero regulatory coefficients for the problem of Example 1.1.

In general, by adopting an optimization-based description of the problem, significant versatility is afforded in tailoring the solution to the specifics of the problem and/or exploring various trade-offs. For example, certain regulatory interactions can be excluded or pre-postulated (e.g., the interaction of known transcription factors with known genes) by setting the related binary variable y_{ij} to zero or one, respectively. Similarly, the total number of genes affecting the expression of a gene i can be restricted to a pre-specified number M using the following constraint:

$$\sum_{j \in I} y_{ij} \leq M \tag{1.7}$$

Alternate representations of this problem can be explored to account for nonlinear interactions, noise in gene expression data, time delay in regulatory interactions, and more. Interested readers are referred to the related articles [1–7] for details. □

The purpose of this illustrative example was to provide an introduction to the iterative process of formulating optimization problems, assessing their output and modifying their structure to address the follow-up questions when dealing with a real-life problem. Next, basic definitions and concepts necessary for correctly describing optimization models and assessing the existence of local and/or global optimal solution points are introduced.

1.2 BASIC CONCEPTS AND DEFINITIONS

We start by introducing basic properties of sets and functions necessary for establishing conditions for the (i) existence and (ii) uniqueness of a global optimum value. These definitions also introduce formal mathematical language and reasoning used in optimization textbooks and articles.

Let S be an arbitrary subset of $\mathbb{R}^N$. The concepts and properties for S are defined in the text.

1.2.1 Neighborhood of a Point

Given a point $\boldsymbol{x}$ in set S (i.e., $\boldsymbol{x} \in S$), an *ε-neighborhood* around x (denoted by $B(\boldsymbol{x},\varepsilon)$) is defined as follows (see Fig. 1.3):

$$B(\boldsymbol{x},\varepsilon) = \{ \boldsymbol{y} \mid \boldsymbol{y} \in S \text{ and } \|\boldsymbol{x} - \boldsymbol{y}\| < \varepsilon \}$$

$B(\boldsymbol{x},\varepsilon)$ is in essence the set of points in set S within an N-dimensional sphere centered at point $\boldsymbol{x}$ with a radius of ε.

1.2.2 Interior of a Set

A point $\boldsymbol{x}$ is in the interior of a set S (denoted by int(S)) if and only if there exists an ε-neighborhood around $\boldsymbol{x}$ with $B(\boldsymbol{x},\varepsilon) \subseteq S$ for some $\varepsilon > 0$.

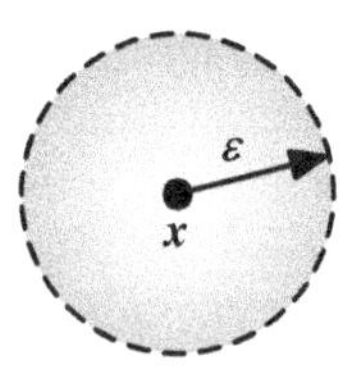

FIGURE 1.3 Schematic representation of the ε-neighborhood of a point $\boldsymbol{x} \in S$ in a two-dimensional space.

This qualitatively means that a "ball" of nonzero size can be constructed around point $\boldsymbol{x}$ so as all points within it belong to set S. This implies that interior of sets exclude "boundary" points.

1.2.3 Open Set

A set S is open if and only if $\text{int}(S) = S$.

This implies that for an open set, nonzero neighborhoods can be constructed around each point so as every point in the neighborhood belongs to the original set. Thus, open sets are identical to their interior as no boundary points are included. For example, set $(0,1)$ is an open set as neighborhoods can be constructed for every point in it that are fully contained within the set by making ε appropriately small.

1.2.4 Closure of a Set

A point $\boldsymbol{x}$ is in the closure of set S (denoted by $cl(S)$) if and only if $S \cap B(\mathbf{x}, \varepsilon) \neq \varnothing$ for every $\varepsilon > 0$.

The closure of a set can be thought of as all the points in a set and all adjacent boundary points irrespective of whether they are part of the original set.

1.2.5 Closed Set

A set S is closed if and only if $cl(S) = S$.

In essence, a closed set contains all of its boundary points, and therefore its closure is identical with the original set. For example, set $[1,2]$ is a closed set, whereas set $(1,2]$ is neither open nor closed.

1.2.6 Bounded Set

A set S is bounded if and only if for every two points $\boldsymbol{x}_1, \boldsymbol{x}_2 \in S$ there exists $M > 0$ such that $\|\boldsymbol{x}_1 - \boldsymbol{x}_2\| < M$.

A set is bounded if any two points within it are only a finite distance apart. This implies that the set of all real numbers $\mathbb{R}$ is unbounded.

1.2.7 Compact Set

The set S is said to be compact if and only if it is both closed and bounded.

Set compactness is important because it guarantees the existence of an optimum solution point when an objective function (that is continuous) is optimized over it. Next, we transition from set properties to function properties underlying optimality conditions for an optimization problem.

1.2.8 Continuous Functions

Function $f: S \to \mathbb{R}$ is continuous at a point $x_0 \in S$, if for every $\varepsilon > 0$ there exists $\delta > 0$ such that $\|\boldsymbol{x} - \boldsymbol{x}_0\| < \delta$ for $\boldsymbol{x} \in S$ implies that $|f(\boldsymbol{x}) - f(x_0)| < \varepsilon$.

The definition of function continuity implies that no matter how small an arbitrary number ε is chosen, a point $\boldsymbol{x}$ close to $\boldsymbol{x}_0$ (i.e., in the δ-neighborhood of $\boldsymbol{x}_0$) can be found such that the difference between the function values at $\boldsymbol{x}_0$ and $\boldsymbol{x}$ is less than ε. Function continuity can be thought of as the absence of any "breaks" in the line that plots the function. As mentioned before, continuous functions optimized over compact, nonempty sets are guaranteed to have an optimal solution value and point.

1.2.9 Global and Local Minima

Let $f: S \to \mathbb{R}$, where S is a nonempty subset of $\mathbb{R}^N$.

- A point $\boldsymbol{x}^* \in S$ is a *global minimum point* of f if for every point $\boldsymbol{x} \in S, f(\boldsymbol{x}) \geq f(\boldsymbol{x}^*)$. $f(\boldsymbol{x}^*)$ is referred to as the *global minimum value* of f over set S.
- A point $\boldsymbol{x}^* \in S$ is a *local minimum point* of f, if there exists $\varepsilon > 0$ such that for every point $\boldsymbol{x} \in N(\boldsymbol{x}^*, \varepsilon) \cap S$ we have $f(\boldsymbol{x}) \geq f(\boldsymbol{x}^*)$.

Therefore, local minimality applies only around a neighborhood $N(\boldsymbol{x}^*, \varepsilon) \cap S$ of the minimum point, whereas global minimality applies over the entire set S. It is possible to have multiple local optimum points and values; however, there is a unique global minimum value. This value may be attainable at multiple points (alternate global minimum points). If we have strict inequalities in the earlier definitions, the point $\boldsymbol{x}^*$ is referred to as a *strict* global or local minimum, respectively. Note that *global and local maxima* are defined in a similar manner.

1.2.10 Existence of an Optimal Solution

After introducing concepts related to set compactness, function continuity and definitions of optimality, the following optimum solution existence criterion can be formally stated. Consider the following general unconstrained optimization problem:

$$\min_{x \in S} \text{ or } \max f(x)$$

where $S \subseteq \mathbb{R}^N$ and $f: S \to \mathbb{R}$ is continuous on S. An optimal solution x^* exists if S is a nonempty and compact set.

This implies that unconstrained optimization problems over compact sets are guaranteed to have an optimal solution. In practice, many (constrained or unconstrained) optimization problems are originally described over unbounded or open sets with sometimes discontinuities in the objective function. It is a good practice to set finite lower and upper bounds on all variables (i.e., set compactness) and eliminate discontinuities in the objective function to ensure the existence of an optimal solution.

The derivation of uniqueness criteria for an optimum solution value (i.e., a single local optimum that is also a global optimum) hinges upon the concept of convexity. Testing for convexity is facilitated by the establishment of differentiability properties for the objective function (and constraints).

1.3 CONVEX ANALYSIS

The concept of convexity is central in optimization because it provides the means for proving the existence of a global optimal solution value (or point). Here, we provide a brief description of convexity of (i) a set, (ii) a function at a point and (iii) a function over an entire set. Interested readers are encouraged to refer to optimization textbooks such as Refs. [8.14] for more details.

1.3.1 Convex Sets and Their Properties

Convex Combination of Two Points Let $\boldsymbol{x}_1, \boldsymbol{x}_2 \in S$. Any point $(1-\lambda)\boldsymbol{x}_1 + \lambda \boldsymbol{x}_2$ with $\lambda \in [0,1]$ is referred to as a *convex combination* of $\boldsymbol{x}_1$ and $\boldsymbol{x}_2$. If $\lambda \in (0,1)$, then it is a *strict* convex combination.

Convex Set Set S is convex if the line segment connecting any two points in the set also lies completely within the set. In mathematical language, S is a convex set if and only if for every two points $\boldsymbol{x}_1, \boldsymbol{x}_2 \in S$ their convex combinations $(1-\lambda)\boldsymbol{x}_1 + \lambda \boldsymbol{x}_2$ for every $\lambda \in [0,1]$ is also within S. If this condition holds for every strict convex combination of $\boldsymbol{x}_1$ and $\boldsymbol{x}_2$, then the set S is a *strict* convex set. Any set not satisfying these requirements is a nonconvex set. Examples of convex and nonconvex sets are shown in Figure 1.4.

Special cases of convex sets frequently arise in the treatment of LP problems (i.e., extreme points, hyperplanes, half-spaces, rays, extreme directions and convex cones). The following provides their definitions.

Extreme Points A point $\boldsymbol{x} \in S$ where S is convex, is an *extreme* point of S if it cannot be represented as the strict convex combination of two distinct points in S. Therefore, if $\boldsymbol{x} = (1-\lambda)\boldsymbol{x}_1 + \lambda \boldsymbol{x}_2$ for $\lambda \in (0,1)$ and $\boldsymbol{x}_1, \boldsymbol{x}_2 \in S$, then $\boldsymbol{x} = \boldsymbol{x}_1 = \boldsymbol{x}_2$ [8].

Hyperplanes Hyperplanes are an extension of straight lines in $\mathbb{R}^2$. A hyperplane H in $\mathbb{R}^N$ is defined as $H = \{\boldsymbol{x} \mid \boldsymbol{x} \in \mathbb{R}^N, \boldsymbol{a}^{\mathrm{T}}\boldsymbol{x} = k\}$, where $\boldsymbol{a} \neq 0$ and $k \in \mathbb{R}$. The vector $\boldsymbol{a}$ is

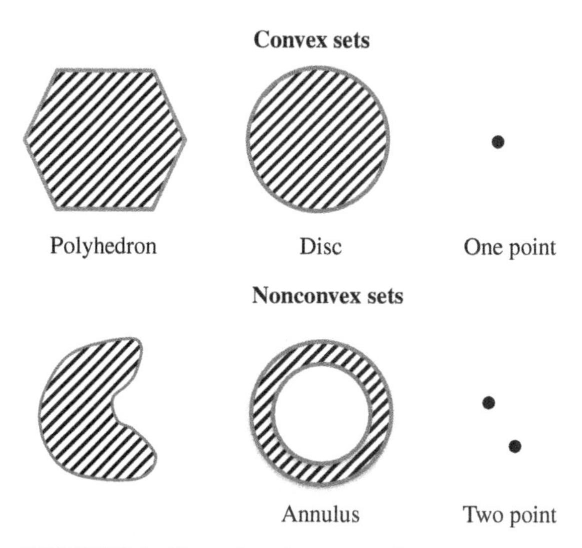

FIGURE 1.4 Examples of convex and nonconvex sets.

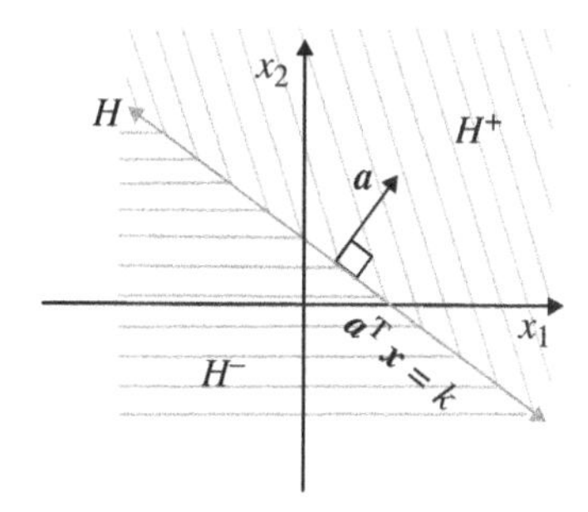

FIGURE 1.5 A schematic representation of a hyperplane and its corresponding half-spaces in a two-dimensional space.

called the *normal* of H as it is the gradient of the linear function $f(\boldsymbol{x}) = \boldsymbol{a}^{\mathrm{T}}\boldsymbol{x}$ and is thus normal to the hyperplane (see Fig. 1.5). Hyperplanes are a central concept in LP (Chapter 2) and in the analysis of metabolic networks arising in metabolite balances under steady state (Chapter 6).

Half-Spaces Each hyperplane divides $\mathbb{R}^N$ into two half-spaces. If H is a hyperplane, as defined earlier, then sets $H^- = \{\boldsymbol{x} \mid \boldsymbol{x} \in \mathbb{R}^N,\ \boldsymbol{a}^{\mathrm{T}}\boldsymbol{x} \leq k\}$ and $H^+ = \{\boldsymbol{x} \mid \boldsymbol{x} \in \mathbb{R}^N,\ \boldsymbol{a}^{\mathrm{T}}\boldsymbol{x} \geq k\}$ are the corresponding half-spaces and are convex (see Fig. 1.5). Sets $H^- = \{\boldsymbol{x} \mid \boldsymbol{x} \in \mathbb{R}^N,\ \boldsymbol{a}^{\mathrm{T}}\boldsymbol{x} < k\}$ and $H^+ = \{\boldsymbol{x} \mid \boldsymbol{x} \in \mathbb{R}^N,\ \boldsymbol{a}^{\mathrm{T}}\boldsymbol{x} > k\}$ are *open* half-spaces. The imposition of any bounds (lower or upper) on the total metabolic flow through a metabolite (see Chapter 6) gives rise to a half-space constraint.

Example 1.2

H as defined in the following is an example of a hyperplane:

$$H = \left\{ \begin{pmatrix} x_1 \\ x_2 \\ x_3 \end{pmatrix} \middle| \begin{pmatrix} x_1 \\ x_2 \\ x_3 \end{pmatrix} \in \mathbb{R}^3, x_1 + x_2 + x_3 = 10 \equiv \begin{pmatrix} 1 & 1 & 1 \end{pmatrix} \begin{pmatrix} x_1 \\ x_2 \\ x_3 \end{pmatrix} = 10 \right\}$$

The normal for H is as follows:

$$\boldsymbol{a} = \begin{pmatrix} 1 \\ 1 \\ 1 \end{pmatrix}$$

The half-spaces defined by H are as follows:

$$H^+ = \left\{ \begin{pmatrix} x_1 \\ x_2 \\ x_3 \end{pmatrix} \middle| \begin{pmatrix} x_1 \\ x_2 \\ x_3 \end{pmatrix} \in \mathbb{R}^3, x_1 + x_2 + x_3 \geq 10 \equiv \begin{pmatrix} 1 & 1 & 1 \end{pmatrix} \begin{pmatrix} x_1 \\ x_2 \\ x_3 \end{pmatrix} \geq 10 \right\}$$

$$H^- = \left\{ \begin{pmatrix} x_1 \\ x_2 \\ x_3 \end{pmatrix} \middle| \begin{pmatrix} x_1 \\ x_2 \\ x_3 \end{pmatrix} \in \mathbb{R}^3, x_1 + x_2 + x_3 \leq 10 \equiv \begin{pmatrix} 1 & 1 & 1 \end{pmatrix} \begin{pmatrix} x_1 \\ x_2 \\ x_3 \end{pmatrix} \leq 10 \right\}$$

□

Rays A ray is a set of points on a line defined as $\{\boldsymbol{y} \mid \boldsymbol{y} = \boldsymbol{x}_0 + \lambda \boldsymbol{d}, \lambda \geq 0, \boldsymbol{d} \neq 0\}$, where $\boldsymbol{x}_0$ and $\boldsymbol{d}$ are the *vertex* and *direction* of the ray, respectively. It is easy to verify that the set describing a ray is convex.

Direction of a Convex Set A nonzero vector $\boldsymbol{d}$ is called a direction of the convex set S, if a ray with vertex $\boldsymbol{x}_0$ and direction $\boldsymbol{d}$ is contained in S for every $x_0 \in S$. Obviously, if the set S is bounded, it has no directions.

Extreme Direction and Extreme Ray of a Convex Set The concept of an extreme direction is similar to that of an extreme point. A direction of a convex set S is called an *extreme* direction if it cannot be represented as a positive combination of any two distinct directions of S, that is, if $\boldsymbol{d} = \lambda_1 \boldsymbol{d}_1 + \lambda_2 \boldsymbol{d}_2$, $\lambda_1, \lambda_2 \geq 0$ then $\boldsymbol{d} = \boldsymbol{d}_1 = \boldsymbol{d}_2$ (see Fig. 1.6). Any ray whose vertex is an extreme point and direction is an extreme direction is called an extreme ray.

Convex Cone A convex set C is called a convex cone if $\lambda \boldsymbol{x} \in C$ for each $\boldsymbol{x} \in C$ for all $\lambda \geq 0$ (see Fig. 1.7). Each cone contains the origin (for $\lambda = 0$) and at least one ray with vertex at the origin. A convex cone can be viewed as a convex set where all points on a line linking the origin and any point in the set also belongs to the set.

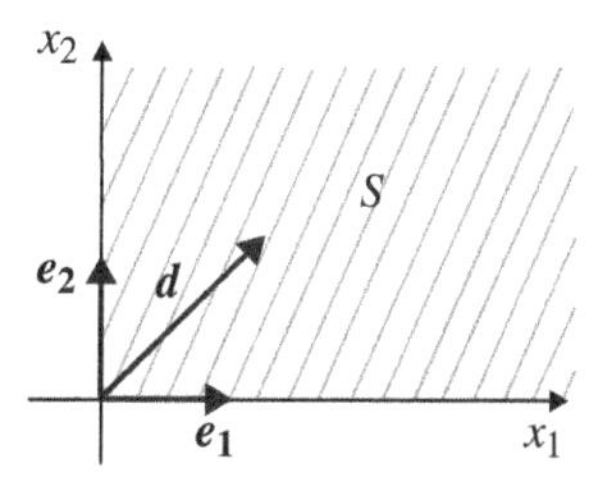

FIGURE 1.6 Examples of a direction (i.e., $\boldsymbol{d}$) and extreme directions (i.e., $\boldsymbol{e}_1$ and $\boldsymbol{e}_2$) of a convex set S.

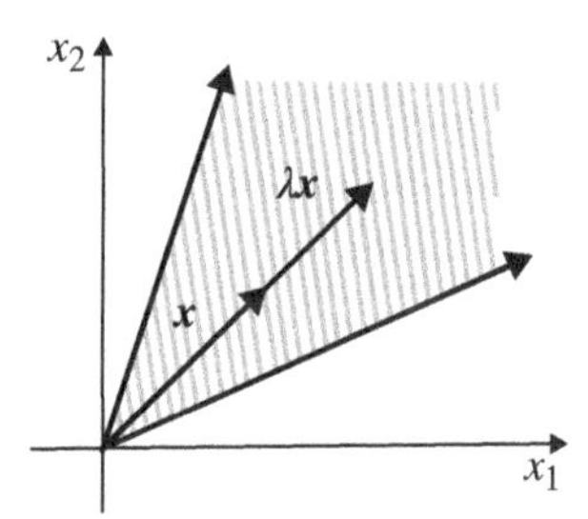

FIGURE 1.7 An example of a convex cone. For any $\boldsymbol{x} \in \boldsymbol{C}$ and $\lambda \geq 0$, we have $\lambda \boldsymbol{x} \in \boldsymbol{C}$.

The concepts of the extreme direction/ray of a convex cone are cornerstones in the extreme pathway analysis of metabolic networks [9–13].

Polyhedral Set and Polyhedral Cone A polyhedral set is the intersection of a finite number of half-spaces. A polyhedral cone is a polyhedral set, whose half-spaces pass through the origin.

For example, set $S = \{\boldsymbol{x} \mid \boldsymbol{x} \in \mathbb{R}^N, \boldsymbol{Ax} \leq \boldsymbol{b}, \boldsymbol{x} \geq 0\}$ is a polyhedral set as it is the intersection of N half-spaces defined by $\boldsymbol{Ax} \leq \boldsymbol{b}$ and the half-space defined by $\boldsymbol{x} \geq 0$. Extreme points of this polyhedral set are the intersections of the half-spaces [8]. The feasible regions of LP problems correspond to polyhedral sets. As we will see in Chapter 2, the solution of LP problems always lies on an extreme point of this polyhedral set.

1.3.2 Convex Functions and Their Properties

Set convexity is important as it ensures that every point within the set (except extreme points) is reachable as a linear combination of others. This has implications for the design of algorithms that search for the optimum within convex sets. Proving set convexity is cumbersome as every two point combination must be tested. Functions provide a way of circumventing this challenge by testing for set convexity through an equivalent function convexity criterion.

Convex and Concave Function Definitions A function defined over a convex set S is *convex over set* S if a line connecting any two points on the function lies above the function. Stated formally, function $f:S \to \mathbb{R}$ is convex in S if and only if for every two points $\boldsymbol{x}_1, \boldsymbol{x}_2 \in S$ and every $\lambda \in [0,1]$ we have:

$$f\left(\lambda \boldsymbol{x}_1 + (1-\lambda)\boldsymbol{x}_2\right) \le \lambda f\left(\boldsymbol{x}_1\right) + (1-\lambda) f\left(\boldsymbol{x}_2\right) \tag{1.8}$$

- Function f is *strictly* convex if we have a strict inequality in Equation 1.8.
- A function $f:S \to \mathbb{R}$ is concave if and only if $-f$ is convex.
- A function f is *convex at the point* $\bar{\boldsymbol{x}} \in S$, if $f(\lambda \bar{\boldsymbol{x}} + (1-\lambda)\boldsymbol{x}) \le \lambda f(\bar{\boldsymbol{x}}) + (1-\lambda) f(\boldsymbol{x})$ for each $\lambda \in (0,1)$ and each $\boldsymbol{x} \in S$.

Figure 1.8 illustrates some examples of convex and nonconvex functions. It is possible for a function to be nonconvex over a set but convex within a defined subset. For example, the function shown in Figure 1.9 is nonconvex in set S_2, but is convex within set S_1. Another example is $f(x) = x^3$, which is nonconvex in $\mathbb{R}$, but is convex for $x \ge 0$.

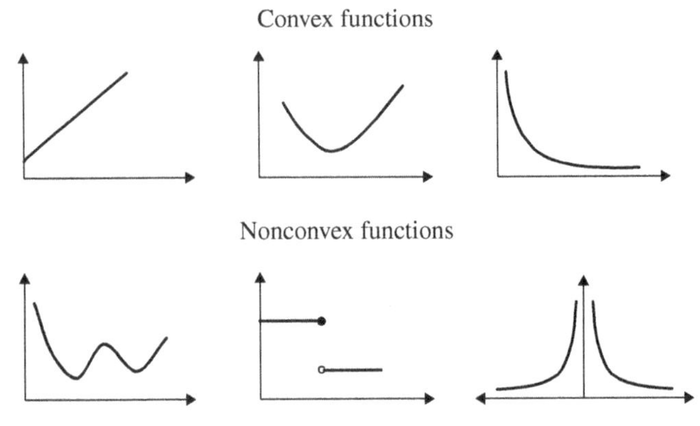

FIGURE 1.8 Examples of convex and nonconvex functions.

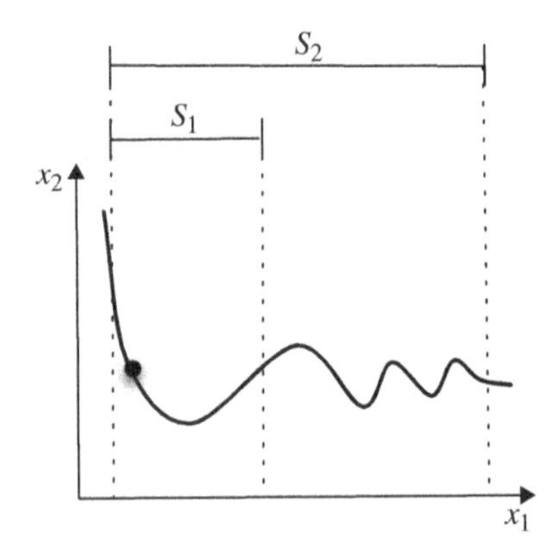

FIGURE 1.9 Convexity of a function within different sets. This function is nonconvex in S_2 but is convex within S_1 and at the point indicated on the graph.

Properties of Convex and Concave Functions If $f, g : S \to \mathbb{R}$ are convex functions in S:

- $f + g$ is convex in S.
- λf is convex in S if $\lambda > 0$.
- max $(f(\boldsymbol{x}), g(\boldsymbol{x}))$ is convex.
- min $(f(\boldsymbol{x}), g(\boldsymbol{x}))$ is generally nonconvex.

Connection of Set Convexity with Function Convexity Let $f(\boldsymbol{x})$ be a convex function in set S, then set $S_c = \{\boldsymbol{x} \in S \mid f(\boldsymbol{x}) \leq c\}$, where c is an arbitrary scalar, is a convex set. The statement is true in the reverse direction as well. Set S_c is also referred to as the *level set* of function f.

Proof: To prove that S_c is convex, we need to show that for any two arbitrary points $\boldsymbol{x}_1$ and $\boldsymbol{x}_2$ in S_c we have $\lambda \boldsymbol{x}_1 + (1-\lambda)\boldsymbol{x}_2 \in S$ for $\lambda \in [0,1]$. In other words, we need to show that $f(\lambda \boldsymbol{x}_1 + (1-\lambda)\boldsymbol{x}_2) \leq c$. Since $\boldsymbol{x}_1, \boldsymbol{x}_2 \in S_c$, we have $f(\boldsymbol{x}_1) \leq c$ and $f(\boldsymbol{x}_2) \leq c$. Therefore, $\lambda f(\boldsymbol{x}_1) + (1-\lambda) f(\boldsymbol{x}_2) \leq \lambda c + (1-\lambda)c = c$. Also, since $\boldsymbol{x}_1, \boldsymbol{x}_2 \in S$ and f is convex, $f(\lambda \boldsymbol{x}_1 + (1-\lambda)\boldsymbol{x}_2) \leq \lambda f(\boldsymbol{x}_1) + (1-\lambda) f(\boldsymbol{x}_2)$. The proof then follows directly from the last two inequalities. The proof for the reverse direction is derived in a similar fashion. □

This is a very important result because it allows testing for set convexity by testing the convexity properties of the functions defining the set as their level set. Note that since the intersection of convex sets is a convex set, the feasible region described by a number of constraints corresponding to convex level sets is also a convex set. This implies that the convexity of the region defined by a set of inequality constraints can be inferred by testing the convexity of each individual function. Figure 1.10 illustrates convex sets arising from level sets associated with convex functions.

Testing for the convexity of a function based on the definitions already provided is often cumbersome. Much more tractable representations of convexity can be drawn by using the partial (first and second-order) derivatives of the function.

Differentiable Functions Let S be a nonempty subset of $\mathbb{R}^N$. A function $f : S \to \mathbb{R}$ is *differentiable* at $\bar{\boldsymbol{x}} \in \text{int}(S)$ if and only if f is continuous at $\bar{\boldsymbol{x}}$ and for each $\Delta \boldsymbol{x}$,

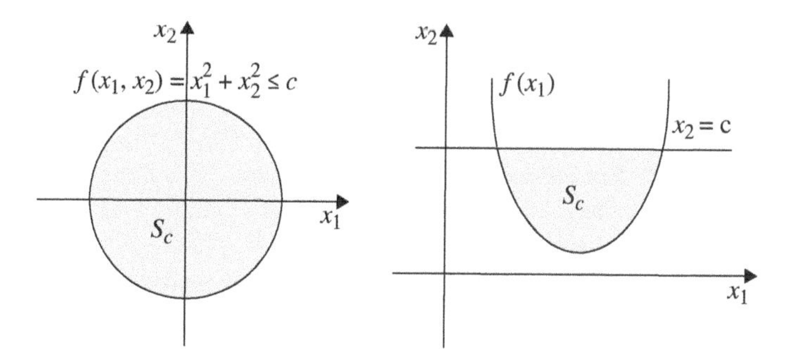

FIGURE 1.10 Examples of convex level sets associated with convex functions.

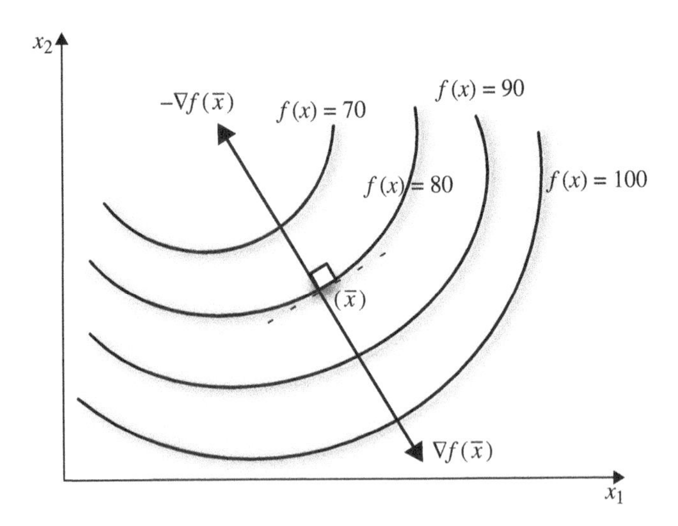

FIGURE 1.11 The gradient vector at any given point is normal to the level set of the function at that point and represents the direction of the steepest ascent.

where $\bar{x} + \Delta x \in S$, there exists a vector $\nabla f(\bar{x})$ (called the *gradient vector*) and a function $\alpha : \mathbb{R}^N \rightarrow \mathbb{R}$ such that

$$f(\bar{x} + \Delta x) = f(\bar{x}) + \nabla^T f(\bar{x}) \Delta x + \alpha(\bar{x} + \Delta x) \|\Delta x\| \tag{1.9}$$

where

$$\nabla f(\bar{x}) = \begin{pmatrix} \dfrac{\partial f(\bar{x})}{\partial x_1} & \dfrac{\partial f(\bar{x})}{\partial x_2} & \cdots & \dfrac{\partial f(\bar{x})}{\partial x_N} \end{pmatrix}^T \tag{1.10}$$

and $\lim_{\Delta x \to 0} \alpha(\bar{x} + \Delta x) = 0$. This definition implies that the linear approximation (or first-order Taylor expansion) of the function f at any point $\bar{x} + \Delta x$ becomes equal to $f(\bar{x} + \Delta x)$ as Δx approaches zero from any direction. Function f is differentiable on an open set $S_0 \subset S$ if it is differentiable for every point in S_0. The gradient of a function points to the direction of greatest increase (*steepest ascent*). Similarly, the negative of the gradient vector represents the direction of the *steepest descent* (see Fig. 1.11). In addition, the gradient of a function at any given point is normal to the level sets of the function at that point.

Twice Differentiable Functions Let $f : S \rightarrow \mathbb{R}$, where S is a nonempty set in $\mathbb{R}^N$. f is *twice differentiable* at $\bar{x} \in int(S)$ if and only if f is continuous at $\bar{x}$ and for each Δx, where $\bar{x} + \Delta x \in S$, there exists a gradient vector $\nabla^T f(\bar{x})$, a $N \times N$ (symmetric) matrix $H(\bar{x})$ (*Hessian* matrix) and a function $a : \mathbb{R}^N \rightarrow \mathbb{R}$ such that

$$f(\bar{x} + \Delta x) = f(\bar{x}) + \nabla^T f(\bar{x}) \Delta x + \frac{1}{2} (\Delta x)^T H(\bar{x}) \Delta x + \alpha(\bar{x} + \Delta x) \|\Delta x\|^2 \tag{1.11}$$

where

$$
\boldsymbol{H}(\bar{\boldsymbol{x}})=\begin{pmatrix}
\frac{\partial}{\partial x_1}\left(\frac{\partial f(\bar{\boldsymbol{x}})}{\partial x_1}\right) & \frac{\partial}{\partial x_1}\left(\frac{\partial f(\bar{\boldsymbol{x}})}{\partial x_2}\right) & \cdots & \frac{\partial}{\partial x_1}\left(\frac{\partial f(\bar{\boldsymbol{x}})}{\partial x_N}\right) \\
\frac{\partial}{\partial x_2}\left(\frac{\partial f(\bar{\boldsymbol{x}})}{\partial x_1}\right) & \frac{\partial}{\partial x_2}\left(\frac{\partial f(\bar{\boldsymbol{x}})}{\partial x_2}\right) & \cdots & \frac{\partial}{\partial x_2}\left(\frac{\partial f(\bar{\boldsymbol{x}})}{\partial x_N}\right) \\
\vdots & \vdots & \ddots & \vdots \\
\frac{\partial}{\partial x_N}\left(\frac{\partial f(\bar{\boldsymbol{x}})}{\partial x_1}\right) & \frac{\partial}{\partial x_N}\left(\frac{\partial f(\bar{\boldsymbol{x}})}{\partial x_2}\right) & \cdots & \frac{\partial}{\partial x_N}\left(\frac{\partial f(\bar{\boldsymbol{x}})}{\partial x_N}\right)
\end{pmatrix} \tag{1.12}
$$

and $\lim_{\Delta x\to 0} \alpha(\bar{\boldsymbol{x}}+\Delta \boldsymbol{x})=0$. As was the case for first-order differentiability, second-order differentiability implies that the quadratic approximation (or second-order Taylor expansion) of the function f at any point $\bar{\boldsymbol{x}}+\Delta\boldsymbol{x}$ becomes equal to $f(\bar{\boldsymbol{x}}+\Delta\boldsymbol{x})$ as $\Delta\boldsymbol{x}$ approaches zero from any direction.

Convexity Check for Differentiable Functions Let $f: S \to \mathbb{R}$, where S is a nonempty open convex set in $\mathbb{R}^N$. If f is differentiable for every point $\boldsymbol{x}$ in S, then it is *convex at a point* $\bar{\boldsymbol{x}} \in S$ if and only if:

$$f(\boldsymbol{x}) \geq f(\bar{\boldsymbol{x}}) + \nabla^{\mathrm{T}} f(\bar{\boldsymbol{x}})(\boldsymbol{x}-\bar{\boldsymbol{x}}), \quad \forall \boldsymbol{x} \in S \tag{1.13}$$

Similarly, f is *concave at a point* $\bar{\boldsymbol{x}} \in S$ if and only if

$$f(\boldsymbol{x}) \leq f(\bar{\boldsymbol{x}}) + \nabla^{\mathrm{T}} f(\bar{\boldsymbol{x}})(\boldsymbol{x}-\bar{\boldsymbol{x}}), \quad \forall \boldsymbol{x} \in S \tag{1.14}$$

In other words, a convex (concave) function always lies above (below) its first-order (linear) approximation at any point $\bar{\boldsymbol{x}} \in S$, respectively. Strict convexity or concavity can be established in a similar manner by converting the inequality signs in Constraints 1.13 and 1.14 to strict inequalities. Convexity or concavity of a function f at a given point $\bar{\boldsymbol{x}} \in S$ can be extended for set S, if Constraints 1.13 or 1.14, respectively, apply for every point $\bar{\boldsymbol{x}} \in S$.

Convexity Check for Twice Differentiable Functions Based on Hessian Matrix If a function f is twice differentiable within a set S, then the information contained within the second-order partial derivatives can be used to test/prove the convexity of function f over the set. Let $f: S \to \mathbb{R}$, where S is a nonempty open convex set in $\mathbb{R}^N$ and f is twice differentiable in S. Function f is convex in S if and only if its Hessian matrix $\boldsymbol{H}(\boldsymbol{x})$ is positive semidefinite (psd) for every point in S. (Note that a matrix $\boldsymbol{M}$ is psd if $\boldsymbol{x}^{\mathrm{T}}\boldsymbol{M}\boldsymbol{x} \geq 0$ for all $\boldsymbol{x} \in S \subseteq \mathbb{R}^N, \boldsymbol{x} \neq \boldsymbol{0}$.)

Proof: (a) We first provide the proof in the forward direction, that is, we show that if f is convex, then $\boldsymbol{H}(\boldsymbol{x})$ is psd for every point in S. Let $\bar{\boldsymbol{x}}$ be an arbitrary point in S.

It follows from the convexity of f that $f(\bar{x}+\lambda x) \geq f(\bar{x})+\nabla^{\mathrm{T}} f(\bar{x})(x-\bar{x}),\ \forall x \in S$ (see Equation 1.13). Also, since f if twice differentiable, we have (see Equation 1.11):

$$f(x)=f(\bar{x})+\nabla^{\mathrm{T}} f(\bar{x})(x-\bar{x})+\frac{1}{2}(x-\bar{x})^{\mathrm{T}} H(\bar{x})(x-\bar{x})+\alpha(x-\bar{x})\|x-\bar{x}\|^2$$

Therefore, by combining the last two expressions, we have $\frac{1}{2}(x-\bar{x})^{\mathrm{T}} H(\bar{x})(x-\bar{x}) +\alpha(x)\|x-\bar{x}\|^2 \geq 0$. As x approaches $\bar{x}$, function $\alpha(x-\bar{x})$ approaches zero implying that $(x-\bar{x})^{\mathrm{T}} H(\bar{x})(x-\bar{x}) \geq 0$ which completes the proof. (b) The proof in the reverse direction proceeds in a similar fashion. □

A concave, strictly convex or strictly concave function is associated with a *negative semidefinite (nsd), positive definite (pd)* or *negative definite (nd)* Hessian matrix, respectively. A matrix is psd, nsd, pd or nd if all of its eigenvalues are nonnegative, nonpositive, positive or negative, respectively. A matrix that is neither psd nor nsd is called *indefinite*. This result enables checking the convexity properties of a function (and consequently of a set) by inspecting the eigenvalues of the corresponding Hessian matrix.

Example 1.3

Check whether the following function is concave or convex:

$$f(x_1,x_2,x_3)=x_1^2+x_2^2+x_3^2+2x_1x_2$$

Solution: This Hessian matrix for this quadratic function is as follows:

$$H=\begin{bmatrix} 2 & 2 & 0 \\ 2 & 2 & 0 \\ 0 & 0 & 2 \end{bmatrix}$$

The eigenvalues of the Hessian matrix can be obtained by solving the characteristic equation $\det(H-\lambda I)=0$, where det represents the determinant of a matrix. The eigenvalues of the Hessian matrix are $\lambda=0,2,4 \geq 0$ implying that it is psd and the function is convex. □

The convexity properties of the following frequently encountered functions can be established by checking the eigenvalues of their Hessian matrix:

- $f(x)=\log(x)$ is concave.
- $f(x,y)=xy$ is neither convex nor concave.
- $f(x,y,z)=xyz$ is neither convex nor concave.
- $f(x,y,z)=\frac{x}{y}$ is neither convex nor concave.
- $f(x,y)=\frac{x^2}{y}$ is convex.
- $f(x,y)=\sqrt{xy}$ is concave.
- $f(x,y)=\frac{1}{xy}$ is convex.

As described earlier, the multiplication of a convex function with a positive scalar and the sum of convex functions yield convex functions. Therefore, complex expressions can be analyzed for their convexity by disassembling them into smaller terms and analyzing each one separately. For example, function $f(x,y)=x^2/y+x\log(x)$ is convex as the sum of two convex functions. In some cases, establishing (or refuting) convexity may require the recombination of various terms. For example, the convexity of function $g(x,y)=(x+y)^2-2xy$ as the sum of a convex $(x+y)^2$ and a nonconvex $(-2xy)$ function cannot be initially determined. However, upon combining the two functions, a single convex function $(x-y)^2$ emerges.

1.3.3 Convex Optimization Problems

Upon establishing a connection between set convexity and function convexity and an eigenvalue-based test, the following optimum solution value uniqueness test can be applied. An optimization problem is convex if the objective function and all inequality constraints are convex and all equality constraints are linear. It can be schematically represented as follows:

$$\begin{aligned}
&\text{minimize}\left(\text{convex function}\right)\\
&\text{subject to}\\
&\left(\text{convex function}\right)\leq 0\\
&\left(\text{linear function}\right)=0\\
&x\in\text{compact set}
\end{aligned}$$

A key attribute of convex optimization problems is that there exists a unique local (and thus global) *minimum value* (see Chapter 9). In addition, if the objective function is strictly convex, then the point associated with the global minimum value is also unique (*global minimum point*). Consequently, LP problems described by linear objective functions and constraints are convex optimization problems. This implies that the optimal solution value of an LP problem (if one exists) is unique (see also Chapter 2).

Example 1.4
The following optimization problem is an example of a convex problem where the objective function is convex and all constraints are linear and thus are convex.

$$\begin{aligned}
&\text{minimize}\quad x^2+y^2\\
&\text{subject to}\\
&x+y\leq 3\\
&y\geq\frac{1}{x}\\
&x\geq 0
\end{aligned}$$

A graphical presentation of the feasible space and the objective function is given in Figure 1.12. □

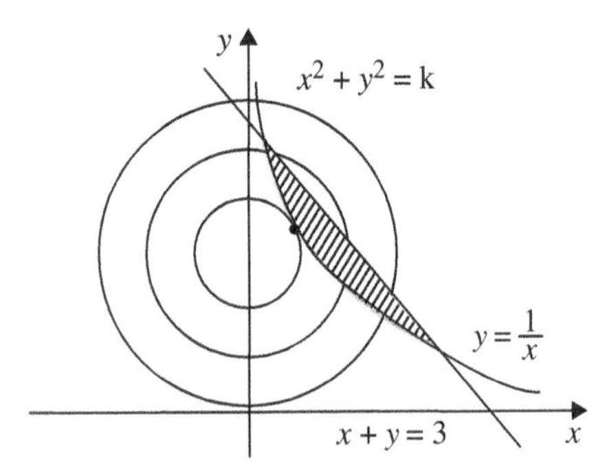

FIGURE 1.12 Graphical representation of the convex optimization problem in Example 1.4. The convexity of the problem implies that we have a global minimum.

1.3.4 Generalization of Convex Functions

Generalized forms of convexity such as quasi- and pseudoconvexity suffice to guarantee global minimality even if the function is nonconvex under certain conditions (see Chapter 9). In the following, we provide only the definitions of these generalized forms. Interested readers are encouraged to refer to standard nonlinear optimization textbooks (e.g., Ref. [14]) for more details.

Quasiconvex Functions f is quasiconvex in S if for every two points $\boldsymbol{x}_1, \boldsymbol{x}_2 \in S$ and each $\lambda \in (0,1)$, we have

$$f\left(\lambda \boldsymbol{x}_1 + (1-\lambda)\boldsymbol{x}_2\right) \leq \max\left\{f(\boldsymbol{x}_1), f(\boldsymbol{x}_2)\right\} \tag{1.15}$$

This definition implies that a function is quasiconvex within a set, if for any two points within the set the value of the function at any of the points in the line segment connecting the two points, is less than or equal to the larger of the values that the function attains at the two examined points.

Pseudoconvex Functions A function f (differentiable within S) is pseudoconvex if for every two points $\boldsymbol{x}_1, \boldsymbol{x}_2 \in S$ with $\nabla^{\mathrm{T}} f(\boldsymbol{x}_2)(\boldsymbol{x}_1 - \boldsymbol{x}_2) > 0$, we have $f(\boldsymbol{x}_1) > f(\boldsymbol{x}_2)$, or equivalently for every two points $\boldsymbol{x}_1, \boldsymbol{x}_2 \in S$ with $f(\boldsymbol{x}_1) \leq f(\boldsymbol{x}_2)$, we have $\nabla^{\mathrm{T}} f(\boldsymbol{x}_2)(\boldsymbol{x}_1 - \boldsymbol{x}_2) \leq 0$.

EXERCISES

1.1 Show whether the following functions are convex, concave or neither:

(a) $f(x) = \sin(x), x \in [0, \pi]$.
(b) $f(x) = x^k, x \geq 0, k \geq 0$.
(c) $f(x) = x \ln(x), x > 0$.

1.2 Consider a single-period inventory model where demand is a random variable with density f; that is, $P(\text{demand} \le x) = \int_0^x f(x')dx'$. Also, all stock-outs are lost sales, C is the unit cost of each item, p is the loss due to inability to fulfill orders (includes loss of revenue and customer goodwill), r is the selling price per unit and l is the salvage value of each unsold item at the end of the period. The problem is to determine the optimal order quantity Q that will maximize the expected net revenue for the season. The expected net revenue, denoted by $\Pi(Q)$, is given by the following equation:

$$\prod(Q) = r\mu + l\int_0^Q (Q-x)f(x)dx - (p+r)\int_Q^x (x-Q)f(x)dx - CQ$$

where μ is the expected demand as follows:

$$\mu = \int_0^x xf(x)dx$$

(a) Show that $\Pi(Q)$ is a concave function in $Q(\ge 0)$.

(b) Based on the result of (a) explain how you will find the optimal ordering policy.

(c) Compute the optimal policy for the following data:

$$C = \$2.50 \quad r = \$5.00 \quad l = 0 \quad p = \$2.50$$

$$f(x) = \begin{cases} \dfrac{1}{400} & 100 \le x \le 500 \\ 0 & \text{otherwise} \end{cases}$$

Hint: Use Leibniz rule for differentiation under the integral sign.

1.3 Identify the relations that constants a and b must satisfy so that the function

$$f(x,y) = x^a y^b \quad x, y \ge 0$$

is (a) convex and (b) concave for every $x, y \ge 0$. Generalize the result for the n-dimensional case:

$$f(x) = \prod_{i=1}^{N} x_i^{a_i} \quad x_i \ge 0$$

1.4 Let $g : \mathbb{R}^N \to \mathbb{R}$ be a concave function, and let function f be defined by $f(x) = \dfrac{1}{g(x)}$. Show that f is convex over $S = [x \mid g(x) > 0]$.

REFERENCES

1. Chemmangattuvalappil N, Task K, Banerjee I: An integer optimization algorithm for robust identification of non-linear gene regulatory networks. *BMC Syst Biol* 2012, **6**:119.
2. Dasika M, Gupta A, Maranas C, Altman R, Dunker A, Hunter L, Jung T, Klein T: A mixed integer linear programming (MILP) framework for inferring time delay in gene regulatory networks. Pacific Symposium on Biocomputing 2004, 2003: pp. 474–485.
3. Foteinou P, Yang E, Saharidis G, Ierapetritou M, Androulakis I: A mixed-integer optimization framework for the synthesis and analysis of regulatory networks. *J Global Optim* 2009, **43**(2–3):263–276.
4. Gupta A, Varner J, Maranas C: Large-scale inference of the transcriptional regulation of Bacillus subtilis. *Comput Chem Eng* 2005, **29**(3):565–576.
5. Lee W, Tzou W: Computational methods for discovering gene networks from expression data. *Briefings Bioinformat* 2009, **10**(4):408–423.
6. Thomas R, Paredes CJ, Mehrotra S, Hatzimanikatis V, Papoutsakis ET: A model-based optimization framework for the inference of regulatory interactions using time-course DNA microarray expression data. *BMC Bioinformat* 2007, **8**:228.
7. Xuan NV, Chetty M, Coppel R, Wangikar PP: Gene regulatory network modeling via global optimization of high-order dynamic Bayesian network. *BMC Bioinformat* 2012, **13**:131.
8. Bazaraa MS, Jarvis JJ, Sherali HD: *Linear programming and network flows*, 4th edn. Hoboken, N.J.: John Wiley & Sons; 2010.
9. Papin JA, Price ND, Edwards JS, Palsson BB: The genome-scale metabolic extreme pathway structure in Haemophilus influenzae shows significant network redundancy. *J Theor Biol* 2002, **215**(1):67–82.
10. Wiback SJ, Palsson BO: Extreme pathway analysis of human red blood cell metabolism. *Biophys J* 2002, **83**(2):808–818.
11. Schilling CH, Edwards JS, Letscher D, Palsson B: Combining pathway analysis with flux balance analysis for the comprehensive study of metabolic systems. *Biotechnol Bioeng* 2000, **71**(4):286–306.
12. Schilling CH, Letscher D, Palsson BO: Theory for the systemic definition of metabolic pathways and their use in interpreting metabolic function from a pathway-oriented perspective. *J Theor Biol* 2000, **203**(3):229–248.
13. Papin JA, Price ND, Wiback SJ, Fell DA, Palsson BO: Metabolic pathways in the post-genome era. *Trends Biochem Sci* 2003, **28**(5):250–258.
14. Bazaraa MS, Sherali HD, Shetty CM: *Nonlinear programming: theory and algorithms*, 3rd edn. Hoboken, N.J.: Wiley-Interscience; 2006.

2

LP AND DUALITY THEORY

Linear programming (LP) involves the minimization or maximization of the weighted sum of a number of variables satisfying a set of linear equality and/or inequality constraints. The term "programming" reflects historical precedence as some of the early LP problems modeled optimal logistics schedules or programs. The Simplex method by Dantzig [1] was the first algorithm for efficiently solving LP problems to optimality. In this chapter, we introduce the basic concepts in LP, discuss the logic of the Simplex method, and briefly introduce the duality theory concepts that will be relied upon for the solution of bilevel optimization problems in metabolic networks discussed in subsequent chapters. Finally, we introduce useful transformation techniques to convert a number of nonlinear optimization problems to an equivalent linear programming representation. Comprehensive treatments of LP can be found in Refs. [2–4]. In addition to dedicated mathematical programming and optimization modeling software such as AMPL [5], GAMS (GAMS Development Corp. Washington DC, USA), and LINGO (LINDO Systems Inc., Chicago, IL, USA), a number of other popular software environments such as MATLAB (Mathworks, Natick, MA, USA) and Microsoft Excel provide easy access to LP solvers. Appendix A provides a brief description of the GAMS optimization modeling environment.

2.1 CANONICAL AND STANDARD FORMS OF AN LP PROBLEM

The same LP problem can be formulated and represented in many different forms. The most widely used forms of an LP problem are *canonical* and *standard* forms. These two representations are in fact equivalent and can be converted to one another.

Optimization Methods in Metabolic Networks, First Edition. Costas D. Maranas and Ali R. Zomorrodi.

2.1.1 Canonical Form

The canonical form of an LP is frequently used to introduce duality relationships. A minimization problem in canonical form contains only inequality constraints of greater than or equal form with all variables restricted to be nonnegative. Similarly, a maximization problem in canonical form consists of only inequality constraints of the form less than or equal with all variables restricted to be nonnegative:

$$\begin{array}{ll} \text{minimize } z = \boldsymbol{c}^{\mathrm{T}}\boldsymbol{x} & \text{maximize } z = \boldsymbol{c}^{\mathrm{T}}\boldsymbol{x} \\ \text{subject to} & \text{subject to} \\ \boldsymbol{Ax} \geq \boldsymbol{b} & \boldsymbol{Ax} \leq \boldsymbol{b} \\ \boldsymbol{x} \geq 0 & \boldsymbol{x} \geq 0 \end{array}$$

where

$\boldsymbol{x}$ is an $N \times 1$ vector called the *vector of decision variables*. Each element x_j of this vector is a decision variable for the optimization problem

$\boldsymbol{c}$ is an $N \times 1$ vector referred to as the *cost vector* for a minimization problem and *profit vector* for a maximization problem. Each element c_j of this vector represents the cost (or profit) associated with each variable x_j

$\boldsymbol{A}$ is an $M \times N$ matrix called the *matrix of coefficients*. Each element a_{ij} of this matrix is the coefficient of variable x_j in constraint i

$\boldsymbol{b}$ is an $M \times 1$ vector denoted as the *right-hand side vector*. For minimization problems with greater than or equal constraints, this vector is sometimes referred to as the *demand vector*

The same LP representations can be also shown in the expanded form as follows:

$$\begin{array}{ll} \text{minimize } z = \sum_{j=1}^{N} c_j x_j & \text{maximize } z = \sum_{j=1}^{N} c_j x_j \\ \text{subject to} & \text{subject to} \\ \sum_{j=1}^{N} a_{ij} x_j \geq b_i, \quad i = 1,2,\ldots,M & \sum_{j=1}^{N} a_{ij} x_j \leq b_i, \quad i = 1,2,\ldots,M \\ x_j \geq 0, \quad j = 1,2,\ldots,N & x_j \geq 0, \quad j = 1,2,\ldots,N \end{array}$$

2.1.2 Standard Form

The standard form of an LP problem is a maximization problem consisting of only equality constraints with all variables restricted to be nonnegative:

$$\begin{array}{l} \text{maximize } z = \boldsymbol{cx} \\ \text{subject to} \\ \boldsymbol{Ax} = \boldsymbol{b} \\ \boldsymbol{x} \geq 0 \end{array}$$

or in expanded form:

$$\text{maximize}\, z = \sum_{j=1}^{N} c_j x_j$$

subject to

$$\sum_{j=1}^{N} a_{ij} x_j = b_i, \quad i = 1,2,\ldots,M$$

$$x_j \geq 0, \quad j = 1,2,\ldots,N$$

When dealing with equality constraints, the elements b_i of vector $\boldsymbol{b}$ are referred to as the *requirement coefficients*.

Note that in both standard and canonical forms the objective function and constraints are linear functions of variables x_j. The standard form of an LP problem is of particular importance as the Simplex method is designed to solve LPs only after they have been converted to their standard form.

Any LP problem can be converted to either canonical or standard form using the following transformations:

- Inequality constraints can be converted into equality constraints by introducing slack variables $\left(s_i^+ \geq 0\right)$ as follows:

$$\sum_{j=1}^{N} a_{ij} x_j \leq b_i \equiv \sum_{j=1}^{N} a_{ij} x_j + s_i^+ = b_i$$

- Variables unrestricted in sign $\left(x_j \in \mathbb{R}\right)$ can be expressed as the difference of two nonnegative real variables:

$$x_j = x_j^+ - x_j^- \ \text{ where}\, x_j^+, x_j^- \geq 0$$

- A minimization problem can be converted to its equivalent problem of maximizing the negative of the objective function:

$$\text{minimize}(z) = -\text{maximize}(-z)$$

For an LP in standard form, the number of variables N is typically larger than the number of equalities M (i.e., $N > M$) leading to an underdetermined system of equations with $N - M$ degrees of freedom if the system of equalities is of full rank.

The following example pictorially illustrates the topology of the feasible region of an LP problem and the unique properties of the optimum solution point that the simplex method hinges upon.

2.2 GEOMETRIC INTERPRETATION OF AN LP PROBLEM

Consider the following LP problem:

$$\begin{aligned}&\text{maximize}\, x_1 + x_2 \quad [\text{LP}_{2.2}]\\&\text{subject to}\\&x_1 + 2x_2 \leq 2\\&2x_1 + x_2 \leq 2\\&x_1, x_2 \geq 0\end{aligned}$$

The values of variables x_1 and x_2 are sought that satisfy all the constraints and maximize the objective function $(x_1 + x_2)$. By plotting all the constraints as equalities and determining the region where each inequality holds, we can identify the feasible solution space of the problem (shaded region in Fig. 2.1). As shown in Chapter 1, this feasible space defines a polyhedral set. The aim here is to find a point in the feasible space with the maximum value for the objective function. To this end, one can draw level sets (also known as contour lines) of the objective function. These are lines of the form $x_1 + x_2 = c$ with increasing values of $c \in \mathbb{R}$ (dotted lines in Fig. 2.1). The last point, where the objective function level set "touches" the feasible region before seizing to intersect with it, corresponds to the optimal solution. Applying this test to this problem reveals that point $\left(\frac{2}{3}, \frac{2}{3}\right)$ is the optimal solution with a maximum objective function value of $\frac{4}{3}$.

In general, for LPs with three variables the objective function level sets form two-dimensional planes while for N-dimensional problems the level sets generally correspond to $(N-1)$-dimensional hyperplanes. Therefore, an optimal solution point of a bounded LP problem will be at one of the vertices of the polyhedral set defining the feasible region (see Fig. 2.1).

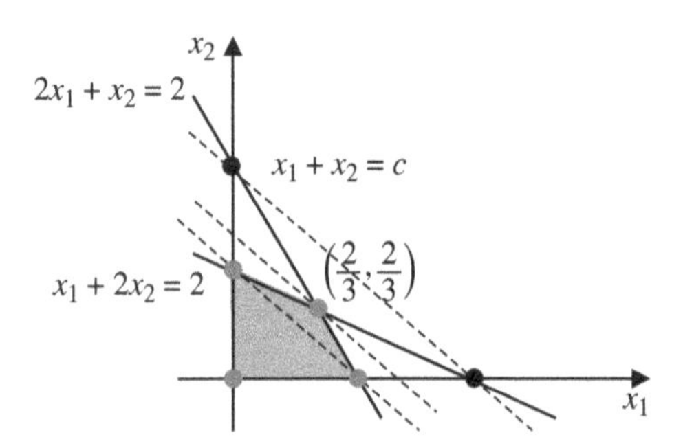

FIGURE 2.1 The geometric representation of the solution of a sample LP problem ($[LP_{2.2}]$). The shaded region indicates the feasible region defined by the constraints. Dashed lines represent the objective function level sets, gray points represent the feasible region vertices, and black points represent infeasible intersection points. The optimal solution to the problem lies at the vertex (2/3,2/3).

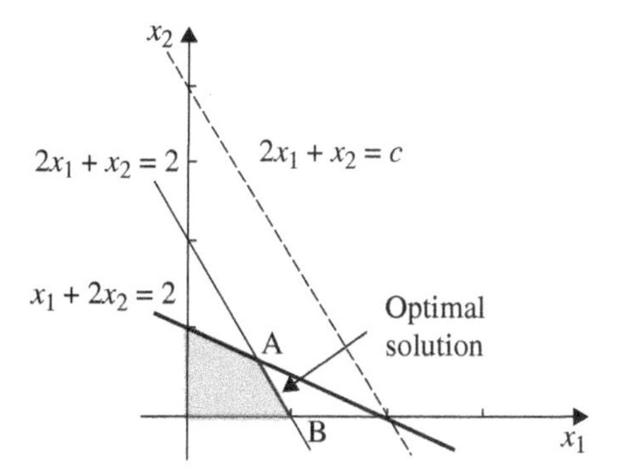

FIGURE 2.2 Multiplicity of optimal solution points for the LP problem. An infinite number of optimal solution points are defined by line segment AB.

Multiplicity of Optimal Solutions In the previous example, the LP problem had a unique optimum solution point as the objective function level set intersects with a single point (i.e., vertex) as it leaves the feasible region. An LP problem may also accept an infinite number of optimum solution points if the objective function level set overlaps with an entire facet of the feasible region polyhedron. The dimensionality of the polyhedron facet composed of all the optimum points may vary depending on the orientation of the objective function level sets and placement of the feasible region. In the above example, if the objective function level set is co-planar to one of the constraints (e.g., $2x_1 + x_2$), then there exists an infinite number of optimum points along the entire line segment AB not just points A and B (see Fig. 2.2).

Infeasible and Unbounded Solutions LP problems can become infeasible or unbounded. Infeasibility implies that the feasible space is empty while unboundedness means that one can arbitrarily increase the value of the objective function (for a maximization problem) without violating any of the constraints. If we add constraint $x_1 + x_2 \geq 2$ to the example problem, then no feasible solution exists and the problem is infeasible (Fig. 2.3a). However, if we replace the constraints of the original example problem with

$$x_1 + 2x_2 \geq 2$$
$$2x_1 + x_2 \geq 2$$

then the problem becomes unbounded (Fig. 2.3b).

This illustrative example demonstrates that an optimum solution point for an LP problem always lies on one of the vertices of the polyhedron. Each one of the vertices of a polyhedron corresponds to the intersection of M hyperplanes containing N variables. Therefore, the task of finding an optimum solution point can be reduced to testing only such vertices. These vertices are referred to as basic feasible solutions (BFSs).

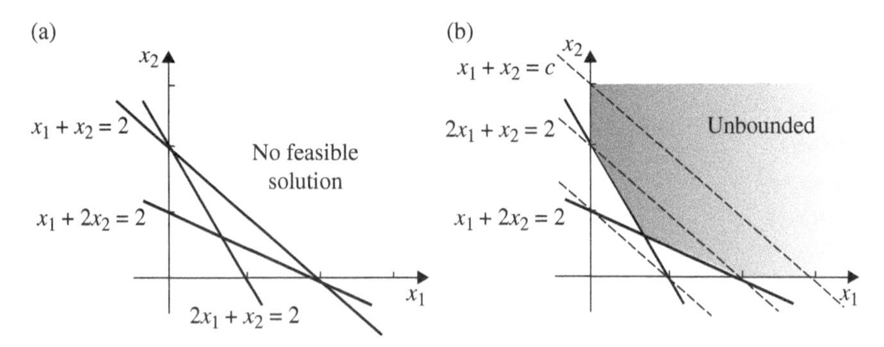

FIGURE 2.3 Examples of (a) infeasible and (b) unbounded LP problems.

2.3 BASIC FEASIBLE SOLUTIONS

Consider the system of equations $\{\boldsymbol{Ax} = \boldsymbol{b}, \boldsymbol{x} \geq \boldsymbol{0}\}$ with $\boldsymbol{A}$ being an $M \times N$ matrix of rank M and $\boldsymbol{b}$ an $M \times 1$ vector. Matrix $\boldsymbol{A}$ can be rewritten as $\boldsymbol{A} = [\boldsymbol{B} \;\; \boldsymbol{N}]$ by permuting columns where $\boldsymbol{B}$ is an $M \times M$ invertible matrix (referred to as *basic matrix* or *basis*) and $\boldsymbol{N}$ is an $M \times (N - M)$ matrix (*nonbasic matrix*).

- A point $\boldsymbol{x} = [\boldsymbol{x}_B \;\; \boldsymbol{x}_N]$ is called a *basic solution* of the system if $\boldsymbol{x}_B = \boldsymbol{B}^{-1}\boldsymbol{b}$ and $\boldsymbol{x}_N = \boldsymbol{0}$. Here, $\boldsymbol{x}_B$ and $\boldsymbol{x}_N$ are the variables multiplying columns of $\boldsymbol{A}$ in $\boldsymbol{B}$ and $\boldsymbol{N}$, respectively.
- Elements of $\boldsymbol{x}_B$ and $\boldsymbol{x}_N$ are called *basic* and *nonbasic variables*, respectively.
- If $\boldsymbol{x}_B \geq \boldsymbol{0}$, then it is called a *BFS*.
- If all elements of $\boldsymbol{x}_B$ are strictly greater than zero, it is referred to as a *non-degenerate* BFS. If at least one element is zero, then it is defined as a *degenerate* BFS.
- Each BFS corresponds to an extreme point (i.e., vertex) of the feasible region and the total number of BFSs is bounded by $\binom{N}{M} = \frac{N!}{(N-M)!M!}$ [2].
- It can be shown that every nonempty polyhedral set defined by $\{\boldsymbol{Ax} = \boldsymbol{b}, \boldsymbol{x} \geq \boldsymbol{0}\}$ has at least one extreme point (or BFS) [2].

Example 2.1
Identify all BFSs for problem $[LP_{2.2}]$ in Section 2.2.

Solution: Constraints are first converted into the standard form by introducing slack variables, x_3 and x_4:

$$x_1 + 2x_2 + x_3 = 2$$
$$2x_1 + x_2 + x_4 = 2$$
$$x_1, x_2, x_3, x_4 \geq 0$$

The matrix of coefficients $\boldsymbol{A}$ for the standard form of the problem is thus

$$\boldsymbol{A} = [\boldsymbol{a}_1, \boldsymbol{a}_2, \boldsymbol{a}_3, \boldsymbol{a}_4] = \begin{bmatrix} 1 & 2 & 1 & 0 \\ 2 & 1 & 0 & 1 \end{bmatrix}$$

Finding BFSs involves identifying the basis $\boldsymbol{B}$ which is a 2×2 matrix with $\boldsymbol{B}^{-1}\boldsymbol{b} \geq \boldsymbol{0}$. The total number of BFSs in this example is bounded by $\binom{4}{2} = 6$. The basis can be constructed by choosing any two columns of $\boldsymbol{A}$ at a time. The following are all possible combinations:

(i)

$$\boldsymbol{B} = [\boldsymbol{a}_1, \boldsymbol{a}_2] = \begin{bmatrix} 1 & 2 \\ 2 & 1 \end{bmatrix}$$

$$\boldsymbol{x}_B = \begin{bmatrix} x_1 \\ x_2 \end{bmatrix} = \boldsymbol{B}^{-1}\boldsymbol{b} = \begin{bmatrix} 2/3 \\ 2/3 \end{bmatrix} > 0, \quad \boldsymbol{x}_N = \begin{bmatrix} x_3 \\ x_4 \end{bmatrix} = \begin{bmatrix} 0 \\ 0 \end{bmatrix}$$

(ii)

$$\boldsymbol{B} = [\boldsymbol{a}_1, \boldsymbol{a}_3] = \begin{bmatrix} 1 & 1 \\ 2 & 0 \end{bmatrix}$$

$$\boldsymbol{x}_B = \begin{bmatrix} x_1 \\ x_3 \end{bmatrix} = \boldsymbol{B}^{-1}\boldsymbol{b} = \begin{bmatrix} 1 \\ 1 \end{bmatrix} > 0, \quad \boldsymbol{x}_N = \begin{bmatrix} x_2 \\ x_4 \end{bmatrix} = \begin{bmatrix} 0 \\ 0 \end{bmatrix}$$

(iii)

$$\boldsymbol{B} = [\boldsymbol{a}_1, \boldsymbol{a}_4] = \begin{bmatrix} 1 & 0 \\ 2 & 1 \end{bmatrix}$$

$$\boldsymbol{x}_B = \begin{bmatrix} x_1 \\ x_4 \end{bmatrix} = \boldsymbol{B}^{-1}\boldsymbol{b} = \begin{bmatrix} 2 \\ -2 \end{bmatrix}, \quad \boldsymbol{x}_N = \begin{bmatrix} x_2 \\ x_3 \end{bmatrix} = \begin{bmatrix} 0 \\ 0 \end{bmatrix}$$

(iv)

$$\boldsymbol{B} = [\boldsymbol{a}_2, \boldsymbol{a}_3] = \begin{bmatrix} 2 & 1 \\ 1 & 0 \end{bmatrix}$$

$$\boldsymbol{x}_B = \begin{bmatrix} x_2 \\ x_3 \end{bmatrix} = \boldsymbol{B}^{-1}\boldsymbol{b} = \begin{bmatrix} 2 \\ -2 \end{bmatrix}, \quad \boldsymbol{x}_N = \begin{bmatrix} x_1 \\ x_4 \end{bmatrix} = \begin{bmatrix} 0 \\ 0 \end{bmatrix}$$

(v)

$$\boldsymbol{B} = [\boldsymbol{a}_2, \boldsymbol{a}_4] = \begin{bmatrix} 2 & 0 \\ 1 & 1 \end{bmatrix}$$

$$\boldsymbol{x}_B = \begin{bmatrix} x_2 \\ x_4 \end{bmatrix} = \boldsymbol{B}^{-1}\boldsymbol{b} = \begin{bmatrix} 1 \\ 1 \end{bmatrix} > 0, \quad \boldsymbol{x}_N = \begin{bmatrix} x_1 \\ x_3 \end{bmatrix} = \begin{bmatrix} 0 \\ 0 \end{bmatrix}$$

(vi)

$$\boldsymbol{B} = [\boldsymbol{a}_3, \boldsymbol{a}_4] = \begin{bmatrix} 1 & 0 \\ 0 & 1 \end{bmatrix}$$

$$\boldsymbol{x}_B = \begin{bmatrix} x_3 \\ x_4 \end{bmatrix} = \boldsymbol{B}^{-1}\boldsymbol{b} = \begin{bmatrix} 2 \\ 2 \end{bmatrix} > 0, \quad \boldsymbol{x}_N = \begin{bmatrix} x_1 \\ x_2 \end{bmatrix} = \begin{bmatrix} 0 \\ 0 \end{bmatrix}$$

As we can see combinations (i), (ii), (v), and (vi) generate the following BFSs:

$$\begin{bmatrix} 2/3 \\ 2/3 \\ 0 \\ 0 \end{bmatrix}, \begin{bmatrix} 1 \\ 0 \\ 1 \\ 0 \end{bmatrix}, \begin{bmatrix} 0 \\ 1 \\ 0 \\ 1 \end{bmatrix}, \begin{bmatrix} 0 \\ 0 \\ 2 \\ 2 \end{bmatrix}$$

while combinations (iii) and (iv) result in basic solutions that are infeasible. When considering only the original variables x_1 and x_2 of the problem without the slack variables x_3 and x_4, the BFSs give rise to the following four points corresponding to the extreme points of the feasible region (see Fig. 2.1):

$$\begin{bmatrix} 2/3 \\ 2/3 \end{bmatrix}, \begin{bmatrix} 1 \\ 0 \end{bmatrix}, \begin{bmatrix} 0 \\ 1 \end{bmatrix}, \begin{bmatrix} 0 \\ 0 \end{bmatrix}$$

□

2.4 SIMPLEX METHOD

Because a solution of an LP always lies at an extreme point of the polyhedral set defining the feasible region, one may suggest that solving an LP amounts to examining all BFSs and reporting the one with the largest objective function value. The problem with this search strategy is that the number of BFSs grows exponentially with the number of variables and constraints, quickly becoming computationally intractable even for moderate values of M and N. For example, an LP problem with 100 variables and 10 constraints (in standard form) can potentially have a maximum of $\binom{100}{10} = 1.7\times10^{13}$ BFSs. Therefore, a systematic approach is required to search for the optimal solution among the set of BFSs without relying on an exhaustive enumeration. The Simplex method solves LP problems in an efficient manner by examining only a generally tiny fraction of the total number of BFSs while guaranteeing convergence to an optimum solution point.

The Simplex method works by iteratively "hopping" from one BFS to an *adjacent* one that improves the objective function. Adjacency between two BFSs implies that they differ in the composition of their bases by only a single variable. The key contribution of the Simplex method is that it provides a provably optimal termination criterion if no improving adjacent BFS can be found. This requires that the problem be expressed in standard form.

Solving for a BFS for an LP Problem in Standard Form Let $\boldsymbol{x} = \begin{bmatrix} \boldsymbol{x}_B \\ \boldsymbol{x}_N \end{bmatrix}$ and $\boldsymbol{A} = [\boldsymbol{B}\ \boldsymbol{N}]$ where $\boldsymbol{x}_B$ and $\boldsymbol{x}_N$ are $M\times1$ and $(N-M)\times1$ vectors and $\boldsymbol{B}$ and $\boldsymbol{N}$ are the $M\times M$ and $M\times(N-M)$ basic and nonbasic matrices, respectively. The system of equations $\boldsymbol{Ax} = \boldsymbol{b}$ can be rewritten as follows:

$$[\boldsymbol{B}\ \boldsymbol{N}]\begin{bmatrix} \boldsymbol{x}_B \\ \boldsymbol{x}_N \end{bmatrix} = \boldsymbol{b} \text{ or } \boldsymbol{B}\boldsymbol{x}_B + \boldsymbol{N}\boldsymbol{x}_N = \boldsymbol{b}$$

Solving for $\boldsymbol{x}_B$, we have

$$\boldsymbol{x}_B + \boldsymbol{B}^{-1}\boldsymbol{N}\boldsymbol{x}_N = \bar{\boldsymbol{b}} \quad \text{where } \bar{\boldsymbol{b}} = \boldsymbol{B}^{-1}\boldsymbol{b} \tag{2.1}$$

For example, if $x_1, x_2, \ldots, x_M$ are the basic variables (or in other words are in the basis), Equation 2.1 can be written in expanded form as follows:

$$\begin{array}{ccccccccccccc} x_1 & + & \cdots & + & 0 & + & \bar{a}_{1,M+1}x_{M+1} & + & \cdots & + & \bar{a}_{1N}x_N & = & \bar{b}_1 \\ \vdots & & \ddots & & \vdots & & \vdots & & & & \vdots & & \vdots \\ 0 & + & \cdots & + & x_M & + & \bar{a}_{M,M+1}x_{M+1} & + & \cdots & + & \bar{a}_{MN}x_N & = & \bar{b}_M \end{array}$$

The objective function can be rewritten as follows by replacing $\boldsymbol{x}_B$ using Equation 2.1:

$$z = \begin{bmatrix} \boldsymbol{c}_B & \boldsymbol{c}_N \end{bmatrix} \begin{bmatrix} \boldsymbol{x}_B \\ \boldsymbol{x}_N \end{bmatrix} = \begin{bmatrix} \boldsymbol{c}_B & \boldsymbol{c}_N \end{bmatrix} \begin{bmatrix} \bar{\boldsymbol{b}} - \boldsymbol{B}^{-1}\boldsymbol{N}\boldsymbol{x}_N \\ \boldsymbol{x}_N \end{bmatrix} = \boldsymbol{c}_B\bar{\boldsymbol{b}} - \left(\boldsymbol{c}_B\boldsymbol{B}^{-1}\boldsymbol{N} - \boldsymbol{c}_N\right)\boldsymbol{x}_N \tag{2.2}$$

Upon setting the value of all nonbasic variables $\boldsymbol{x}_N$ to zero, it follows from Equations 2.1 and 2.2 that

$$\begin{cases} \boldsymbol{x}_B = \bar{\boldsymbol{b}} \\ \boldsymbol{x}_N = \boldsymbol{0} \\ z_{\text{BFS}} = \boldsymbol{c}_B\bar{\boldsymbol{b}} \end{cases} \quad \text{or} \quad \begin{cases} x_j = \bar{b}_j \quad \forall j \in J^B \\ x_j = 0 \quad \forall j \in J^N \\ z_{\text{BFS}} = \sum_{j \in J^B} c_j \bar{b}_j \end{cases} \tag{2.3}$$

where J^B is the set of variables in the basis and J^N is the set of variables not in the basis, respectively. The key advantage of using the standard form is that the feasibility of this basic solution can be quickly ascertained by simply checking whether $\bar{b}_j \geq 0, \forall j \in J^B$. This would not have been possible if inequalities and/or negative variables were left in the problem formulation.

Moving to an Adjacent BFS that Improves the Objective Function The task of finding an improving adjacent BFS can be split into two separate decisions (1) which nonbasic variable to enter the basic set by assuming a nonzero value and simultaneously (2) which basic solution to leave the basis and thus assume a value of zero. We can first explore decision (1) by assessing the sensitivity of the objective function to the increase of a nonbasic variable value from zero to one. For ease of presentation we temporarily assume that the first M variables are basic while the remaining $N-M$ are nonbasic. Let nonbasic variable x_s, where $s \in \{M+1, \ldots, N\}$, be a candidate for inclusion into the basis. We are interested to quantify the change in the objective function value per unit increase in the value of x_s. If we increase the value of nonbasic variable x_s from zero to one, then the new Standard representation of the LP becomes

$$\begin{array}{ccccccccc} x_1 & + & \cdots & + & 0 & + & \bar{a}_{1s}(1) & = & \bar{b}_1 \\ \vdots & & \ddots & & \vdots & & \vdots & & \vdots \\ 0 & + & \cdots & + & x_M & + & \bar{a}_{Ms}(1) & = & \bar{b}_M \end{array}$$

The new solution to the LP is thus

$$x_j = \bar{b}_j - \bar{a}_{js} \quad j = 1,2,\ldots,M$$
$$x_s = 1$$
$$x_j = 0 \qquad j = M+1, M+2,\ldots,N, j \neq s$$

The new objective function value becomes

$$z = \sum_{j=1}^{M} c_j \left(\bar{b}_j - \bar{a}_{js}\right) + c_s \tag{2.4}$$

Therefore, the change in the objective function value per unit change of x_s is

$$\Delta z = \left(\sum_{j=1}^{M} c_j \left(\bar{b}_j - \bar{a}_{js}\right) + c_s\right) - \sum_{j=1}^{M} c_j \bar{b}_j = c_s - \sum_{j=1}^{M} c_j \bar{a}_{js} = \bar{c}_s \tag{2.5}$$

Term $\bar{c}_s$ is called the *relative payoff* or *sensitivity* of the nonbasic variable s and represents the net change in the objective function value for a unit increase in the value of the nonbasic variable j (i.e., $x_s = 1$). Observe that if $\bar{c}_s > 0$, we can improve the objective function by increasing the value of a nonbasic variable from its current value of zero. The maximum value that the nonbasic variable ultimately assumes is determined by the nonnegativity requirement for all variables. Because BFSs correspond to vertices that are the intersection of M hyperplanes in an M-dimensional space, entering a nonbasic variable into the basis implies that one current basic variable must leave the basis (i.e., become nonbasic by setting it equal to zero). Therefore, moving from a BFS to an adjacent one involves swapping a basic with a nonbasic variable.

Selecting the Nonbasic Variable That Enters the Basis The nonbasic variable x_s that is chosen to enter the basis is the one with the maximum relative payoff. This yields the maximum increase in the value of the objective function per unit of nonbasic variable value change:

$$\bar{c}_s = \max_{\substack{j \in J^N \\ \bar{c}_j > 0}} \bar{c}_j \quad \text{or} \quad s = \arg\max_{\substack{j \in J^N \\ \bar{c}_j > 0}} \bar{c}_j \tag{2.6}$$

Here J^N denotes the set of variables that are not in the basis.

Selecting the Basic Variable that Leaves the Basis Upon having the nonbasic variable x_s enter the basis, the new solution for the basic variables x_j has to be modified according to Equation 2.1:

$$\boldsymbol{x}_B = \bar{\boldsymbol{b}} - \boldsymbol{B}^{-1}\boldsymbol{a}_s x_s = \bar{\boldsymbol{b}} - \bar{\boldsymbol{a}}_s x_s \quad \text{or} \quad x_j = \bar{b}_j - \bar{a}_{js} x_s, \quad \forall j \in J^B \tag{2.7}$$

where $\bar{\boldsymbol{b}} = \boldsymbol{B}^{-1}\boldsymbol{b}$ and $\bar{\boldsymbol{a}}_s = \boldsymbol{B}^{-1}\boldsymbol{a}_s$ with $\boldsymbol{a}_s$ being column of matrix $\boldsymbol{N}$ (or $\boldsymbol{A}$) corresponding to x_s. Furthermore, J^B denotes the set of variables that are in the basis. To maintain feasibility of the modified basic solution we need to have

$$x_j = \bar{b}_j - \bar{a}_{js}x_s \geq 0, \quad \forall j \in J^B$$

Because $\bar{b}_j \geq 0$, $\forall j \in J^B$ as the original basic solution was feasible, the maximum possible value of x_s depends on the sign and magnitude of $\bar{a}_{js}$:

- If for all $j \in J^B$, $\bar{a}_{js} \leq 0$, then x_s can increase arbitrarily without making any other basic variables negative. This means that the original LP problem is unbounded.
- If for some $j \in J^B$, $\bar{a}_{js} > 0$, then x_j decreases in value as x_s increases. In this case, the first x_j to reach zero as we increase the value of x_s would be the variable that needs to leave the basis. It is easy to see that the basic variable x_r to leave the basis is the one with a subscript r that minimizes the ratio $\frac{\bar{b}_j}{\bar{a}_{js}}$. Therefore,

$$\frac{\bar{b}_r}{\bar{a}_{rs}} = \min_{\substack{j \in J^B \\ \bar{a}_{js} \geq 0}} \left(\frac{\bar{b}_j}{\bar{a}_{js}} \right) = x_s \quad \text{or} \quad r = \arg\min_{\substack{j \in J^B \\ \bar{a}_{js} \geq 0}} \left(\frac{\bar{b}_j}{\bar{a}_{js}} \right) \tag{2.8}$$

  This is referred to as the *minimum ratio rule* for determining the basic variable x_r that must leave the basis. It also determines the maximum value that the non-basic variable entering the basis (x_s) can assume.

By identifying the entering and leaving variables, we have constructed a new BFS. The values of the variables in the new basis are as follows:

- Nonbasic variable $x_s = 0$ enters the basis with the new value of $x_s = (\bar{b}_r / \bar{a}_{rs})$
- Basic variable $x_r = \bar{b}_r$ becomes a nonbasic variable $x_r = 0$
- The values of all other basic variables (other than x_r) are updated as $x_j = \bar{b}_j - \bar{a}_{js}\left(\frac{\bar{b}_r}{\bar{a}_{rs}}\right), \quad \forall j \in J^B, j \neq r$
- The values of all nonbasic variables other than x_s remain at zero $x_j = 0, \quad \forall j \in J^N, j \neq s$
- The change in the value of the objective function is equal to the relative profit of the nonbasic variable that enters the basis times the value it assumes based on the minimum ratio rule:

$$\Delta z = \bar{c}_s \left(\frac{\bar{b}_r}{\bar{a}_{rs}} \right)$$

  Note that because $\bar{a}_{rs}, \bar{c}_s > 0$, if $\bar{b}_r$ is strictly greater than zero ($\bar{b}_r > 0$), then $\Delta z > 0$ (i.e., the objective function strictly improves).

This procedure of moving to another adjacent BFS is repeated until the optimality criterion of the Simplex method is satisfied (see the following section) or the problem is found to be unbounded.

Optimality Conditions Based on Equation 2.5, a BFS is optimal if the relative profits of all nonbasic variables are non-positive (i.e., $\bar{c}_j \le 0,\ \forall j \in J^N$). It is a unique optimal solution point (vertex) if the relative profits of all nonbasic variables are strictly negative. If the relative benefit for some nonbasic variables is equal to zero, then they can enter the basis without changing the value of the objective function implying that there exist alternate optima. The number of nonbasic variables with zero relative payoff values defines the number of alternate optimum vertices and the dimensionality of the facet of the polyhedron composed of alternate optimum points.

Degeneracy and Cycling A BFS with one or more basic variables equal to zero is called a *degenerate* BFS. This can arise from a constraint with a right hand-side equal to zero ($b_j = 0$) or can be introduced during row operations ($\bar{b}_j = 0$). In this case, the value of the next nonbasic variable entering the basis, x_s will also be zero (*degenerate solution*) due to the minimum ratio rule (see Eq. 2.8). Cycling between several degenerate solutions could result in costly iterations with no improvement in the objective function. Modern Simplex implementations (e.g., CPLEX [6] and Gurobi [7]) have devised ways to circumvent this problem.

Summary of the Simplex Method The basic steps of the Simplex method are summarized as follows:

Step 0. Pick an initial BFS. There are several methods for finding an initial BFS described in LP textbooks [2, 3].

Step 1. Compute the value of all basic variables $\boldsymbol{x_B} = \bar{\boldsymbol{b}}$, where $\bar{\boldsymbol{b}} = \boldsymbol{B}^{-1}\boldsymbol{b}$ and set the value of all nonbasic variables to zero $\boldsymbol{x_N} = 0$.

Step 2. Calculate the relative profit for all nonbasic variables (using Eq. 2.5):

$$\bar{c}_j = c_j - \boldsymbol{c_B}\boldsymbol{B}^{-1}\boldsymbol{a}_j,\ \forall j \in J^N$$

Step 3. If all $\bar{c}_j \le 0$, then the current BFS is optimal. Otherwise, select the nonbasic variable s with the largest relative profit $\bar{c}_j$ to enter the basis (see Eq. 2.6):

$$\bar{c}_s = \max_{\substack{j \in J^N \\ \bar{c}_j > 0}} \bar{c}_j$$

Step 4. If for all $j \in J^B$ $\bar{a}_{js} \le 0$, then the problem is unbounded. Otherwise, find the index of the basic variable x_r which should be removed from the basis using the minimum ratio rule (see Eq. 2.8):

$$r = \arg\min_{\substack{j \in J^B \\ \bar{a}_{js} \ge 0}} \left(\frac{\bar{b}_j}{\bar{a}_{js}} \right)$$

Step 5. Set the value of the entering variable in the basis to $x_s = \left(\bar{b}_r / \bar{a}_{rs}\right)$.

Step 6. Update the values of the basic variables using Equation 2.7 and matrices $\boldsymbol{B}$ and $\boldsymbol{N}$ by swapping columns $\boldsymbol{a}_r$ and $\boldsymbol{a}_s$. Return to Step 1.

The Simplex tableau is one of the most popular ways of systematizing the implementation of the Simplex method. It is simply the tabular representation of the coefficients of Equations 2.1 and 2.2 (after some rearrangements). The tableau is updated at each iteration of the Simplex method (Step 7) by performing pivoting and elementary row operations [2, 3]. In addition to the Simplex method, modern LP solvers such as CPLEX [6] and Gurobi [7] provide additional algorithms for solving LP problems including the primal-dual and interior point or barrier methods as well as network flow algorithms [2, 3, 8, 9]. Given the diversity of the LP problems and their applications, no single algorithm dominates the others in all applications and their efficiency must be assessed on a case-by-case basis.

2.5 DUALITY IN LINEAR PROGRAMMING

Every LP problem with N variables and M constraints is associated with another LP problem called the *dual* which contains M variables and N constraints. The designation of primal and dual is arbitrary as the dual of the dual problem is the primal problem (see Example 2.2). Therefore, they are more properly referred to as the primal-dual pair. The term primal generally denotes the "original" LP problem. The primal-dual pair shares a number of important properties, which will be elaborated in the subsequent sections.

2.5.1 Formulation of the Dual Problem

Consider the following LP in canonical form:

$$\begin{aligned}
&\text{minimize}\, z = \boldsymbol{c}^{\mathrm{T}}\boldsymbol{x} \\
&\text{subject to} \\
&\boldsymbol{A}\boldsymbol{x} \geq \boldsymbol{b} \\
&\boldsymbol{x} \geq 0
\end{aligned}$$

where $\boldsymbol{x}, \boldsymbol{c}$ and $\boldsymbol{b}$ are $N \times 1$, $N \times 1$ and $M \times 1$ vectors, respectively and $\boldsymbol{A}$ is an $M \times N$ matrix. The dual linear problem associated with the primal is defined as follows:

$$\begin{aligned}
&\text{maximize}\, z = \boldsymbol{b}^{\mathrm{T}}\boldsymbol{y} \\
&\text{subject to} \\
&\boldsymbol{A}^{\mathrm{T}}\boldsymbol{y} \leq \boldsymbol{c} \\
&\boldsymbol{y} \geq 0
\end{aligned}$$

where $\boldsymbol{y}$ is an $M \times 1$ vector. Through inspection of the primal and dual problems, several key connections can be gleaned (see Fig. 2.4):

- While the primal involves a minimization, the dual involves a maximization.
- The primal minimization problem contains "≥" constraints while the dual maximization problem contains "≤" constraints (i.e., the direction of inequalities is reversed).
- Each constraint in the dual problem corresponds to a variable in the primal problem and vice-versa. This means that the primal problem has N variables and M constraints whereas the dual problem has M variables and N constraints.
- The matrix of the coefficients for the dual is the transpose of the matrix of coefficients for the primal and vice-versa.
- The cost vector $\boldsymbol{c}$ in the primal serves as the right-hand side vector in the dual and vice-versa.

The dual pair for any LP problem, including a mixture of inequality and equality constraints as well as real and positive variables, can be derived by following the links shown in Figure 2.5. If the original LP problem involves a maximization operator, then one can start from the right column of Figure 2.5 to create the corresponding dual problem using the entries in the left-most column. Inequalities are restricted to "≥" for minimization problems and "≤" for maximization problems by reversing directionality (i.e., multiply by −1) whenever needed.

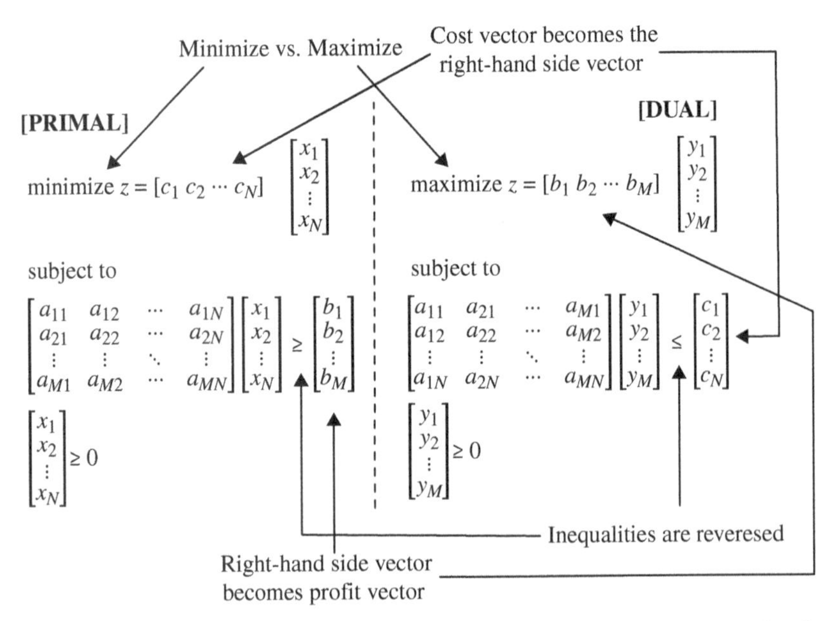

FIGURE 2.4 Connections between a primal problem and its dual for an LP in canonical form.

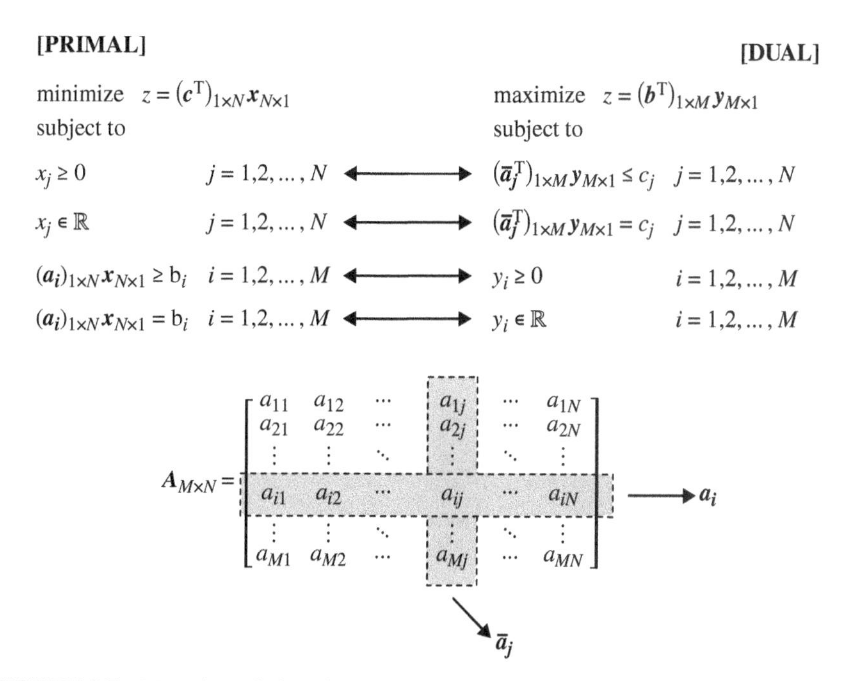

FIGURE 2.5 Matching relations between primal and dual variables and constraints for an LP in the general form.

Example 2.2

Show that the dual of the dual is the primal.

Solution: Let the LP minimization problem in canonical form be the primal problem and consider its dual as noted earlier:

$$\text{maximize } z = \boldsymbol{b}^{\mathrm{T}} \boldsymbol{y}$$
$$\text{subject to}$$
$$\boldsymbol{A}^{\mathrm{T}} \boldsymbol{y} \leq \boldsymbol{c}$$
$$\boldsymbol{y} \geq 0$$

Let $\boldsymbol{w}$ denote the variables vector for the dual of this problem. According to the rules of Figures 2.4 and 2.5: (i) the right-hand side vector of this problem serves as the vector of the cost coefficients in its dual and vice-versa, (ii) the coefficient matrix is transposed, and (iii) the dual variables $\boldsymbol{w}$ are declared as nonnegative as all constraints in the original problem are "≤" inequalities. Therefore, the dual of the dual can thus be written as follows:

$$\text{minimize } z = \boldsymbol{c}^{T} \boldsymbol{w}$$
$$\text{subject to}$$
$$\boldsymbol{A}\boldsymbol{w} \geq \boldsymbol{b}$$
$$\boldsymbol{w} \geq \boldsymbol{0}$$

which is identical to the original primal problem in canonical form. □

2.5.2 Primal-Dual Relations

A number of important properties underpin a primal-dual pair. In the following, we briefly review these connections skipping most mathematical proofs [2, 3].

Weak Duality Theorem The weak duality theorem establishes bounding relations between feasible solutions for the primal and dual problems. Consider an LP minimization problem in canonical form and let $\boldsymbol{x}_0$ and $\boldsymbol{y}_0$ be feasible solutions to the primal and dual problems, respectively. The objective function of the primal at $\boldsymbol{x}_0$ is greater than or equal to the objective function of the dual at $\boldsymbol{y}_0 : \boldsymbol{c}^{\mathrm{T}}\boldsymbol{x}_0 \geq \boldsymbol{b}^{\mathrm{T}}\boldsymbol{y}_0$.

Proof

From the feasibility of $\boldsymbol{x}_0$ and $\boldsymbol{y}_0$ for the primal and dual, we have $\boldsymbol{A}\boldsymbol{x}_0 \geq \boldsymbol{b}, \boldsymbol{x}_0 \geq 0$ and $\boldsymbol{A}^{\mathrm{T}}\boldsymbol{y}_0 \leq \boldsymbol{c}, \boldsymbol{y}_0 \geq 0$. By multiplying both sides of $\boldsymbol{A}\boldsymbol{x}_0 \geq \boldsymbol{b}$ with $\boldsymbol{y}_0$, and multiplying both sides of $\boldsymbol{A}^{\mathrm{T}}\boldsymbol{y}_0 \leq \boldsymbol{c}$ with $\boldsymbol{x}_0$ we obtain, respectively:

$$\left(\boldsymbol{A}\boldsymbol{x}_0\right)^{\mathrm{T}}\boldsymbol{y}_0 \geq \boldsymbol{b}^{\mathrm{T}}\boldsymbol{y}_0 \quad \text{and} \quad \left(\boldsymbol{A}^{\mathrm{T}}\boldsymbol{y}_0\right)^{\mathrm{T}}\boldsymbol{x}_0 \leq \boldsymbol{c}^{\mathrm{T}}\boldsymbol{x}_0$$

However, $\left(\boldsymbol{A}^{\mathrm{T}}\boldsymbol{y}_0\right)^{\mathrm{T}}\boldsymbol{x}_0 = \left(\boldsymbol{y}_0^{\mathrm{T}}\boldsymbol{A}\right)\boldsymbol{x}_0 = \boldsymbol{y}_0^{\mathrm{T}}\left(\boldsymbol{A}\boldsymbol{x}_0\right) = \left(\boldsymbol{A}\boldsymbol{x}_0\right)^{\mathrm{T}}\boldsymbol{y}_0$ which implies $\boldsymbol{c}^{\mathrm{T}}\boldsymbol{x}_0 \geq \boldsymbol{b}^{\mathrm{T}}\boldsymbol{y}_0$. □

Note that since the choice of primal and dual is arbitrary, the weak duality theorem essentially implies that the objective function for any feasible solution to the minimization problem is always greater than or equal to the objective function for a feasible solution to the maximization problem. In other words, the objective function to the minimization problem provides an upper bound to that of the maximization problem and vice-versa. The following results can be discerned from the weak duality theorem:

<u>Result 1:</u> If $\boldsymbol{x}_0$ and $\boldsymbol{y}_0$ are finite feasible solutions of the primal and dual problems, respectively, and $\boldsymbol{c}^{\mathrm{T}}\boldsymbol{x}_0 = \boldsymbol{b}^{\mathrm{T}}\boldsymbol{y}_0$, then $\boldsymbol{x}_0$ is the optimal solution to the primal (minimization problem) and $\boldsymbol{y}_0$ is the optimal solution to the dual (maximization problem).

<u>Result 2:</u> If either the primal or the dual is unbounded, then the other problem is infeasible. Note that the reverse is not necessarily true because if one problem is infeasible it is possible for the other to be infeasible too (see the subsequent property).

<u>Result 3:</u> If either the primal or dual is infeasible, the other is either infeasible or unbounded. An example of an infeasible pair is as follows:

minimize $z = -x$	maximize $z = y$
subject to	subject to
$0x \geq 1$	$0y \leq -1$
$x \geq 0$	$y \geq 0$

Strong Duality Theorem Building upon the weak duality theorem that establishes only a bounding relation between the primal and the dual objective function values, the *strong duality theorem* provides matching relations between the respective optima (if they exist). Specifically, if the primal or dual problem has a finite optimal solution so does the other one and the optimal objective function values are equal to one another.

An alternative way of imposing optimality conditions for an LP problem without equating the objective functions of the primal and dual relies on the Karush–Kuhn–Tucker (KKT) conditions (see below).

2.5.3 The Karush-Kuhn-Tucker (KKT) Optimality Conditions

Consider the following primal-dual pair:

$$\begin{array}{ll} \text{minimize}\, z = \boldsymbol{c}^{\mathrm{T}}\boldsymbol{x} \quad [P] & \quad \text{maximize}\, z = \boldsymbol{b}^{\mathrm{T}}\boldsymbol{y} \quad [D] \\ \text{subject to} & \quad \text{subject to} \\ \boldsymbol{A}\boldsymbol{x} \geq \boldsymbol{b} & \quad \boldsymbol{A}^{\mathrm{T}}\boldsymbol{y} \leq \boldsymbol{c} \\ \boldsymbol{x} \geq 0 & \quad \boldsymbol{y} \geq 0 \end{array}$$

The following conditions are referred to as the *complementary slackness* (CS) conditions:

$$\begin{cases} (\boldsymbol{A}\boldsymbol{x}-\boldsymbol{b})^{\mathrm{T}}\boldsymbol{y}=0 \\ (\boldsymbol{c}-\boldsymbol{A}^{\mathrm{T}}\boldsymbol{y})^{\mathrm{T}}\boldsymbol{x}=0 \end{cases} \quad \text{or} \quad \begin{cases} (\boldsymbol{a}_i\boldsymbol{x}-b_i)y_i=0, & i \in 1,2,\ldots,M \\ (c_j-\overline{\boldsymbol{a}}_j^{\mathrm{T}}\boldsymbol{y})x_j=0, & j \in 1,2,\ldots,N \end{cases} \tag{2.9}$$

where $\boldsymbol{a}_i$ and $\overline{\boldsymbol{a}}_j$ represent a row i and column j of matrix $\boldsymbol{A}$, respectively. The CS conditions, in essence ensure that if a constraint is non-binding (i.e., holds as inequality: $\boldsymbol{a}_i\boldsymbol{x} > b_i$ or $\overline{\boldsymbol{a}}_j^{\mathrm{T}}\boldsymbol{y} < c_j$), then the corresponding dual variable must be equal to zero. Therefore, if the dual variable associated with a constraint is positive, then the constraint must be binding (i.e., holds as equality: $\boldsymbol{a}_i\boldsymbol{x} = b_i$ or $\overline{\boldsymbol{a}}_j^{\mathrm{T}}\boldsymbol{y} = c_j$).

Theorem 2.1

Let $\boldsymbol{x}^*$ and $\boldsymbol{y}^*$ be feasible solutions to $[P]$ and $[D]$, respectively. Point $\boldsymbol{x}^*$ is an optimal solution to $[P]$ and $\boldsymbol{y}^*$ is an optimal solution to $[D]$, if and only if the *CS conditions* hold.

Proof: Because points $\boldsymbol{x}^*$ and $\boldsymbol{y}^*$ are feasible solutions to $[P]$ and $[D]$, respectively we have

$$\begin{array}{lll} \boldsymbol{A}\boldsymbol{x}^* \geq \boldsymbol{b} & \text{and} & \boldsymbol{x}^* \geq 0 \\ \boldsymbol{A}^{\mathrm{T}}\boldsymbol{y}^* \leq \boldsymbol{c} & \text{and} & \boldsymbol{y}^* \geq 0 \end{array}$$

By multiplying $\boldsymbol{A}\boldsymbol{x}^* \geq \boldsymbol{b}$ by $\boldsymbol{y}^* \geq 0$ and $\boldsymbol{A}^{\mathrm{T}}\boldsymbol{y}^* \leq \boldsymbol{c}$ by $\boldsymbol{x}^* \geq 0$, we obtain $(\boldsymbol{A}\boldsymbol{x}^*)^{\mathrm{T}}\boldsymbol{y}^* \geq \boldsymbol{b}^{\mathrm{T}}\boldsymbol{y}^*$ and $(\boldsymbol{A}^{\mathrm{T}}\boldsymbol{y}^*)^{\mathrm{T}}\boldsymbol{x}^* \leq \boldsymbol{c}^{\mathrm{T}}\boldsymbol{x}^*$, respectively. Using the CS condition definitions (Eq. 2.9), $(\boldsymbol{A}\boldsymbol{x}^*)^{\mathrm{T}}\boldsymbol{y}^*$ and $(\boldsymbol{A}^{\mathrm{T}}\boldsymbol{y}^*)^{\mathrm{T}}\boldsymbol{x}^*$ in the first and second inequality can be replaced with $\boldsymbol{b}^{\mathrm{T}}\boldsymbol{y}^*$ and $\boldsymbol{c}^{\mathrm{T}}\boldsymbol{x}^*$, respectively. This yields the following two matching inequalities:

$$\boldsymbol{c}^{\mathrm{T}}\boldsymbol{x}^* \geq \boldsymbol{b}^{\mathrm{T}}\boldsymbol{y}^*$$
$$\boldsymbol{b}^{\mathrm{T}}\boldsymbol{y}^* \leq \boldsymbol{c}^{\mathrm{T}}\boldsymbol{x}^*$$

which imply equality of $\boldsymbol{c}^{\mathrm{T}}\boldsymbol{x}^*$ with $\boldsymbol{b}^{\mathrm{T}}\boldsymbol{y}^*$. Therefore, the objective function values of the primal and dual problems are equal if and only if the CS conditions hold (the proof in the reverse direction is straightforward). It then follows from Result 1 of the weak duality theorem mentioned earlier that $\boldsymbol{x}^*$ is optimal to $[P]$ and $\boldsymbol{y}^*$ is optimal to $[D]$ completing the proof. □

Considering the earlier theorem, the optimality conditions for a general LP problem $[P]$ consist of three constraint sets:

Primal feasibility:

$$\boldsymbol{A}\boldsymbol{x} \geq \boldsymbol{b}, \quad \boldsymbol{x} \geq 0$$

Dual feasibility:

$$\boldsymbol{A}^{\mathrm{T}}\boldsymbol{y} \leq \boldsymbol{c}, \quad \boldsymbol{y} \geq 0$$

Complementary slackness:

$$(\boldsymbol{A}\boldsymbol{x} - \boldsymbol{b})^{\mathrm{T}}\boldsymbol{y} = 0, \; (\boldsymbol{c} - \boldsymbol{A}^{\mathrm{T}}\boldsymbol{y})^{\mathrm{T}}\boldsymbol{x} = 0$$

These three conditions together are referred to as the *KKT optimality conditions*. A point $\boldsymbol{x}^*$ is an optimal solution to $[P]$, if and only if the KKT conditions hold. It is important to note that imposing the CS condition gives rise to the emergence of bilinear (nonconvex) terms. The CS condition can be alternatively recast as a set of linear constraints but only at the cost of introducing binary variables. The KKT and duality conditions are used extensively to convert bilevel optimization problems that commonly arise in metabolic engineering applications to single-level optimization problems with explicit constraints (see Chapter 8). As we will see in Chapter 9, the KKT optimality conditions can be extended for nonlinear programming problems.

2.5.4 Economic Interpretation of the Dual Variables

Consider again the canonical minimization problem [P] and its dual [D]:

minimize $z = \boldsymbol{c}^{\mathrm{T}}\boldsymbol{x}$ $[P]$	maximize $z = \boldsymbol{b}^{\mathrm{T}}\boldsymbol{y}$ $[D]$
subject to	subject to
$\boldsymbol{A}\boldsymbol{x} \geq \boldsymbol{b}$	$\boldsymbol{A}^{\mathrm{T}}\boldsymbol{y} \leq \boldsymbol{c}$
$\boldsymbol{x} \geq 0$	$\boldsymbol{y} \geq 0$

Let $\boldsymbol{x}^*$ and $\boldsymbol{y}^*$ represent the optimal solutions to the primal $[P]$ and dual $[D]$ problems, respectively. If $\boldsymbol{B}$ represents the basis in the optimal solution for the primal, we have $\boldsymbol{x}^*_B = \boldsymbol{B}^{-1}\boldsymbol{b}$. From the duality theorems we have

$$z^* = \begin{bmatrix} \boldsymbol{B}^{-1}\boldsymbol{b} \\ 0 \end{bmatrix} \begin{bmatrix} \boldsymbol{c}_B & \boldsymbol{c}_N \end{bmatrix} = \boldsymbol{b}^{\mathrm{T}}\boldsymbol{y}^*$$

or,

$$z^* = \boldsymbol{c}_B\boldsymbol{B}^{-1}\boldsymbol{b} = \boldsymbol{b}^{\mathrm{T}}\boldsymbol{y}^*$$

Therefore,

$$\frac{\partial z^*}{\partial \boldsymbol{b}} = \boldsymbol{c}_B\boldsymbol{B}^{-1} = \boldsymbol{y}^*$$

or

$$\frac{\partial z^*}{\partial b_i} = y_i^*, \quad i = 1,2,\ldots,M$$

This implies that the optimal value of each dual variable y_i^* is equal to the rate of change in the objective function value with respect to a unit increase in the right-hand side value of the ith constraint. In economic terms, $\boldsymbol{y}^*$ can be interpreted as the vector of *shadow prices* for the demand vector $\boldsymbol{b}$ that is the incremental cost of producing one more unit of the product i (i.e., an increase in demand for product i). Note that since $y_i^* \geq 0$, we have $(\partial z^* / \partial b_i) \geq 0$.

Example 2.3

Generate the dual of the following LP problem and verify that the objective functions of the primal and dual are equal. In addition, verify that the KKT conditions hold at the optimal solution.

$$\begin{aligned}
&\text{maximize } 50x_1 + 100x_2 \\
&\text{subject to} \\
&2x_1 + x_2 \leq 1250 \\
&2x_1 + 5x_2 \leq 1000 \\
&2x_1 + 3x_2 \leq 900 \\
&x_2 \leq 150 \\
&x_1, x_2 \geq 0
\end{aligned}$$

Solution: Let y_1, y_2, y_3 and y_4 denote the dual variables associated with $2x_1 + x_2 \leq 1250$, $2x_1 + 5x_2 \leq 1000$, $2x_1 + 3x_2 \leq 900$ and $x_2 \leq 150$, respectively. By making use of the rules from Figure 2.5, the dual can be constructed in a straightforward manner.

$$\begin{aligned}&\text{minimize } z = 1250y_1 + 1000y_2 + 900y_3 + 150y_4\\&\text{subject to}\\&2y_1 + 2y_2 + 2y_3 + 0y_4 \geq 50 \rightarrow x_1\\&y_1 + 5y_2 + 3y_3 + y_4 \geq 100 \rightarrow x_2\\&y_1, y_2, y_3, y_4 \geq 0\end{aligned}$$

The dual problem involves four variables and two constraints corresponding to the four constraints and two variables in the primal problem, respectively. In contrast to the primal, it involves the minimization of an objective function whose coefficients match the entries of the right-hand side vector in the primal problem. The two constraints of the dual problem are of the form "≥" because $x_1, x_2 \geq 0$. Furthermore, the coefficients of the first and second constraints in the dual are the coefficients of variables x_1 and x_2 respectively, in the four constraints of the primal problem. Dual variables y_1, y_2, y_3 and y_4 are all nonnegative as the corresponding constraints in the primal problem are all "≤" inequalities. The reader can verify that the optimal solution and the optimal objective function values to the primal and dual solutions are as follows:

$$x_1^* = 375.0, \quad x_2^* = 50.0, \quad z_{\text{primal}}^* = 23750.0$$

$$y_1^* = 0, \quad y_2^* = 12.5, \quad y_3^* = 12.5, \quad y_4^* = 0, \quad z_{\text{dual}}^* = 23750.0$$

In accordance with the duality theorems, we have $z_{\text{primal}}^* = z_{\text{dual}}^*$. Next we check whether the KKT conditions hold at the optimal solution. Primal and dual feasibility obviously hold at their respective optimal solution points. Testing for the CS conditions requires checking which constraints of the primal problem are binding at the optimal solution and comparing with the corresponding dual variables values:

Constraint $2x_1 + x_2 \leq 1250$:

$$2x_1^* + x_2^* = 800 < 1250, \quad y_1^* = 0$$

Constraint $2x_1 + 5x_2 \leq 1000$:

$$2x_1^* + 5x_2^* = 1000, \quad y_2^* = 12.5 > 0$$

Constraint $2x_1 + 3x_2 \leq 900$:

$$2x_1^* + 3x_2^* = 900, \quad y_3^* = 12.5 > 0$$

Constraint $x_2 \leq 150$:

$$x_2^* = 50 < 150, \quad y_4^* = 0$$

Therefore, the CS conditions hold. □

Example 2.4

Consider the following LP problem (primal) which corresponds to the mathematical structure of a flux balance analysis (FBA) problem in metabolic networks (see Chapter 3):

$$\begin{aligned} &\text{maximize } z = x_b \quad \text{for some } b \in J \\ &\text{subject to} \\ &\sum_{j \in J} S_{ij} x_j = 0, \qquad \forall i \in I \\ &LB_j \le x_j \le UB_j, \quad \forall j \in J \\ &x_j \in \mathbb{R}, \qquad \forall j \in J \end{aligned}$$

where

$J = \{1,2,\ldots,N\}$: Set of variables.
$I = \{1,2,\ldots,M\}$: Set of equality constraints.
x_j: Decision variables.
S_{ij}: Coefficient of contribution of variable x_j in equality constraint i.
LB_j, UB_j: The lower and upper variable bounds, respectively.

Solution: Observe that the primal problem has N variables and $M+2N$ constraints (M equalities and $2N$ inequalities). Hence, the dual problem will have $M+2N$ variables and N constraints. Dual variables corresponding to primal constraints are defined as follows (see Table 2.1):

$$\begin{aligned} &\sum_{j \in J} S_{ij} x_j = 0, \quad \forall i \in I \quad \rightarrow \quad \lambda_i \in \mathbb{R} \\ &-x_j \le -LB_j, \quad \forall j \in J \quad \rightarrow \quad \mu_j^{\text{LB}} \ge 0 \\ &x_j \le UB_j, \quad \forall j \in J \quad \rightarrow \quad \mu_j^{\text{UB}} \ge 0 \end{aligned}$$

In addition, associated with each variable in the primal is a constraint in the dual. The coefficient of each variable x_j in any primal constraint serves as the coefficient of the respective dual variables in the dual constraint associated with x_j (see Table 2.1A). Furthermore, x_b is the only variable that appears in the primal problem's objective function. Therefore, the right-hand side of the dual constraints associated with all $x_j, j \ne b$ is zero, whereas that for x_b is equal to one (see Table 2.1B). Finally, note that since the right-hand side of the primal constraints $\sum_{j \in J} S_{ij} x_j = 0$ is zero, the coefficients of the associated dual variables λ_i in the dual objective function will also be zero. Following a similar line of reasoning, the coefficients of the dual variables μ_j^{LB} and μ_j^{UB} in the dual objective function

TABLE 2.1 Constructing the Dual Constraints Corresponding to Each Primal Variable x_j in Example 2.4

(A) Determining the Terms Appearing in the Left-Hand Side of Each Dual Constraint

Primal Constraint	Associated Dual Variable	Coefficient of x_j in This Constraint	Respective Term(s) in the Dual Constraint
$\sum_{j \in J} S_{ij} x_j = 0, \quad \forall i \in I$	λ_i	S_{ij}	$\sum_{i \in I} S_{is} \lambda_i$
$-x_j \le -LB_j, \ \forall j \in J$	μ_j^{LB}	-1	$-\mu_j^{\mathrm{LB}}$
$x_j \le UB_j, \quad \forall j \in J$	μ_j^{UB}	1	μ_j^{UB}

(B) Determining the Right-Hand Side of Each Dual Constraint

Primal Variable(s)	Coefficient in the Primal Objective Function	Right-Hand Side Value of the Dual Constraint Associated with This Primal Variable
$x_j, \ \forall j \in J - \{b\}$	0	0
x_b	1	1

(C) Determining the Coefficient of Each Dual Variable in the Dual Objective Function

Primal Constraint(s)	Associated Dual Variable	Right-Hand Side Value of This Primal Constraint	Respective Terms in the Dual Objective Function
$\sum_{j \in J} S_{ij} x_j = 0, \quad \forall i \in I$	λ_i	0	$0\lambda_i$
$-x_j \le -LB_j, \ \forall j \in J$	μ_j^{LB}	$-LB_j$	$-LB_j \mu_j^{\mathrm{LB}}$
$x_j \le UB_j, \quad \forall j \in J$	μ_j^{UB}	UB_j	$UB_j \mu_j^{\mathrm{UB}}$

will be $-LB_j$ and UB_j, respectively (see Table 2.1C). The dual problem can thus be formulated as follows:

$$\text{minimize } z = \sum_{i \in I} 0\lambda_i + \sum_{j \in J} \left(-LB_j\right) \mu_j^{\mathrm{LB}} + \sum_{j \in J} UB_j \mu_j^{\mathrm{UB}}$$

subject to

$$\sum_{i \in I} S_{ij} \lambda_i + \mu_j^{\mathrm{UB}} - \mu_j^{\mathrm{LB}} = 0, \quad \forall j \in J - \{b\} \quad \rightarrow \quad x_j \in \mathbb{R}\left(j \ne b\right)$$

$$\sum_{i \in I} S_{ib} \lambda_i + \mu_b^{\mathrm{UB}} - \mu_b^{\mathrm{LB}} = 1 \quad \rightarrow \quad x_b \in \mathbb{R}$$

$$\mu_j^{\mathrm{LB}}, \mu_j^{\mathrm{UB}} \ge 0, \quad \forall j \in J, \quad \lambda_i \in \mathbb{R}, \quad \forall i \in I$$

According to the Result 1 of the weak duality theorem, if the primal and dual problems are for a given pair of primal and dual variable values feasible and their objective function values equal, then this pair of points are optimal solutions to the primal and dual problems. Therefore, one can solve the primal and dual problems simultaneously by collecting all the primal and dual constraints together, setting the primal objective equal to the dual objective in the form of a constraint and by maximizing or minimizing a dummy objective function such as $z = 0$ as shown in the following:

$$\underset{x_j, \lambda_i, \mu_j^{\mathrm{LB}}, \mu_j^{\mathrm{UB}}}{\text{minimize}} \quad z = 0$$

subject to

$$\sum_{j \in J} S_{ij} x_j = 0, \quad \forall i \in I$$

$$LB_j \le x_j \le UB_j, \quad \forall j \in J$$

$$\sum_{i \in I} S_{ij} \lambda_i + \mu_j^{\mathrm{UB}} - \mu_j^{\mathrm{LB}} = 0, \quad \forall j \in J - \{b\}$$

$$\sum_{i \in I} S_{ib} \lambda_i + \mu_b^{\mathrm{UB}} - \mu_b^{\mathrm{LB}} = 1$$

$$x_b = \sum_{j \in J} \left(-LB_j\right) \mu_j^{\mathrm{LB}} + \sum_{j \in J} UB_j \mu_j^{\mathrm{UB}}$$

$$x_j \in \mathbb{R}, \quad \forall j \in J, \quad \mu_j^{\mathrm{LB}}, \mu_j^{\mathrm{UB}} \ge 0, \quad \forall j \in J, \quad \lambda_i \in \mathbb{R}, \quad \forall i \in I$$

Note that instead of the duality condition (i.e., primal objective function = dual objective function), one can alternatively enforce the CS condition to solve this problem. However, here the duality condition is preferable as the equivalent CS condition gives rise to nonlinear constraints. The utility of this problem structure will be highlighted in Chapter 8 when addressing bilevel optimization problems. Note that infeasibility or unboundedness of the primal (or dual) will render the entire optimization problem above infeasible. □

2.6 NONLINEAR OPTIMIZATION PROBLEMS THAT CAN BE TRANSFORMED INTO LP PROBLEMS

In this section, we briefly describe a number of nonlinear optimization problems that can be directly solved using a LP formulation by means of a transformation.

2.6.1 Absolute Values in the Objective Function

Consider the following minimization problem:

$$\text{minimize } z = \sum_{i \in \{1,2,\ldots,N_y\}} |y_i|$$

subject to

$$\boldsymbol{Ax} + \boldsymbol{y} = \boldsymbol{b}$$

$$\boldsymbol{x} \ge 0, \boldsymbol{y} \in \mathbb{R}$$

where $\boldsymbol{x}, \boldsymbol{y}$ and $\boldsymbol{b}$ are $N_x \times 1$, $N_y \times 1$ and $M \times 1$ vectors, respectively, and $\boldsymbol{A}$ is an $M \times (N_x + N_y)$ matrix. Note that even though the constraints are linear, the objective function is nonlinear due to the presence of absolute values thereby rendering the entire optimization problem nonlinear. This problem can be linearized by using the following transformation:

$$|y_i| = y_i^+ \quad \text{where} \quad y_i^+ \geq 0, \quad \forall i \in \{1,2,\ldots,N_y\}$$

and addition of the following constraints to the problem:

$$y_i^+ \geq y_i \quad \text{and} \quad y_i^+ \geq -y_i$$

Therefore, the original problem can be recast as the following LP problem:

$$\begin{aligned}
&\text{minimize}\, z = \sum_{i \in \{1,2,\ldots,N_y\}} y_i^+ \\
&\text{subject to} \\
&\boldsymbol{Ax} + \boldsymbol{y} = \boldsymbol{b} \\
&\boldsymbol{y}^+ \geq \boldsymbol{y} \\
&\boldsymbol{y}^+ \geq -\boldsymbol{y} \\
&\boldsymbol{x}, \boldsymbol{y}^+ \geq 0, \quad \boldsymbol{y} \in \mathbb{R}
\end{aligned}$$

Note that this transformation technique is a viable linearization procedure only if the sum of the absolute values are minimized (not maximized) in the objective function of the original problem.

2.6.2 Minmax Optimization Problems with Linear Constraints

Consider the following minmax optimization problem:

$$\begin{aligned}
&\text{minimize} \left\{ \underset{j \in \{1,2,\ldots,N_y\}}{\text{maximize}} \quad y_j \right\} \\
&\text{subject to} \\
&\boldsymbol{Ax} + \boldsymbol{y} = \boldsymbol{b} \\
&\boldsymbol{x} \geq \boldsymbol{0}, \boldsymbol{y} \in \mathbb{R}
\end{aligned}$$

This problem is nonlinear due to the presence of the minmax operator in the objective function. If z denotes the maximum of $\{y_j\}$ over $j \in \{1,2,\ldots,N_y\}$

$$z = \underset{j \in \{1,2,\ldots,N_y\}}{\text{maximize}} \quad \{y_j\}$$

then we can impose $z \geq y_j$, $\forall j \in \{1,2,..,N_y\}$. The optimization problem can thus be linearized as follows:

$$\begin{aligned}&\text{minimize } z\\&\text{subject to}\\&z \geq y_j, \quad j = 1,2,\ldots,N_y\\&\boldsymbol{Ax} + \boldsymbol{y} = \boldsymbol{b}\\&\boldsymbol{x} \geq \boldsymbol{0},\ \boldsymbol{y} \in \mathbb{R}\end{aligned}$$

The minimization operator will ensure that at least one of y_j variables becomes equal to z. It is easy to verify that maxmin optimization problems can be linearized in a similar way. However, neither maxmax nor minmin problems can be linearized as described above. They generally require the introduction of binary variables to ensure that at least one of the constraints $z \geq y_j$ (or $z \leq y_j$) is satisfied as an equality.

2.6.3 Linear Fractional Programming

A linear fractional programming problem consists of a fractional objective function with linear numerator and denominator and linear constraints:

$$\begin{aligned}&\text{maximize}\left(\text{or minimize}\right) z = \frac{\boldsymbol{cx} + p}{\boldsymbol{dx} + q}\\&\text{subject to}\\&\boldsymbol{Ax} \leq \boldsymbol{b}\\&\boldsymbol{x} \geq 0\end{aligned}$$

where p and q are scalars. This optimization problem is nonlinear due to the presence of a fractional term in the objective function. This problem can be linearized as follows [10]:

Assume that the feasible region $\{\boldsymbol{Ax} \leq \boldsymbol{b}, \boldsymbol{x} \geq 0\}$ is bounded and the denominator of the objective function is positive: $\boldsymbol{dx} + q > 0$. Let r be a new variable defined as follows:

$$r = \frac{1}{\boldsymbol{dx} + q}$$

The objective function can thus be written as $z = \boldsymbol{c}(r\boldsymbol{x}) + rp$. The original vector of variables $\boldsymbol{x}$ is mapped one-to-one to a new vector of variables $\boldsymbol{y}$ scaled by scalar variable r:

$$\boldsymbol{y} = r\boldsymbol{x} \geq 0$$

By introducing r and replacing variable $\boldsymbol{x}$ with variable $\boldsymbol{y}$, the original optimization problem can be rewritten as follows:

$$\begin{aligned}
&\text{maximize}\left(\text{or minimize}\right) z = \boldsymbol{cy} + rp \\
&\text{subject to} \\
&\boldsymbol{A}\left(\frac{\boldsymbol{y}}{r}\right) \le \boldsymbol{b} \\
&r = \frac{1}{\boldsymbol{dx} + q} \\
&\boldsymbol{x}, \boldsymbol{y} \ge 0, r > 0
\end{aligned}$$

or

$$\begin{aligned}
&\text{maximize}\left(\text{or minimize}\right) z = \boldsymbol{cy} + rp \\
&\text{subject to} \\
&\boldsymbol{Ay} \le r\boldsymbol{b} \\
&\boldsymbol{dy} + qr = 1 \\
&\boldsymbol{y} \ge 0, r > 0
\end{aligned}$$

which is a linear program with respect to the new variables $\boldsymbol{y}$. In essence, this transformation rescales both numerator and denominator at the same time (with r) so as the denominator becomes equal to one. The scaling factor remains a variable until the problem is solved.

Example 2.5

Transform the following linear fractional programming problem to an equivalent linear programming representation:

$$\begin{aligned}
&\text{maximize } z = \frac{2x_1 + x_2 + 3x_3 + 6}{x_1 + 3x_3 + 4} \\
&\text{subject to} \\
&x_1 + 3x_2 + x_3 \le 10 \\
&3x_1 + 2x_2 \le 7 \\
&x_1, x_2, x_3 \ge 0
\end{aligned}$$

Solution: Considering the general form of a linear fractional programming problem, we have

$$p = 6, \quad q = 4, \quad \boldsymbol{c} = \begin{bmatrix} 2 & 1 & 3 \end{bmatrix} \quad d = \begin{bmatrix} 1 & 0 & 3 \end{bmatrix}, \quad \boldsymbol{b} = \begin{bmatrix} 10 \\ 7 \end{bmatrix}$$

Also, following the transformation described earlier, we obtain

$$r = \frac{1}{x_1 + 3x_3 + 4}, \quad \begin{bmatrix} y_1 \\ y_2 \\ y_3 \end{bmatrix} = r \begin{bmatrix} x_1 \\ x_2 \\ x_3 \end{bmatrix}$$

The original problem can thus be rewritten as follows:

$$\begin{aligned}
&\text{maximize } z = 2y_1 + y_2 + 3y_3 + 6r \\
&\text{subject to} \\
&y_1 + 3y_2 + y_3 \le 10r \\
&3y_1 + 2y_2 \le 7r \\
&y_1 + 3y_3 + 4r = 1 \\
&y_1, y_2, y_3 \ge 0, \quad r > 0
\end{aligned}$$

If y^*_1, y^*_2, y^*_3 are the optimal solutions to the transformed linear problem, the optimal solutions to the original problem are $x^*_1 = \frac{y^*_1}{r}, x^*_2 = \frac{y^*_2}{r}, x^*_3 = \frac{y^*_3}{r}$. □

EXERCISES

2.1 Solve the following LP problem:

$$\begin{aligned}
&\text{maximize } 4x_1 + 3x_2 \\
&\text{subject to} \\
&x_1 + 2x_2 \le 7 \\
&-2x_2 - x_2 \le 5 \\
&5x_1 + 2x_2 \le 16 \\
&x_1, x_2 \ge 0
\end{aligned}$$

(a) Using graphical means.

(b) By applying the Simplex method for two iterations.

2.2 Consider the following LP problem:

$$\begin{aligned}
&\text{minimize } 3x_1 + x_2 + 31x_3 \\
&\text{subject to} \\
&6x_1 + 9x_2 + 26x_3 \le 26 \\
&3x_1 + 2x_2 - 26x_3 \le 20 \\
&x_1, x_2, x_3 \ge 0
\end{aligned}$$

(a) Generate its dual problem.

(b) Apply the duality theorems to generate a set of conditions that include both primal and dual variables guaranteeing optimality for both problems if a feasible point exists.

2.3 Reformulate the following problem as an LP and generate its dual problem:

$$\text{minimize} \quad \sum_{j \in J} \left| x_j - a_j \right|$$

subject to

$$\sum_{j \in J} S_{ij} x_j = 0, \qquad \forall i \in I$$

$$LB_j \leq x_j \leq UB_j, \quad \forall j \in J$$

$$x_j \in \mathbb{R}, \qquad \forall j \in J$$

where

$J = \{1,2,\ldots,N\}$: Set spanned by variables.
$I = \{1,2,\ldots,M\}$: Set spanned by equality constraints.
x_j: Decision variables.
a_j: A constant
S_{ij}: Coefficient of contribution of variable x_j in equality constraint i.
LB_j, UB_j: The lower and upper variable bounds, respectively.

2.4 Transform the following problem to an LP:

$$\text{minimize} \frac{x_1}{x_2}$$

subject to

$$\sum_{j \in J} S_{ij} x_j = 0, \qquad \forall i \in I$$

$$LB_j \leq x_j \leq UB_j, \quad \forall j \in J$$

$$x_j \geq 0, \qquad \forall j \in J$$

where all sets, variables and parameters are defined as in Exercise 2.3 and $LB_j > 0$ and $UB_j > 0$.

REFERENCES

1. Dantzig GB, Orden AWP: The generalized simplex method for minimizing a linear form under linear inequality restraints. *Pacific J Mathemat* 1955, **5**(2):187–195.
2. Bazaraa MS, Jarvis JJ, Sherali HD: *Linear programming and network flows*, 4th edn. Hoboken, N.J.: John Wiley & Sons; 2010.

3. Ignizio JP, Cavalier TM: *Linear programming*. Englewood Cliffs, N.J.: Prentice Hall; 1994.
4. Rardin RL: *Optimization in operations research*. Upper Saddle River, N.J.: Prentice Hall; 1998.
5. Fourer R, Gay DM, Kernighan BW: *AMPL: a modeling language for mathematical programming*, 2nd edn. Pacific Grove, CA: Thomson/Brooks/Cole; 2003.
6. IBM: IBM ILOG CPLEX Optimization Studio v12.5.1 documentation; 2015. Available at http://www-03.ibm.com/software/products/en/ibmilogcpleoptistud/ (accessed September 1, 2015).
7. Gurobi Optimization Inc.: Gurobi Optimizer Reference Manual; 2015. Available at http://www.gurobi.com (accessed September 1, 2015).
8. Wright SJ: *Primal-dual interior-point methods*. Philadelphia, P.A.: Society for Industrial and Applied Mathematics; 1997.
9. Bertsimas D, Tsitsiklis JN: *Introduction to linear optimization*. Belmont, M.A.: Athena Scientific; 1997.
10. Charnes A, Cooper WW: Programming with linear fractional functions *Naval Res Logist Q* 1962, **9**:182–186.

3

FLUX BALANCE ANALYSIS AND LP PROBLEMS

This chapter starts with a brief introduction of metabolism and highlights basic concepts underpinning mathematical modeling of metabolism along with a description of genome-scale metabolic (GSM) models. Subsequently, flux balance analysis (FBA) for metabolic networks is presented. Finally, a number of popular applications of FBA based on linear programming (LP) including the modeling of gene/reaction knockouts, calculation of maximum theoretical yield for a product of interest, flux variability analysis and flux coupling analysis of metabolic networks are discussed.

Metabolism encompasses the complete set of reactions allowing a living organism to convert nutrient molecules such as sugars into energy and the building blocks of life (DNA, RNA, etc.). Compounds participating in metabolic reactions as reactants or products are referred to as *metabolites*. Metabolic reactions are mediated by a series of biological catalysts called enzymes: a class of specialized proteins that generally have a high degree of specificity for their substrates. The rates of enzymatically catalyzed metabolic reactions are dramatically higher than those in the absence of a catalyst (typically 10^6–10^{12} times greater). In contrast to chemical catalysts often requiring elevated temperatures and pressures as well as extreme pH conditions to enable efficient chemical transformations, enzymes are capable of carrying out biochemical transformations with high specificity and yield under mild physiological conditions. In the absence of enzymes, metabolic reactions typically occur at such slow rates that they can be considered inactive. The detailed description of metabolism and relevant biochemistry is beyond the scope of this book. The reader is encouraged to refer to dedicated biochemistry textbooks such as Refs. 1, 2, 3 for thorough treatments.

Optimization Methods in Metabolic Networks, First Edition. Costas D. Maranas and Ali R. Zomorrodi.

3.1 MATHEMATICAL MODELING OF METABOLISM

The incredible functional diversity of enzymes in living organisms enables them to catalyze thousands of different metabolic reactions interconverting hundreds of metabolites [4–6]. The richness of enzymatic functions gives rise to a complex network of reactions that synthesize all biomass precursors, generate ATP to power biochemical transformations, balance redox, and respond to externally imposed perturbations. Mathematical modeling and computational analyses play a key role in assessing the range of possible metabolic flux distributions (i.e., allowable metabolic phenotypes) that can be traversed by an organism under different environmental conditions and genetic backgrounds. In addition, these tools can be used to identify systematic ways of engineering metabolism by removing or adding genes or by modulating their expression levels for biotechnological and biomedical applications.

3.1.1 Kinetic Modeling of Metabolism

Kinetic modeling allows for the quantification of reaction fluxes as functions of metabolite concentrations, enzyme levels and parameters capturing enzyme turnover, saturation and allosteric regulation. Different forms of mechanistic expressions such as Michaelis–Menten or Hill-type kinetics can be used to describe reaction rates. Upon assembling all kinetic expressions, a system of (nonlinear) ordinary differential equations (ODEs) representing the conservation of mass for each metabolite is solved to obtain the time-dependent profile of metabolite concentrations and reaction fluxes. Kinetic models are currently available for only a limited number of well-studied microorganisms and cover generally only a limited portion of metabolism [7–12]. This is because expanding kinetic models is hindered by the current paucity of experimental data on concentrations and fluxes at a scale needed to support unambiguous kinetic parameter value identification [13, 14].

3.1.2 Stoichiometric-Based Modeling of Metabolism

In contrast to kinetic modeling, where the conservation of mass is captured by a system of nonlinear ODEs, stoichiometric-based modeling relies on a system of linear equations describing the conservation of mass for each metabolite at steady-state using primarily stoichiometric information. Over the past decade, stoichiometric-based genome-scale metabolic (GSM) models have been reconstructed for a growing list of organisms from microbial to multicompartment and multicellular [15]. The global nature of these models enables the assessment of theoretical limits of metabolic performance and the identification of plausible engineering strategies using constraint-based methods such as flux balance analysis (FBA). A key advantage of this modeling formalism lies in the minimal amount of biological knowledge and data it requires to assess the possible range of potentially feasible metabolic phenotypes. Furthermore, the linearity of the underlying mathematical models affords significant computational savings and tractability (compared to kinetic models) even for GSM models. This simplicity, however, comes at the expense of not being able to capture metabolite concentration information, enzyme saturation or nonlinearities due to kinetic and regulatory effects. In the subsequent sections, we briefly introduce genome-scale stoichiometric-based models of metabolism and describe the basic elements of FBA.

3.2 GENOME-SCALE STOICHIOMETRIC MODELS OF METABOLISM

A stoichiometric GSM model is a global inventory of the metabolic reactions for a given organism and a link to the corresponding genes (see Fig. 3.1). Such a model can, in principle, be reconstructed for any organism with an annotated genome. The decision to include a metabolic reaction in the model is made based on whether the enzyme(s) catalyzing the reaction is encoded in the genome of the organism of interest. In addition to the list of reactions, GSM models contain information regarding reaction reversibility based on experimental evidence and/or thermodynamic information. Some GSM models also contain condition-dependent accessibility to different reactions. For example, some reactions cannot carry flux in the presence of glucose or oxygen in the *Escherichia coli* *i*AF1260 model [16]. A number of important features of GSM models are discussed next.

3.2.1 Gene–Protein–Reaction Associations

GSM models contain links between metabolic reactions and genes encoding the relevant enzymes. This information is stored in gene to protein to reaction (GPR) associations. GPRs can be broadly classified into four different categories (see Fig. 3.2):

(i) **One-to-one mapping:** A reaction R is catalyzed by a single enzyme (protein) E which is encoded by a unique gene G.

(ii) **Isozymes:** A reaction R is catalyzed by either of two enzymes $E1$ or $E2$ encoded by genes $G1$ and $G2$, respectively. Either one of the two (or more) genes is sufficient to enable this reaction. They are thus denoted with an OR relation in the corresponding GPR association ($G1$ *OR* $G2$).

Reaction	Reaction equation	Genes
HEX1	[c] : atp + glc-D $\longrightarrow$ adp + g6p + h	*glk*
PGI	[c] : g6p $\rightleftarrows$ f6p	*pgi*
PFK	[c] : atp + f6p $\longrightarrow$ adp + fdp + h	*pfkA* OR *pfkB*
FBA	[c] : fdp $\rightleftarrows$ dhap + g3p	*fbaA* OR *fbaB*
TPI	[c] : dhap $\rightleftarrows$ g3p	*tpiA*
GAPD	[c] : g3p + nad + pi $\rightleftarrows$ 13dpg + h + nadh	*gapA* OR ($gapC_1$ AND $gapC_2$)
PGK	[c] : 3pg + atp $\rightleftarrows$ 13dpg + adp	*pgk*
PGM	[c] : 2pg $\rightleftarrows$ 3pg	*gpmA* OR *gpmB* OR *gpmM*
ENO	[c] : 2pg $\rightleftarrows$ h2o + pep	*eno*
PYK	[c] : adp + h + pep $\longrightarrow$ atp + pyr	*pykA* OR *pykF*
PDH	[c] : coa + nad + pyr $\longrightarrow$ accoa + co2 + nadh	*aceE* AND *aceF* AND *lpdA*
GLCtex	glc-D[e] $\rightleftarrows$ glc-D[p]	*phoE* OR *ompF* OR *ompN* OR *ompC*
GLCt2pp	glc-D[p] + h[p] $\longrightarrow$ glc-D[c] + h[c]	*galP*
EX_glc(e)	[e] : glc-D $\rightleftarrows$	None
⋮	⋮	⋮
Biomass	[c]: $S_1X_1 + S_2X_2 + \cdots + S_{atp}X_{atp} \rightarrow$ adp + h + pi	None

FIGURE 3.1 A schematic representation of a stoichiometric GSM model.

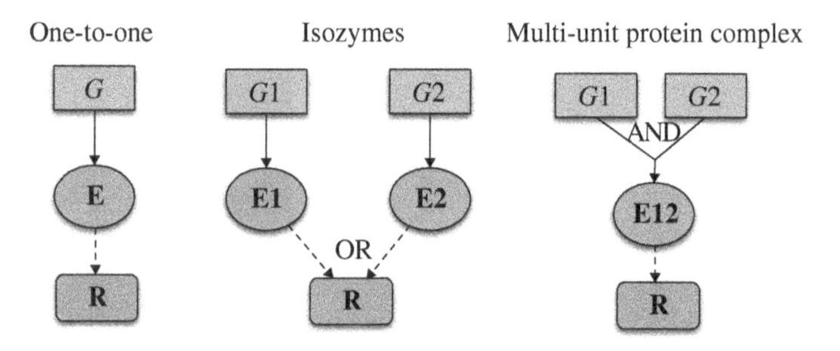

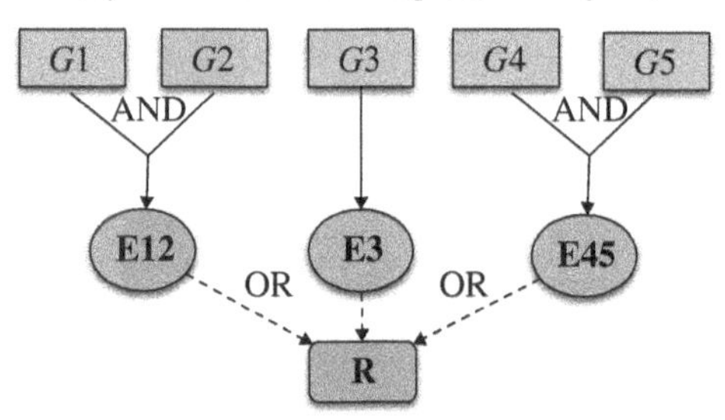

FIGURE 3.2 Different types of GPR relationships. *G*, E and R denote genes, enzymes and reactions, respectively. In addition, solid arrows represent enzyme production by one or more genes while dashed arrows denote catalyzing a reaction by one or more enzymes.

(iii) **Multiunit protein complex:** A reaction R is catalyzed by a multiunit protein complex composed of two (or more) proteins $E1$ and $E2$ coded by genes $G1$ and $G2$, respectively. Both genes are needed to enable this reaction and thus are denoted with an AND relationship ($G1$ AND $G2$) in the GPR association.

(iv) **Combination of isozymes and multiunit protein complexes**: Infrequently, a reaction R is encoded by multiple isozymes some of which are multiunit protein complexes. The relevant genes are linked using a combination of AND and OR relations in the GPR association. For example, a GPR relation of the form ($G1$ AND $G2$) OR $G3$ OR ($G4$ AND $G5$) means that reaction R is encoded by three isozymes, two of which are multiunit protein complexes encoded by ($G1$ AND $G2$) and ($G4$ AND $G5$), respectively, while the third isozyme is coded by gene $G3$ (see Fig. 3.2).

3.2.2 The Biomass Reaction

A fundamental component of a GSM model is the presence of an *ad hoc biomass reaction* that provides a convenient way to inventory all components that make up biomass. The biomass reaction serves as a drain on all biomass precursors at their appropriate biological ratios and can be abstracted as follows:

$$\sum_{i\in I^{\text{biomass}}} S_{i,\text{biomass}} X_i + S_{\text{ATP,biomass}}\text{ATP} \xrightarrow{v_{\text{biomass}}} \text{biomass} + \text{ADP} + \text{H}^+ + \text{P}$$

where X_i is a biomass precursor (e.g., amino acids, lipids and carbohydrates) and $S_{i,\text{biomass}}$ is its corresponding stoichiometric coefficient of biomass precursor i. These coefficients are determined according to the fraction of biomass contributed by metabolite X_i measured on a molar basis. It has units of $\frac{\text{mmol}}{\text{gDW}}$ where gDW denotes grams of dry cell weight. The amount of ATP needed for growth $S_{\text{ATP,biomass}}$ per unit of biomass produced is referred to as *growth-associated maintenance* (GAM) *ATP*. The biomass composition and thus biomass reaction vary from one organism to another with chlorophyll and various pigments present in photosynthetic organisms and specialized cofactors in archaea. The biomass reaction flux is indicative of the cell's growth per amount of the limiting resource(s) taken up from the extracellular environment. This limiting resource is typically glucose but other resources such as oxygen, light or ammonium can become limiting under different conditions. Interested readers are encouraged to refer to available textbooks and literature [17, 18] to find out more details about stoichiometric-based models and how they are reconstructed. It is worth noting that the *ad hoc* metabolite referred to as "biomass" on the right-hand side of the biomass reaction is usually omitted in GSM model representations.

3.2.3 Metabolite Compartments

Each metabolite in a GSM model may have one or more designations denoting the relevant cellular compartment. For example, "A[c]" refers to metabolite A in the cytosol, whereas "A[e]" refers to A in the extracellular space. Designations [c] and [e] are only two out of many compartment designations typically encountered in GSM models including [m] for mitochondria, [p] for periplasmic membrane or plastid, [l] for lumen, etc.

3.2.4 Scope and Applications

GSM models contain a few hundred to over a thousand metabolites and reactions. For example, the latest *i*JO1366 metabolic model of *E. coli* accounts for 1136 metabolites, 2251 reactions, and 1366 genes [19]; while the most up-to-date *Saccharomyces cerevisiae* model (Yeast 7.0) consists of 2218 metabolites, 3493 reactions, and 916 genes [20]. GSM models can be viewed as mathematically structured databases of the metabolic repertoire of a given organism. These models can be used to elucidate the performance limits or to identify ways to engineer a microbial host toward an overproduction goal. This requires a mathematical procedure or algorithm to operate on the GSM model to extract the desired output. Frequently, these procedures rely on the solution of an optimization problem.

3.3 FLUX BALANCE ANALYSIS (FBA)

FBA is the most widely used computational method for the analysis of stoichiometric-based GSM models. In the following sections, we enumerate all elements needed to perform an FBA calculation.

3.3.1 Cellular Inputs, Outputs and Metabolic Sinks

Cells convert carbon substrates into energy currency metabolites (e.g., ATP), biomass precursors and by-products that are transported to the extracellular space. FBA relies on the application of mass and energy conservation laws along with thermodynamic feasibility considerations. The application of conservation principles requires the assessment of all inputs and outputs as well as sources and sinks for the system. Metabolites can be exported out of the cell or be taken up from the extracellular environment. Some of these input molecules are available in excess (e.g., water), whereas others are limited by their extracellular concentration (e.g., carbon source) and/or diffusion rate (e.g., oxygen). Terminal products of metabolism that are exported out of the cell (e.g., acetic acid or CO_2) are system outputs. All these inputs and outputs are modeled using *exchange reactions*, which are artificial reactions carrying metabolites across the system's boundary (see Fig. 3.3). All exchange reactions are treated as reversible with the forward direction representing export and the backward direction denoting import. For example, the exchange reaction for a metabolite A is defined as follows:

$$\text{EX\_A(e)}:\ \text{A}[\text{e}] \leftrightharpoons$$

Unlike exchange reactions which are artificial constructs modeling the transfer of a metabolite between the extracellular environment [e] and the system's boundary, transport reactions model the actual transport of a metabolite between the cytosol [c] and the extracellular environment [e] (see Fig. 3.3), or between the cytosol and other compartments such as the mitochondria (in metabolic models of eukaryotes). A transport reaction can be passive relying solely on diffusion or active involving the consumption of ATP or the simultaneous translocation of protons and/or ions. For example, the passive transport of metabolite A using transport reaction At is modeled as follows:

$$\text{At}:\ \text{A}[\text{c}] \leftrightharpoons \text{A}[\text{e}]$$

Metabolites can also be sequestered within sinks in a metabolic model. The two most commonly used sinks are the biomass reaction and non-growth-associated maintenance ATP. The biomass reaction in essence acts as a sink that funnels all necessary

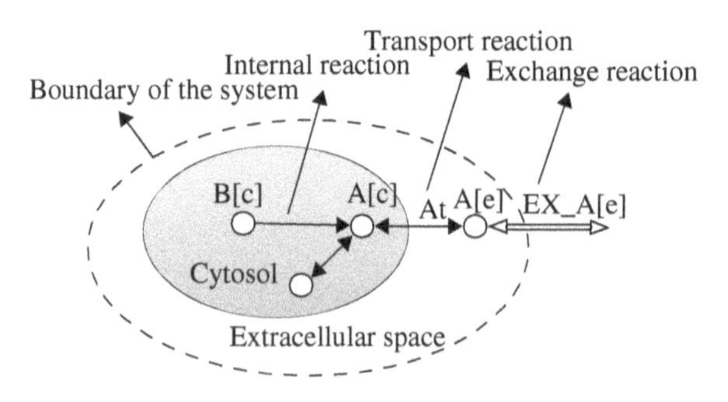

FIGURE 3.3 Exchange, transport and internal reactions.

biomass constituents at their physiologically relevant ratios. Most flux-carrying metabolic reactions in a model operate so as a dedicated or multiple biomass components are produced. In addition to growth, metabolic resources are also diverted for housekeeping and cellular maintenance purposes that are not described in detail in GSM models. Instead, only their energy demands are imposed in the form of an ATP-equivalent sink referred to as *non-growth maintenance* (NGAM) *ATP*. The required amount of NGAM ATP varies between organisms. For example, the NGAM ATP for *E. coli* is $8.39\frac{\text{mmol}}{\text{gDW}\cdot\text{h}}$ [16], while that for *S. cerevisiae* is $1.00\frac{\text{mmol}}{\text{gDW}\cdot\text{h}}$ [21].

3.3.2 Component Balances

The conservation of a metabolite i on a molar basis participating in multiple reactions in a metabolic network can be expressed as follows:

$$\frac{dC_i}{dt} = \sum_{j\in J} S_{ij} v_j, \quad \forall i \in I \tag{3.1}$$

where $I = \{1,\ldots,M\}$ and $J = \{1,\ldots,N\}$ are the sets of metabolites and reactions, respectively, C_i is the concentration of a metabolite i, v_j is the flux of reaction j and S_{ij} is the stoichiometric coefficient of metabolite i in reaction j. The value of S_{ij} is negative, positive or zero if metabolite i serves as a reactant, product or does not participate in reaction j, respectively. This system of ODEs describes both the steady-state and dynamic behavior of the metabolic network. In kinetic models, fluxes v_j in Equation 3.1 are replaced by kinetic expressions resulting in a system of nonlinear ODEs with metabolite concentrations C_i as variables. In contrast, in only stoichiometric-based GSM models, reaction fluxes are not substituted by kinetic expressions but instead are treated as free variables. In addition, a pseudo state-steady condition is assumed for all intracellular metabolites given that the time constants for metabolic reactions (milliseconds to tens of seconds) are typically much smaller than those of other cellular processes such as transcriptional regulation (minutes) or cellular growth (several minutes or hours). Equation 3.1 thus reduces to

$$\sum_{j\in J} S_{ij} v_j = 0, \quad \forall i \in I \tag{3.2}$$

By recasting this equation in matrix form, we obtain:

$$\begin{bmatrix} S_{11} & S_{12} & \cdots & S_{1N} \\ S_{21} & S_{22} & \cdots & S_{2N} \\ \vdots & \vdots & \ddots & \vdots \\ S_{M1} & S_{M2} & \cdots & S_{MN} \end{bmatrix} \begin{bmatrix} v_1 \\ v_2 \\ \vdots \\ v_N \end{bmatrix} = \begin{bmatrix} 0 \\ 0 \\ \vdots \\ 0 \end{bmatrix}$$

or equivalently

$$\boldsymbol{S} \cdot \boldsymbol{v} = 0 \tag{3.3}$$

where $\boldsymbol{S}$ and $\boldsymbol{v}$ are referred to as the stoichiometric matrix and flux vector, respectively. Matrix $\boldsymbol{S}$ captures all structural properties of the metabolic network.

Given that most metabolites participate in many metabolic reactions, the number of metabolites is smaller than the number of reactions in a metabolic network (i.e., $M < N$). This renders the system of equations shown in Equation 3.2 or 3.3 underdetermined implying that there is an infinite number of flux vectors that can satisfy the steady-state conservation of mass equation. Any flux vector that satisfies Equation 3.3 is said to be in the *null space* of matrix $\boldsymbol{S}$. Therefore, Equation 3.3 captures the range of all allowable metabolic phenotypes of the system. However, not all these metabolic states are physiologically relevant. *Constraint-based analysis and reconstruction* (COBRA) methods (such as FBA) aim to constrain the feasible space of fluxes defined by Equation 3.3 as much as possible in order to eliminate physiologically irrelevant metabolic phenotypes.

3.3.3 Thermodynamic and Capacity Constraints

Both thermodynamic and flux capacity constraints are manifested as lower bounds (LBs) and/or upper bounds (UBs) on reaction fluxes:

$$LB_j \leq v_j \leq UB_j \tag{3.4}$$

The reversibility of an enzymatically catalyzed reaction can be deduced based on the change in the Gibbs free energy (ΔG) of the reaction. The exact value of (ΔG) within the crowded cellular environment is difficult to assess due to uncertainties in metabolite concentrations and departures from standard conditions. A reaction is generally treated as *irreversible* if (ΔG) remains only negative (or only positive) even upon considering these uncertainties. In contrast, a reaction is treated as *reversible* if (ΔG) can switch signs for different physiologically relevant concentration ranges. This gives rise to three cases for reaction bounds:

(i) Reaction occurs only in the forward direction $(LB_j = 0, UB_j = M)$
(ii) Reaction occurs only in the reverse direction $(LB_j = -M, UB_j = 0)$
(iii) Reaction is reversible $(LB_j = -M, UB_j = M)$

where M is a large positive number (e.g., 1000) in comparison to the limiting resource transport flux (e.g., 10 $\frac{\text{mmol}}{\text{gDW} \cdot \text{h}}$ of glucose) used to scale the fluxes in the network. Typically, if a reaction is found to be irreversible in the reverse direction, it is rewritten in the GSM model so as the thermodynamically feasible direction becomes the forward direction (i.e., by swapping reactants and products). When no Gibbs free energy data is available, reaction reversibility is inferred by comparing with previously reconstructed models of related species or according to literature sources. In the absence of any corroborating evidence, the reaction is generally assumed to be reversible. Chapter 5 addresses the thermodynamic analysis of metabolic networks.

In addition to thermodynamic considerations, a number of other factors can limit reaction flux. For example, the lower bound of all exchange reactions is set to zero, except for those corresponding to metabolites that are available in the growth medium. The lower bound is set to $-M$ for metabolites that are available in the growth medium in excess (i.e., water), whereas it is set to a finite negative value for limiting resources (such as the carbon source). A common practice is to set the limiting resource (i.e., generally the carbon substrate) to $10\frac{\text{mmol}}{\text{gDW}\cdot\text{h}}$ thus providing convenient scaling for all fluxes in the network and a basis for the calculated biomass or product yield. The upper bound for all exchange fluxes is set to M implying that all metabolites with an exchange reaction in the model can be exported. Flux bounds can also be derived based on the measurements of external fluxes or metabolic flux analysis (MFA) measurements of internal metabolic fluxes (see Chapter 10). Capacity constraints could also be derived based on the enzyme activity and turnover [22]. Great care must be exercised to set appropriate bounds for all reactions. Errors in exchange reaction bounds could lead to either a metabolic model with no feasible flux distributions or models that can generate an unbounded amount of ATP through the unrestricted flow of protons across the membrane. In addition, overly permissive bounds for internal reactions could lead to the creation of thermodynamically infeasible cycles [23, 24] (see Chapter 5). A summary of how the lower and upper bounds are commonly set is given in Table 3.1.

3.3.4 Objective Function

Even upon the addition of thermodynamic and capacity constraints, mass balances alone cannot pinpoint the specific metabolic state of the system under a given environmental or genetic condition as there exists an infinite number of flux distributions satisfying these constraints. We thus need to choose from all these allowable metabolic phenotypes the one that is physiologically relevant under the current environmental condition. In FBA, this is addressed by invoking a fitness function acting as a surrogate for the most plausible physiological state of the system. Metabolic networks have evolved to ensure the efficient

TABLE 3.1 Typical Lower (LB_j) and Upper (UB_j) Bounds for Different Types of Reactions. M is a large positive number (e.g., 1000).

Type of Reaction	(LB_j, UB_j)
Irreversible	$(0, M)$
Reversible	$(-M, M)$
Exchange reaction for metabolites not in the growth medium	$(0, M)$
Exchange reactions for metabolites available in excess in the growth medium	$(-M, M)$
Exchange reactions for limiting substrates in the growth medium[1]	$(-c, M)$ $c > 0 \& c \ll M$
Non-growth ATP maintenance (NGAM)	$(c, c), c > 0 \ \& c \ll M$
Reactions with experimental flux measurements	$\left(v_j^{\text{min,exp}}, v_j^{\text{max,exp}}\right)$

[1] The value of c depends on the limiting substrate and organism.

conversion of resources into cellular components and energy equivalents supporting growth in response to selection pressures. Therefore, metabolic fluxes tend to assume values in support of the overall cellular fitness function. Even though a universal fitness function cannot always be put forth, maximization of biomass production yield provides a reasonable surrogate. Maximum biomass yield implies that metabolic fluxes are apportioned so as all biomass constituents are produced with maximum limiting resource usage efficiency in the exact ratios needed in the biomass reaction. Other such surrogates include the minimization or maximization of ATP generation or the minimization of the sum of all fluxes in the network. Nonlinear objective functions such as the minimization of metabolic adjustment (MOMA) have also been put forth to characterize the metabolic state of a knockout mutant strain [25] (see Chapter 10 for a detailed description of MOMA). A comparison of the accuracy of various objective functions under various genetic and environmental backgrounds can be found in Refs. 26–30.

3.3.5 FBA Optimization Formulation

Maximizing or minimizing a postulated objective function subject to a number of constraints gives rise to an optimization problem used to identify the metabolic flux distributions. Constrained optimization provides a systematic framework for imposing constraints and elucidating various optimality trade-offs in FBA. In its most general form, FBA is an LP problem maximizing or minimizing a linear combination of reaction fluxes subject to the conservation of mass, thermodynamic and capacity constraints as shown in the following vector and expanded forms:

$$
\begin{array}{ll}
\text{maximize}(\text{or minimize})\, z = \boldsymbol{c}^{\mathrm{T}}\boldsymbol{v} & \text{maximize}(\text{or minimize})\ z = \sum_{j \in J} c_j v_j \\
\text{subject to} & \text{subject to} \\
\boldsymbol{S}\boldsymbol{v} = 0 & \sum_{j \in J} S_{ij} v_j = 0, \quad \forall i \in I \\
\mathrm{LB} \le \boldsymbol{v} \le \mathrm{UB} & LB_j \le v_j \le UB_j, \quad \forall j \in J \\
\boldsymbol{v} \in \mathbb{R} & v_j \in \mathbb{R}
\end{array}
$$

The most widely used objective function in the FBA of metabolic networks is the maximization of the biomass reaction flux (i.e., $c_{\text{biomass}} = 1,\ c_j = 0,\ \forall j \in J - \{\text{biomass}\}$) built upon the assumption that the cell is striving to maximally allocate all available resources towards growth as noted earlier.

$$
\begin{aligned}
&\text{maximize} \quad z = v_{\text{biomass}} \quad [\text{FBA}] \\
&\text{subject to} \\
&\sum_{j \in J} S_{ij} v_j = 0, \quad \forall i \in I \\
&LB_j \le v_j \le UB_j, \quad \forall j \in J \\
&v_j \in \mathbb{R}
\end{aligned}
$$

Reaction fluxes (except for biomass) in FBA usually have units of $\frac{\text{mmol}}{\text{gDW} \cdot \text{h}}$, whereas the biomass flux has units of $\frac{\text{gDW}}{\text{gDW} \cdot \text{h}} = \text{h}^{-1}$. It is important to stress that maximization of biomass flux in FBA corresponds to the maximization of the *biomass production yield* not biomass production (growth) rate. This is because in FBA, the uptake rate of the limiting carbon source is set to a pre-specified value and the biomass flux is computed with respect to that value as the basis. For example, if the limiting carbon source uptake is set to $10 \frac{\text{mmol}}{\text{gDW} \cdot \text{h}}$, the FBA computed value of the biomass flux is the maximum yield for the chosen basis. The key advantage of FBA is that it can quantitatively assess all possible metabolic phenotypes by making use of only stoichiometry and bounds on reaction fluxes. Ultimately, internal flux measurements based on MFA (see Chapter 10) are needed to test the validity of the employed biomass maximization principle in FBA. The linearity of FBA optimization models implies that a GSM model spanning thousands of reactions and metabolites can be analyzed relatively quickly. The relatively low data overhead and tractability for large GSM models have made FBA an attractive analysis method. The following example provides a step-by-step description of how an FBA calculation for a prototype stoichiometric metabolic model is carried out.

Example 3.1

Prepare the FBA optimization problem formulation for the demonstration metabolic network represented in Figure 3.4. Assume that A is the limiting resource allowed to be taken up with a maximum flux of $10 \frac{\text{mmol}}{\text{gDW} \cdot \text{h}}$, B is an unlimited resource, F is the exported metabolic product and the *Sink* represents the flux diverted towards biomass formation. In this example, a single flux directed towards the biomass sink is maximized.

Solution: Conservation equations on a molar basis at steady state for each metabolite yield:

$$
\begin{aligned}
A[\text{e}]&: \quad -v_{\text{EX\_A(e)}} + v_{\text{At}} = 0 \\
B[\text{e}]&: \quad -v_{\text{EX\_B(e)}} + v_{\text{Bt}} = 0 \\
A[\text{c}]&: \quad -v_{\text{At}} - v_{\text{rxn1}} = 0 \\
B[\text{c}]&: \quad -v_{\text{Bt}} - v_{\text{rxn1}} = 0 \\
C[\text{c}]&: \quad v_{\text{rxn1}} - v_{\text{rxn2}} - v_{\text{rxn3}} = 0 \\
D[\text{c}]&: \quad v_{\text{rxn3}} - v_{\text{Sink}} = 0 \\
E[\text{c}]&: \quad v_{\text{rxn3}} - v_{\text{rxn4}} = 0 \\
F[\text{c}]&: \quad v_{\text{rxn2}} + v_{\text{rxn4}} - v_{\text{Ft}} = 0 \\
F[\text{e}]&: \quad v_{\text{Ft}} - v_{\text{EX\_F(E)}} = 0
\end{aligned}
$$

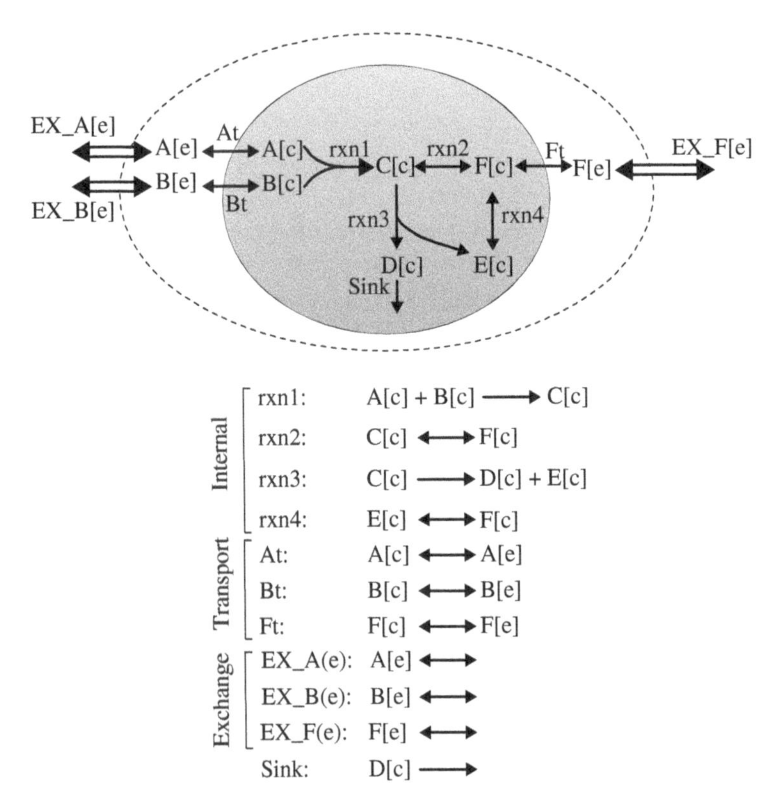

FIGURE 3.4 A demonstration metabolic network for Example 3.1.

These linear balances can be assembled into a stoichiometric matrix as follows:

	rxn1	rxn2	rxn3	rxn4	At	Bt	Ft	EX_A(e)	EX_B(e)	EX_F(e)	Sink
$A[e]$	0	0	0	0	1	0	0	−1	0	0	0
$B[e]$	0	0	0	0	0	1	0	0	−1	0	0
$A[c]$	−1	0	0	0	−1	0	0	0	0	0	0
$B[c]$	−1	0	0	0	0	−1	0	0	0	0	0
$C[c]$	1	−1	−1	0	0	0	0	0	0	0	0
$D[c]$	0	0	1	0	0	0	0	0	0	0	−1
$E[c]$	0	0	1	−1	0	0	0	0	0	0	0
$F[c]$	0	1	0	1	0	0	−1	0	0	0	0
$F[e]$	0	0	0	0	0	0	1	0	0	−1	0

Assuming that A is the limiting resource, bounds for each reaction are set as:

$$\begin{array}{lll}
\text{EX_A(e)}: & \text{LB}_{\text{EX_A(e)}} = -10, & \text{UB}_{\text{EX_A(e)}} = M \\
\text{EX_B(e)}: & \text{LB}_{\text{EX_B(e)}} = -M, & \text{LB}_{\text{EX_B(e)}} = M \\
\text{At}: & \text{LB}_{\text{At}} = -M, & \text{UB}_{\text{At}} = M \\
\text{Bt}: & \text{LB}_{\text{Bt}} = -M, & \text{UB}_{\text{Bt}} = M \\
\text{rxn1}: & \text{LB}_{\text{rxn1}} = 0, & \text{UB}_{\text{rxn1}} = M \\
\text{rxn2}: & \text{LB}_{\text{rxn2}} = -M, & \text{UB}_{\text{rxn2}} = M \\
\text{rxn3}: & \text{LB}_{\text{rxn3}} = 0, & \text{UB}_{\text{rxn3}} = M \\
\text{rxn4}: & \text{LB}_{\text{rxn4}} = -M, & \text{UB}_{\text{rxn4}} = M \\
\text{Sink}: & \text{LB}_{\text{Sink}} = 0, & \text{UB}_{\text{Sink}} = M \\
\text{Ft}: & \text{LB}_{\text{Ft}} = -M, & \text{UB}_{\text{Ft}} = M \\
\text{EX_F(e)}: & \text{LB}_{\text{EX_F(e)}} = 0, & \text{UB}_{\text{EX_F(e)}} = M
\end{array}$$

The FBA optimization formulation is assembled as follows:

$$\begin{array}{l}
\text{maximize} \quad z = v_{\text{Sink}} \\
\text{subject to} \\
-v_{\text{EX_A(e)}} + v_{\text{At}} = 0 \\
-v_{\text{EX_B(e)}} + v_{\text{Bt}} = 0 \\
-v_{\text{At}} - v_{\text{rxn1}} = 0 \\
-v_{\text{Bt}} - v_{\text{rxn1}} = 0 \\
v_{\text{rxn1}} - v_{\text{rxn2}} - v_{\text{rxn3}} = 0 \\
v_{\text{rxn3}} - v_{\text{Sink}} = 0 \\
v_{\text{rxn3}} - v_{\text{rxn4}} = 0 \\
v_{\text{rxn2}} + v_{\text{rxn4}} - v_{\text{Ft}} = 0 \\
v_{\text{Ft}} - v_{\text{EX_F(e)}} = 0 \\
-10 \leq v_{\text{EX_A(e)}} \leq M \\
-M \leq v_{\text{EX_B(e)}} \leq M \\
-M \leq v_{\text{At}} \leq M \\
-M \leq v_{\text{Bt}} \leq M \\
0 \leq v_{\text{rxn1}} \leq M \\
-M \leq v_{\text{rxn2}} \leq M \\
0 \leq v_{\text{rxn3}} \leq M \\
-M \leq v_{\text{rxn4}} \leq M \\
0 \leq v_{\text{Sink}} \leq M \\
-M \leq v_{\text{Ft}} \leq M \\
0 \leq v_{\text{EX_F(e)}} \leq M
\end{array}$$

$$v_{\text{rxn1}}, v_{\text{rxn2}}, v_{\text{rxn3}}, v_{\text{rxn4}}, v_{\text{Sink}}, v_{\text{At}}, v_{\text{At}}, v_{\text{Bt}}, v_{\text{Ft}}, v_{\text{EX\_A(e)}}, v_{\text{EX\_B(e)}}, v_{\text{EX\_F(e)}} \in \mathbb{R} \qquad \square$$

In the next example, we present an example of FBA calculations for a GSM model in order to predict the biomass production flux in a defined growth medium.

Example 3.2
Use the *i*AF1260 metabolic model of *E. coli* [16] to calculate the maximum biomass flux in a minimal medium with glucose as the carbon source under (i) aerobic and (ii) anaerobic conditions. Use a maximum of $10\frac{\text{mmol}}{\text{gDW}\cdot\text{h}}$ of glucose uptake as the basis.

Solution: The list of reactions and metabolites, the stoichiometric matrix, and the bounds on fluxes can be obtained from the supplementary material of Ref. 16. Table 3.2 summarizes the reaction bounds for various types of reactions in the model under aerobic and anaerobic conditions. The flux for a number of reactions was set to zero in accordance with regulatory constraints related to the presence of glucose and/or oxygen. The complete list of switched-off reactions can also be obtained from the supplementary material of Ref. 16.

Upon specifying the reaction bounds, the FBA optimization problem that calculates the maximum biomass flux is assembled. A GAMS implementation is available on the book's webpage. The reader can verify that the maximum value of the biomass flux under the minimal glucose aerobic and anaerobic conditions are 0.929027 and 0.230867 h^{-1}, respectively. The maximum biomass yield under the anaerobic condition is significantly lower as the TCA cycle is nonfunctional and significantly less ATP is produced in the absence of oxygen. To counteract this, *E. coli* growing anaerobically increases glucose uptake rate to restore biomass production rate despite the drop in the biomass yield. Note, however, that the faster consumption of glucose cannot be captured by FBA as here the upper bound on glucose uptake flux is pre-specified at $10\frac{\text{mmol}}{\text{gDW}\cdot\text{h}}$. □

TABLE 3.2 Lower and Upper Bounds on Reaction Fluxes for FBA Simulations Using the *i*AF1260 Metabolic Model of *Escherichia coli* in a Minimal Medium with Glucose as the Carbon Source and under Aerobic and Anaerobic Conditions ($M = 1000$).

Type of Reaction	(LB_j, UB_j)
Irreversible	$(0, M)$
Reversible	$(-M, M)$
Exchange reaction for metabolites not in the growth medium	$(0, M)$
Exchange reactions for metabolites available in excess in the growth medium[1]	$(-M, M)$
Exchange reaction for glucose	$(-10, M)$
Exchange reaction for oxygen (aerobic/anaerobic)	$(-20, M)/(0, M)$
Reactions that cannot carry any flux due to the regulatory constraints	$(0, 0)$
ATP maintenance	$(8.39, 8.39)$

[1] These reactions include EX_ca2(e), EX_cl(e), EX_co2(e), EX_cobalt2(e), EX_cu2(e), EX_fe2(e), EX_fe3(e), EX_h(e), EX_h2o(e), EX_k(e), EX_mg2(e), EX_mn2(e), EX_mobd(e), EX_na1(e), EX_nh4(e), EX_pi(e), EX_so4(e), EX_tungs(e), EX_zn2(e), EX_cbl1(e).

In the subsequent sections, we describe a number of computational methods based on LP that have been very useful in answering important inquiries in metabolic networks.

3.4 SIMULATING GENE KNOCKOUTS

Metabolic gene knockouts can be captured by FBA using the GPR relationships that translate the effect of genetic interventions at the reaction level. It can be captured within the FBA formalism in a straightforward manner with the following constraint:

$$v_j = 0, \quad \forall j \in J^{\mathrm{KO}}$$

where J^{KO} is the set of reactions that need to be inactivated as a result of the gene knockout(s).

Gene/Reaction Essentiality and Synthetic Lethality A gene or reaction whose deletion is lethal (i.e., arrests growth) is called an *essential gene* or *reaction.* "No growth" is captured in FBA as a maximum biomass flux of zero or less than a pre-specified viability threshold. Gene non-essentiality implies that there may exist other genes that provide backup for the loss of function. Errors in prediction of gene essentiality imply that the metabolic model erroneously contains reactions that complement for the lost gene function. In contrast, errors in the prediction of gene non-essentiality are generally indicative of missing annotations in the model (see Chapter 6). Moving to more than one gene deletion at a time, a synthetic lethal gene pair is a pair of genes whose simultaneous deletion is lethal but individual deletions are not (i.e., individual genes are nonessential). For example, two genes providing isozymes for an essential reaction form a synthetic lethal gene pair. The concept of synthetic lethality can be extended to higher orders in a straightforward manner. It is worth noting that the set of essential or synthetic lethal genes (or by extension reactions) is highly dependent on the environmental and growth condition. Interested readers can refer to Ref. [31] for more details.

Example 3.3

Formulate an optimization problem to identify the list of all essential reactions under the aerobic and anaerobic condition using the *i*AF1260 metabolic model of *E. coli* [16]. Use a maximum of $10\,\dfrac{\mathrm{mmol}}{\mathrm{gDW}\cdot\mathrm{h}}$ of glucose uptake as the basis.

Solution: To identify the list of all essential reactions, the following problem needs to be solved for each reaction $j^* \in J$ in the network:

$$\begin{aligned}
&\text{maximize} \quad z = v_{\mathrm{biomass}} \quad [\text{FBA-KO}]\\
&\text{subject to}\\
&\sum_{j\in J} S_{ij} v_j = 0, \quad \forall i \in I\\
&LB_j \le v_j \le UB_j, \quad \forall j \in J\\
&v_{j^*} = 0\\
&v_j \in \mathbb{R}
\end{aligned}$$

The lower and upper bounds on reaction fluxes under the aerobic and anaerobic conditions can be obtained from Table 3.2. A pseudo-code to enumerate the list of all essential reactions is as follows:

Find the maximum biomass flux for the wild-type network $v_{\text{biomass}}^{\text{max,WT}}$ using [FBA]

Set the desired viability threshold $0 \le \varepsilon < 1$

$J^{\text{essential}} = \{\ \}$

For $j^* \in J - J^{\text{exch.}}$

Solve $\left[\text{FBA-KO}\right]$

If $v_{\text{biomass}}^{\text{max,KO}} \le \varepsilon v_{\text{biomass}}^{\text{max,WT}}$

Add j^* to $J^{\text{essential}}$

Here, $J^{\text{exch.}}$ is the set of exchange reactions, which are excluded because they do not represent biological functions. A GAMS implementation of the gene essentiality analysis for this example is available on the Book's webpage. □

3.5 MAXIMUM THEORETICAL YIELD

3.5.1 Maximum Theoretical Yield of Product Formation

Many metabolic engineering applications of FBA involve the overproduction of a target metabolite. The most straightforward way for calculating the maximum product yield [MPY] capacity of the organism is to maximize the flux of the exchange reaction exporting that product metabolite ($v_{\text{EX\_P(e)}}$) (see Fig. 3.5):

$$
\begin{aligned}
&\text{maximize} \quad z = v_{\text{EX\_P(e)}} \quad \left[\text{MPY}\right] \\
&\text{subject to} \\
&\sum_{j \in J} S_{ij} v_j = 0, \quad \forall i \in I \\
&LB_j \le v_j \le UB_j, \quad \forall j \in J \\
&v_j \in \mathbb{R}, \quad \forall j \in J
\end{aligned}
$$

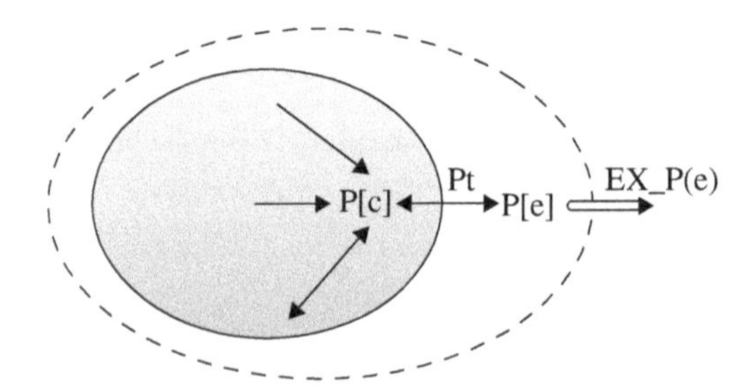

FIGURE 3.5 The maximum product formation yield calculation requires the maximization of the flux of the exchange reaction corresponding to that product.

The solution to this problem provides the *maximum theoretical yield* of the product. This metric is usually used as the basis to gauge the efficiency of different substrates, growth conditions, and gene knockout/knockin combinations toward the targeted overproduction. When calculating the maximum yield, it is very important to check the bounds on potentially limiting resources (e.g., glucose or oxygen). In general, the bound on only a single resource will become active at the optimum solution unless they are preset at values that are consistent with the utilization ratios (e.g., 1:2 ratio of glucose to oxygen uptake rates in *E. coli*). It is possible for the limiting resource to change depending on the exported metabolites, regulation and growth media composition, and concentrations. In addition, the bounds of all transport reactions need to be carefully scrutinized to prevent the formation of thermodynamically infeasible cycles that would shuttle in and out of the cytosol pairs of metabolites leading to unbounded fluxes (see Chapter 6). Maximum yield calculations can be carried out not only for the *wild-type* organism but also for mutants thereof to assess the effect of different genetic backgrounds on yield. Frequently, maximum product yield calculations include lower bounds for the production of biomass providing a quantitative basis for tracing trade-offs between product vs. biomass yield.

3.5.2 Biomass vs. Product Trade-Off

In most cases, the production of a chemical of interest is in direct competition with biomass formation. As a result, maximization of the product formation flux is associated with a zero biomass flux (i.e., no growth). This can be problematic for metabolic engineering applications where a minimum level of growth (biomass formation) is required. This can be ascertained in FBA calculations by the addition of a constraint to the [MPY] formulation requiring the biomass flux in the network to be at least a pre-specified fraction ($0 \le f \le 1$) of the maximum theoretical biomass flux identified beforehand through FBA:

$$v_{\text{biomass}} \ge f\ v_{\text{biomass}}^{\max}.$$

Generally, the value of f is dependent upon the application, but $f = 0.1$ (i.e., 10%) is typically used to impose a minimal level of growth. Alternatively, one may choose to use not just a single value for f but instead assess the effect of different values for f on product yield. The formal trade-off between biomass and target metabolite production can be assessed and graphically illustrated by sequentially maximizing and minimizing the exchange flux of the target product for varying levels of the biomass production:

$$\begin{aligned}
&\text{maximize / minimize} \quad z = v_{\text{EX_P(e)}} \\
&\text{subject to} \\
&\sum_{j \in J} S_{ij} v_j = 0, \quad \forall i \in I \\
&LB_j \le v_j \le UB_j, \quad \forall j \in J \\
&v_{\text{biomass}} \ge f\, v_{\text{biomass}}^{\max} \\
&v_j \in \mathbb{R}
\end{aligned}$$

This optimization problem is solved for different values of the parameter $0 \leq f \leq 1$ and the obtained minimum and maximum product formation fluxes are plotted as a function of the maximum biomass flux (see Example 3.4).

Example 3.4

Use the *i*AF1260 metabolic model of *E. coli* [16] to (a) calculate the maximum theoretical yield of ethanol production in a minimal medium with glucose as the sole carbon source under aerobic and anaerobic conditions. (b) Generate the tradeoff plot for ethanol production under the aerobic and anaerobic growth conditions in (a). Use a maximum of $10 \frac{\text{mmol}}{\text{gDW} \cdot \text{h}}$ of glucose uptake as the basis for both (a) and (b).

Solution:

(a) The maximum theoretical yield for ethanol production can be found by solving [MPY] with $v_{\text{EX\_etoh(e)}}$ (exchange reaction flux for ethanol in *i*AF1260) as the objective function. The bounds for different reactions are again taken from Table 3.2. A GAMS implementation of this problem is available on the book's website. Readers can verify that the maximum theoretical yield for ethanol production is $20 \frac{\text{mmol}}{\text{gDW} \cdot \text{h}}$ under both aerobic and anaerobic condition.

(b) The trade-off plot can be obtained by solving the following optimization problem for varying values of parameter $0 \leq f \leq 1$:

$$\begin{aligned}
&\text{maximize}(\text{and minimize}) \quad z = v_{\text{EX_etoh(e)}} \\
&\text{subject to} \\
&\sum_{j \in J} S_{ij} v_j = 0, \quad \forall i \in I \\
&LB_j \leq v_j \leq UB_j, \quad \forall j \in J \\
&v_{\text{biomass}} = f \, v_{\text{biomass}}^{\text{max,WT}} \\
&v_j \in \mathbb{R}
\end{aligned}$$

A GAMS implementation of this problem is available on the Book's website. The trade-off plots for the aerobic and anaerobic conditions are shown in Figure 3.6.

As evident from the Figure 3.6, ethanol production is decoupled from biomass production under the aerobic condition, while it becomes coupled to biomass formation under the anaerobic condition. This is because the ethanol production pathway is coupled with the regeneration of redox cofactor NAD^+, which is required to drive reaction glyceraldehyde-3-phosphate dehydrogenase (GAPD) in glycolysis under the anaerobic condition for which the electron transport chain is inactive. As glycolysis generates many compounds required for the biosynthesis of biomass precursors, redox balance of NAD^+ makes ethanol an obligatory by-product of biomass formation. □

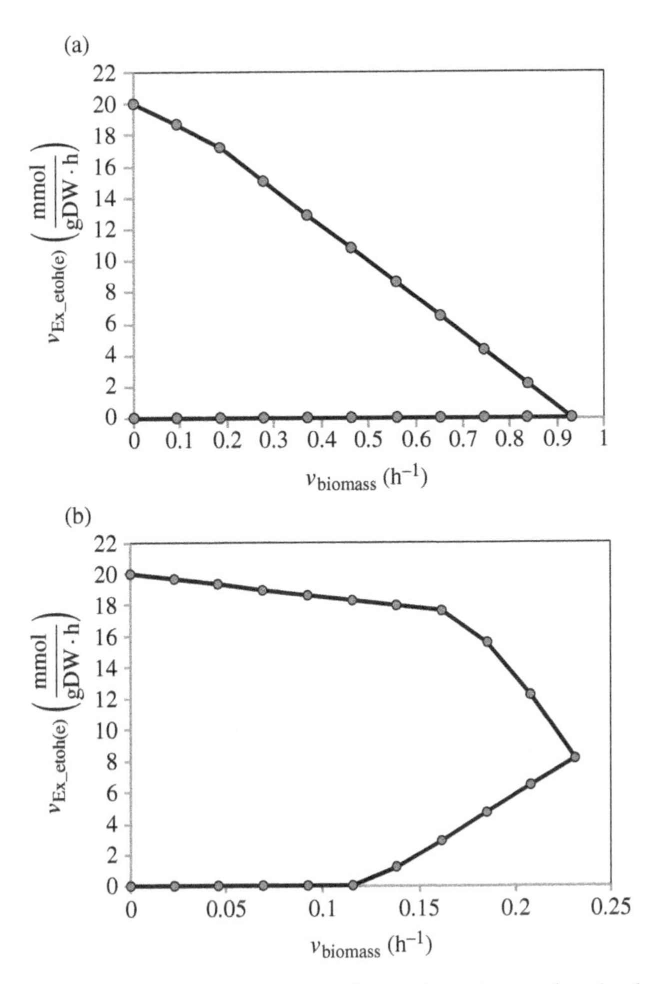

FIGURE 3.6 Trade-off plot for ethanol production in *Escherichia coli* under the (a) aerobic and (b) anaerobic condition in a minimal medium with glucose as the carbon source (see Example 3.4).

3.6 FLUX VARIABILITY ANALYSIS (FVA)

Generally, LP problems arising in FBA involve alternate optima with many flux distributions leading to the same maximum biomass flux value. This means that different nonbasic variables can be brought into the basis without changing the optimal value of the objective function as they have a relative profit of zero (see Chapter 2). The number of such fluxes determines the dimensionality of the alternate optima space. These variables can be identified using flux variability analysis (FVA) [32] where the flux of each reaction in the network is maximized and minimized, one at a time, while fixing the biomass flux at some fraction f of the optimal value obtained from FBA $\left(v_{\text{biomass}}^{\max}\right)$:

$$\begin{aligned}
&\text{maximize}\left(\text{and minimize}\right) \quad z = v_j \quad [\text{FVA}] \\
&\text{subject to} \\
&\sum_{j \in J} S_{ij} v_j = 0, \quad \forall i \in I \\
&LB_j \leq v_j \leq UB_j, \quad \forall j \in J \\
&v_{\text{biomass}} = f\, v_{\text{biomass}}^{\max} \\
&v_j \in \mathbb{R}, \quad \forall j \in J
\end{aligned}$$

Therefore, if there are N reactions in the network, one needs to solve $2N$ optimization problems. Reactions with unequal minimum and maximum fluxes have a range of values consistent with the imposed optimality criterion. FVA provides a convenient way to inscribe within a box the N-dimensional polytope describing the feasible region of the FBA problem. Due to the high dimensionality of the problem, the vast majority of points within the box are in fact outside the feasible region. This implies that results obtained from FVA analysis must be carefully interpreted. Even though a particular flux can vary without affecting the value of the objective function, this typically requires that the remaining fluxes in the network change in concert (see Fig. 3.7). Therefore, FVA does not resolve the polytope describing all alternate optimal solutions but rather inscribes it within the smallest possible box. Note that the basic FVA optimization formulation can be extended with additional constraints to account for reaction removals, imposed over-production targets, or experimental flux measurements. Interested readers can refer to Ref. [32] for more details on the flux variability analysis of metabolic networks.

Blocked Reactions FVA can be used to identify *blocked reactions* in a metabolic network under a given environmental and growth condition. Blocked reactions are

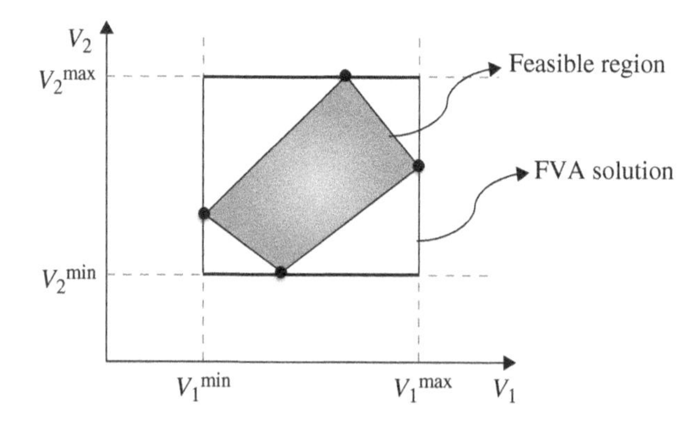

FIGURE 3.7 Flux variability analysis (FVA) provides lower and upper bounds on all reaction fluxes (solid circles on the graph). These reaction bounds are reachable only if the remaining reactions adjust their values accordingly. Source: Adapted from Ref. 32 with permission of Elsevier.

reactions with both a minimum and maximum flux value (identified by FVA) equal to zero under the examined condition. Reactions can be blocked due to the presence of regulatory constraints that are imposed in the FBA simulations or due to the presence of network gaps in the model (i.e., metabolites for which there is no production or consumption pathway in the model). The set of blocked reactions in a metabolic network can be dependent on the growth condition or independent of the growth and environmental conditions if they are caused by network gaps. The importance of identifying blocked reactions is twofold: (i) they can be used as a basis to improve the quality of metabolic network reconstructions by helping to identify dead-end metabolites (see Chapter 6 for more details), or (ii) they can significantly reduce the search space of FBA-based methods by pre-setting their value to zero (see Chapter 4).

Example 3.5

Perform FVA under aerobic minimal glucose using the *i*AF1260 metabolic model [16]. Report the reactions whose flux values are fixed (under max biomass) and those with a different minimum and maximum flux value. In addition, report the list of blocked reactions in the model.

Solution: Problem [FVA] is solved for each reaction in the network. The value of $v_{\text{biomass}}^{\text{max}}$ was already found in Example 3.2 to be $0.929027\,\text{h}^{-1}$. We examine flux variability using the criterion $\left|v_j^{\text{max}} - v_j^{\text{min}}\right| < \varepsilon = 10^{-6}$. Blocked reactions are identified by checking whether $\left|v_j^{\text{min}}\right| = \left|v_j^{\text{max}}\right| \leq \varepsilon$. As many as 674 reactions are found to have different minimum and maximum flux values under maximum biomass formation requirement. Out of the 1709 reactions whose fluxes are fixed, 755 correspond to blocked reactions. A GAMS implementation of the simulations for this example is available on the book's website. □

3.7 FLUX COUPLING ANALYSIS

FVA cannot capture reaction flux dependencies arising from stoichiometric balances between pairs of reactions. To this end, the range of possible values of reaction *flux ratios* rather than individual fluxes needs to be assessed. Flux Coupling Finder (FCF) [33] globally identifies these flux ratio dependencies in GSM networks. FCF identifies the minimum and maximum flux ratio for every reaction pair (j_1, j_2), $\forall j_1, j_2 \in J$ in the network and then assembles results into coupled sets. The FCF optimization problem is formulated as follows:

$$\text{maximize}\left(\text{and minimize}\right) \quad R = \frac{v_{j_1}}{v_{j_2}} \quad [\text{FCF}]$$

subject to

$$\sum_{j \in J} S_{ij} v_j = 0, \quad \forall i \in I$$

$$LB_j \leq v_j \leq UB_j, \quad \forall j \in J$$

$$v_j \geq 0, \quad \forall j \in J$$

Unlike FVA, which corresponds to an LP problem, FCF yields a linear fractional programming problem. In addition, in FCF all reversible reactions are decomposed into a forward and a reverse reaction thus ensuring that all reaction fluxes are non-negative and enabling the transformation of the problem into a LP problem (see Chapter 2). Prior to performing FCF, one needs to perform FVA to identify and exclude all blocked reactions from the flux coupling analysis. As discussed in Chapter 2, the linear fractional program [FCF] can be transformed to an LP problem by using the following variable transformations:

$$t = \frac{1}{v_{j_2}}, \quad \overline{v}_j = tv_j, \quad \forall j \in J$$

This converts [FCF] to the following LP:

$$\begin{aligned}
&\text{maximize (and minimize)} \quad R = \overline{v}_{j_1} \\
&\text{subject to} \\
&\sum_{j \in J} S_{ij}\overline{v}_j = 0, \quad \forall i \in I \\
&tLB_j \le \overline{v}_j \le tUB_j, \quad \forall j \in J \\
&\overline{v}_{j_2} = 1 \\
&\overline{v}_j \ge 0, \quad \forall j \in J \\
&t \ge 0
\end{aligned}$$

As described in Chapter 2, variable t rescales the denominator of the ratio to become equal to one. The value of the scaling factor t is revealed upon solving the optimization problem. Fluxes v_j can be recovered from $\overline{v}_j$ using the scaling factor t. By solving [FCF], a number of different outcomes are possible for the ratio of a pair of arbitrary reaction fluxes v_1 and v_2 [33] (see Fig. 3.8):

(i) *Directional coupling* ($v_1 \rightarrow v_2$): If $R_{min} = 0$ and $R_{max} = c > 0$ where c is finite. A nonzero flux for reaction 1 implies a nonzero flux for reaction 2 ($v_2 \ge v_1 / c$) but not necessarily the reverse.

(ii) *Partial coupling* ($v_1 \leftrightarrow v_2$): If $R_{min} = c_1$ and $R_{max} = c_2$ with $c_1 \ne c_2$ and c_1 and c_2 are finite. A nonzero flux for reaction 1 implies a nonzero but variable flux for reaction 2 and vice-versa.

(iii) *Full coupling* ($v_1 \Leftrightarrow v_2$): If $R_{min} = R_{max} = c$ where c is finite. A nonzero flux for reaction 1 implies a fixed nonzero flux for reaction 2 and vice-versa.

(iv) *Directional coupling* ($v_2 \rightarrow v_1$): If $R_{min} = c > 0$ and $R_{max} = \infty$, where c is finite. A nonzero flux for reaction 2 implies a nonzero flux for reaction 1 ($v_1 \ge cv_2$) but not necessarily the reverse.

Any reaction pair that does not conform to any of the categories described above is classified as an *uncoupled* pair. Given these definitions, the following

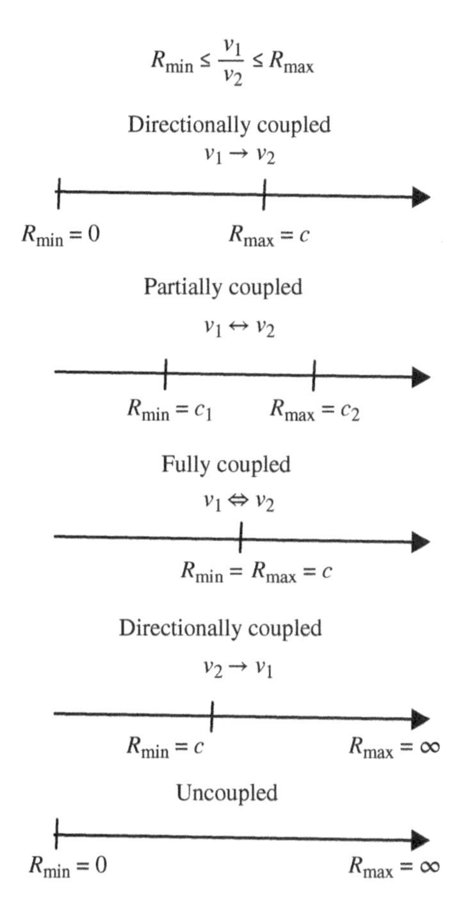

FIGURE 3.8 All possible coupling outcomes in a metabolic network. Source: Adapted from Ref. 33 with permission of Cold Spring Harbor Laboratory Press.

pseudo-code provides a workflow for the identification of all coupled reaction pairs in a genome-scale model through the iterative solution of the [FCF] optimization formulation:

Identify the set of all blocked reactions $J^{blocked}$

$AlreadyCoupled(j) = 0, \quad \forall j \in J$

$For\ j_1 \in J - J^{blocked}$ AND $AlreadyCoupled(j_1) = 0$

$\quad For\ j_2 \in J - J^{blocked}$ AND $ord(j_2) > ord(j_1)$

$\quad\quad$ Find R_{min} and R_{max} by solving $[FCF]$

$\quad\quad If\ R_{min} = 0$ and $R_{max} = c > 0$

$\quad\quad\quad v_{j_1} \rightarrow v_{j_2}$

$$\textit{Elseif } R_{min} = c_1 > 0 \text{ AND } R_{max} = c_2 > 0 \text{ AND } c_1 \neq c_2$$

$$v_{j_1} \leftrightarrow v_{j_2}$$

$$AlreadyCoupled(j_2) = 1$$

$$\textit{Elseif } R_{min} = R_{max} = c > 0$$

$$v_{j_1} \Leftrightarrow v_{j_2}$$

$$AlreadyCoupled(j_2) = 1$$

$$\textit{Elseif } R_{min} = c > 0 \text{ and } R_{max} = \infty$$

$$v_{j_2} \rightarrow v_{j_1}$$

Here, operator *ord* denotes the ordinality of a set member. Parameter $AlreadyCoupled(j) = 1$ implies that this reaction has already been assigned to a coupled reaction set and should not be examined for any additional coupling relations. If reaction 1 is coupled with reactions 2 and 3, then reactions 2 and 3 are also coupled. The recursive application of this property allows for the assembly of *coupled sets* composed of reactions that are coupled with one another. Interested readers are encouraged to explore a number of follow-up efforts to FCF that introduced more efficient computational implementations for enumerating coupling relations between fluxes [34, 35].

A number of important coupling relations that can be gleaned by applying flux coupling analysis are illustrated in Figure 3.9. Reactions v_1, v_2 and v_3 in the figure are directionally coupled to flux v^*, implying that a nonzero flux for either of v_1, v_2 or v_3 guarantees a nonzero flux for v^*. Conversely, if the flux of v^* is set to zero (e.g., as a result of a gene knock-out), then v_1, v_2 and v_3 will not be able to carry any flux in which case they are referred to as *affected reactions*. The effect of a gene knock-out is propagated upstream to all directionally coupled reactions whose flux is switched off. Alternatively, a nonzero flux for v^* implies a nonzero flux for the directionally coupled fluxes v_4, v_5 and v_6. Consequently, the direct elimination of either of v_4, v_5 or v_6 will force the flux through v^* to zero. Therefore, reactions v_4, v_5 or v_6 can be thought of as

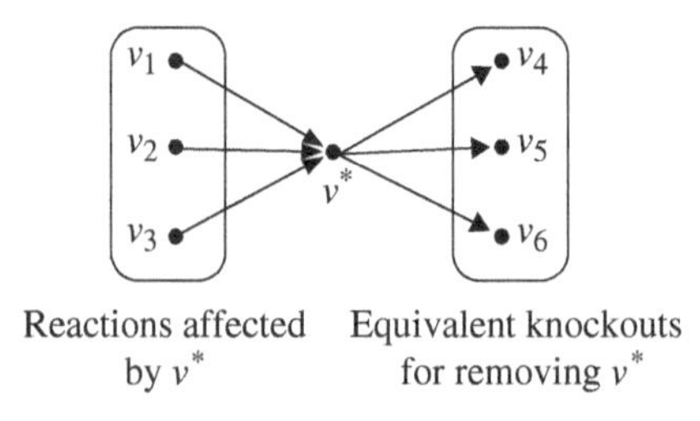

FIGURE 3.9 Affected and equivalent knockout reaction sets for a reaction v^*. Here, solid circles represent reactions and arrows represent coupling relationships (i.e., $v_1, v_2, v_3 \rightarrow v^*$ and $v^* \rightarrow v_4, v_5, v_6$). Deletion of v^* will force the flux of reactions in the affected reaction set (i.e., v_1, v_2 and v_3) to zero. Alternatively, removing either of reactions in the equivalent knockout set (i.e., v_4, v_5 and v_6) will force the flux of reaction v^* to zero. Source: Adapted from Ref. 33 with permission of Cold Spring Harbor Laboratory Press.

equivalent knockout alternatives for reaction v^*. The implication of this analysis is that the set of reactions that are directionally coupled to biomass are necessary for growth thus forming an *essential core* of reactions. This essential core of reactions along with a set of additional reactions that can be drawn from a set of non-essential alternatives form *minimal reaction sets* that support growth. Similarly, the manipulation (up/down-modulation) of only one flux in a fully coupled reaction set will have the same effect as manipulating any other reaction in that set. FCF has thus implications not only in the analysis of the structural properties of a metabolic network but also in guiding genetic manipulations for microbial strain design (see Chapters 4 and 8).

EXERCISES

3.1 Use the *i*AF1260 GSM model of *E. coli* [16] to solve the following problems. Set the lower and upper bounds for the fluxes of each reaction as mentioned in Table 3.2. The reaction regulation information (i.e., inactive reactions under specific environmental conditions) is available in the Supplementary Information of the *i*AF1260 model paper [16].

(a) What is the dimensionality of the subspace spanning metabolic fluxes consistent with maximum biomass formation under aerobic glucose conditions? How does the dimentionality change when biomass is maximized under anaerobic conditions?

(b) Derive the dual of the biomass maximization problem [FBA] and show that its minimum matches the maximum of the primal representation.

(c) Evaluate the maximum and minimum succinate production flux [FVA] in anaerobic conditions for biomass yield ranging from 0 to 100% of the maximum theoretical yield in increments of 10%. Do the same after removing the alcohol dehydrogenase, formate-hydrogen lyase and the lactate dehydrogenase reactions. Plot your results and comment on their implication on the best production mode.

(d) Perform flux coupling analysis [FCF] between the biomass flux and all other fluxes in the network in aerobic glucose minimal condition. What is the number of fluxes that are either partially or fully coupled to the biomass flux? What is the implication for these reactions?

(e) Identify all reactions that are non-essential under aerobic minimal glucose conditions but become essential under anaerobic conditions.

3.2 Consider a proprietary microbial production system whose metabolic model has been reconstructed. It includes $i = 1,\ldots,N$ metabolites and $j = 1,\ldots,M$ reactions with S_{ij} denoting the stoichiometric coefficient of metabolite i in reaction j. Assume aerobic minimal glucose conditions with a maximum glucose uptake rate of 1 mole (as the basis), maximum oxygen uptake rate of $v_{O_2}^{\text{uptake}}$ moles/mole glucose and non-growth-associated ATP maintenance fixed at $v_{\text{ATP}}^{\text{maint.}}$ moles/mole glucose.

(a) This strain needs to be optimized for the production of a protein whose amino acid molar composition is r_k, where $k = 1,\ldots,20$ denotes all amino acids and MW_k represents the molecular weight of each amino acid. Derive the optimization formulation that identifies the maximum theoretical yield (gr protein/mole glucose) for growth-arrested cells.

(b) Economic considerations require that the team look for an alternative carbon substrate than glucose. Let c_i be the price per gram and MW_i be the molecular weight of a list of potential carbon substrates $I^{\text{substrate}}$. All other nutrients are available in excess. Prepare an optimization formulation that solves for the cheapest carbon source per unit of biomass growth. Is it possible for the cheapest solution to involve a mixture of carbon substrates?

REFERENCES

1. Voet D, Voet JG: *Biochemistry*, 4th edn. Hoboken, N.J.: John Wiley & Sons; 2011.
2. Nelson DL, Lehninger AL, Cox MM: *Lehninger principles of biochemistry*, 5th edn. New York: W.H. Freeman; 2008.
3. Palsson BØ: Systems Biology: Constraint-based Reconstruction and Analysis, Cambridge: Cambridge University Press; 2015.
4. Schomburg I, Chang A, Placzek S, Söhngen C, Rother M, Lang M, Munaretto C, Ulas S, Stelzer M, Grote A *et al*: BRENDA in 2013: integrated reactions, kinetic data, enzyme function data, improved disease classification: new options and contents in BRENDA. *Nucleic Acids Res* 2013, **41**(Database issue):D764–772.
5. Caspi R, Altman T, Billington R, Dreher K, Foerster H, Fulcher CA, Holland TA, Keseler IM, Kothari A, Kubo A *et al*: The MetaCyc database of metabolic pathways and enzymes and the BioCyc collection of Pathway/Genome Databases. *Nucleic Acids Res* 2014, **42**(Database issue):D459–471.
6. Kanehisa M, Goto S, Sato Y, Kawashima M, Furumichi M, Tanabe M: Data, information, knowledge and principle: back to metabolism in KEGG. *Nucleic Acids Res* 2014, **42**(Database issue):D199–205.
7. Kadir TA, Mannan AA, Kierzek AM, McFadden J, Shimizu K: Modeling and simulation of the main metabolism in *Escherichia coli* and its several single-gene knockout mutants with experimental verification. *Microb Cell Fact* 2010, **9**:88.
8. Rizzi M, Baltes M, Theobald U, Reuss M: In vivo analysis of metabolic dynamics in Saccharomyces cerevisiae: II. Mathematical model. *Biotechnol Bioeng* 1997, **55**(4):592–608.
9. Chassagnole C, Noisommit-Rizzi N, Schmid JW, Mauch K, Reuss M: Dynamic modeling of the central carbon metabolism of *Escherichia coli*. *Biotechnol Bioeng* 2002, **79**(1):53–73.
10. Khodayari A, Zomorrodi AR, Liao JC, Maranas CD: A kinetic model of *Escherichia coli* core metabolism satisfying multiple sets of mutant flux data. *Metab Eng* 2014, **25C**:50–62.
11. Wright BE, Butler MH, Albe KR: Systems analysis of the tricarboxylic acid cycle in Dictyostelium discoideum. I. The basis for model construction. *J Biol Chem* 1992, **267**(5):3101–3105.
12. Bakker BM, Michels PA, Opperdoes FR, Westerhoff HV: Glycolysis in bloodstream form Trypanosoma brucei can be understood in terms of the kinetics of the glycolytic enzymes. *J Biol Chem* 1997, **272**(6):3207–3215.

13. Zomorrodi AR, Lafontaine Rivera JG, Liao JC, Maranas CD: Optimization-driven identification of genetic perturbations accelerates the convergence of model parameters in ensemble modeling of metabolic networks. *Biotechnol J* 2013, **8**(9):1090–1104.
14. Zamora-Sillero E, Hafner M, Ibig A, Stelling J, Wagner A: Efficient characterization of high-dimensional parameter spaces for systems biology. *BMC Syst Biol* 2011, **5**:142.
15. Kim TY, Sohn SB, Kim YB, Kim WJ, Lee SY: Recent advances in reconstruction and applications of genome-scale metabolic models. *Curr Opin Biotechnol* 2012, **23**(4): 617–623.
16. Feist AM, Henry CS, Reed JL, Krummenacker M, Joyce AR, Karp PD, Broadbelt LJ, Hatzimanikatis V, Palsson B: A genome-scale metabolic reconstruction for *Escherichia coli* K-12 MG1655 that accounts for 1260 ORFs and thermodynamic information. *Mol Syst Biol* 2007, **3**:121.
17. Palsson B: *Systems biology: Properties of reconstructed networks*. In. Cambridge ; New York: Cambridge University Press; 2006: 1 online resource (xii, 322 p.).
18. Thiele I, Palsson B: A protocol for generating a high-quality genome-scale metabolic reconstruction. *Nat Protoc* 2010, **5**(1):93–121.
19. Orth JD, Conrad TM, Na J, Lerman JA, Nam H, Feist AM, Palsson B: A comprehensive genome-scale reconstruction of *Escherichia coli* metabolism--2011. *Mol Syst Biol* 2011, **7**:535.
20. Aung HW, Henry SA, Walker LP: Revising the Representation of Fatty Acid, Glycerolipid, and Glycerophospholipid Metabolism in the Consensus Model of Yeast Metabolism. *Ind Biotechnol (New Rochelle N Y)* 2013, **9**(4):215–228.
21. Famili I, Forster J, Nielsen J, Palsson BO: Saccharomyces cerevisiae phenotypes can be predicted by using constraint-based analysis of a genome-scale reconstructed metabolic network. *Proc Natl Acad Sci U S A* 2003, **100**(23):13134–13139.
22. Cotten C, Reed JL: Mechanistic analysis of multi-omics datasets to generate kinetic parameters for constraint-based metabolic models. *BMC Bioinformat* 2013, **14**:32.
23. De Martino D, Capuani F, Mori M, De Martino A, Marinari E: Counting and correcting thermodynamically infeasible flux cycles in genome-scale metabolic networks. *Metabolites* 2013, **3**(4):946–966.
24. Schellenberger J, Lewis NE, Palsson B: Elimination of thermodynamically infeasible loops in steady-state metabolic models. *Biophys J* 2011, **100**(3):544–553.
25. Segrè D, Vitkup D, Church GM: Analysis of optimality in natural and perturbed metabolic networks. *Proc Natl Acad Sci U S A* 2002, **99**(23):15112–15117.
26. Schuetz R, Kuepfer L, Sauer U: Systematic evaluation of objective functions for predicting intracellular fluxes in *Escherichia coli*. *Mol Syst Biol* 2007, **3**:119.
27. Feist AM, Palsson BO: The biomass objective function. *Curr Opin Microbiol* 2010, **13**(3):344–349.
28. Harcombe WR, Delaney NF, Leiby N, Klitgord N, Marx CJ: The ability of flux balance analysis to predict evolution of central metabolism scales with the initial distance to the optimum. *PLoS Comput Biol* 2013, **9**(6):e1003091.
29. Shlomi T, Berkman O, Ruppin E: Regulatory on/off minimization of metabolic flux changes after genetic perturbations. *Proc Natl Acad Sci U S A* 2005, **102**(21):7695–7700.
30. Lee D, Smallbone K, Dunn WB, Murabito E, Winder CL, Kell DB, Mendes P, Swainston N: Improving metabolic flux predictions using absolute gene expression data. *BMC Syst Biol* 2012, **6**:73.

31. Suthers PF, Zomorrodi A, Maranas CD: Genome-scale gene/reaction essentiality and synthetic lethality analysis. *Mol Syst Biol* 2009, **5**:301.
32. Mahadevan R, Schilling CH: The effects of alternate optimal solutions in constraint-based genome-scale metabolic models. *Metab Eng* 2003, **5**(4):264–276.
33. Burgard AP, Nikolaev EV, Schilling CH, Maranas CD: Flux coupling analysis of genome-scale metabolic network reconstructions. *Genome Res* 2004, **14**(2):301–312.
34. David L, Marashi SA, Larhlimi A, Mieth B, Bockmayr A: FFCA: a feasibility-based method for flux coupling analysis of metabolic networks. *BMC Bioinformat* 2011, **12**:236.
35. Larhlimi A, David L, Selbig J, Bockmayr A: F2C2: a fast tool for the computation of flux coupling in genome-scale metabolic networks. *BMC Bioinformat* 2012, **13**:57.

4

MODELING WITH BINARY VARIABLES AND MILP FUNDAMENTALS

Binary variables are frequently used in optimization formulations in metabolic network analysis for modeling the presence (or absence) of reactions and for imposing restrictions on the total number of active reactions or regulatory interactions. This chapter focuses on modeling tasks using binary variables followed by fundamental concepts and solution procedures for mixed-integer linear programming (MILP) problems.

In general, optimization variables can be continuous or discrete. An *integer variable* is a variable that is restricted to assume only integer (discrete) values. A *binary variable* is an integer variable that can assume a value of only zero or one. An MILP problem is a mathematical optimization problem, where the objective function and all constraints are linear and a subset of the variables is restricted to be integer. A general MILP problem is represented as follows:

$$\begin{aligned}
&\text{minimize} \quad z = \boldsymbol{c}^{\mathrm{T}}\boldsymbol{x} + \boldsymbol{d}^{\mathrm{T}}\boldsymbol{y} \quad [\text{MILP}] \\
&\text{subject to} \\
&\boldsymbol{A}\boldsymbol{x} + \boldsymbol{B}\boldsymbol{y} \geq \boldsymbol{b} \\
&\boldsymbol{x} \in \mathbb{R}^N, \quad \boldsymbol{y} \in \mathbb{Z}^M
\end{aligned}$$

where $\mathbb{R}$ and $\mathbb{Z}$ represent the set of real and integer numbers, whereas $\boldsymbol{x}$ and $\boldsymbol{y}$ are vectors of continuous and integer variables, respectively. Furthermore, $\boldsymbol{c}$, $\boldsymbol{d}$ and $\boldsymbol{b}$

Optimization Methods in Metabolic Networks, First Edition. Costas D. Maranas and Ali R. Zomorrodi.

are $N \times 1$, $M \times 1$ and $L \times 1$ parameter vectors and $\boldsymbol{A}$ and $\boldsymbol{B}$ are $L \times N$ and $L \times M$ matrices of coefficients, respectively. If all variables are restricted to integer values, then the problem is referred to as an integer linear programming (ILP) problem, whereas if all variables can assume real values the problem reverts back to a linear programming (LP). The most widely used form of an MILP is a special case where all integer variables are restricted to be binary ($\boldsymbol{y} \in \{0,1\}^M$). These problems are often referred to as 0–1 MILPs.

Every MILP with bounded integer variables can be transformed into a 0–1 MILP. A straightforward way of replacing an integer variable z with binary variables y_k is as follows:

$$z = \sum_{k=z^L}^{z^U} k y_k \tag{4.1}$$

$$\sum_{k=z^L}^{z^U} y_k = 1 \tag{4.2}$$

$$y_k = \{0,1\}$$

where z^L and z^U are the lower and upper bound on z, respectively. Although this description is the most direct way of translating bounded integer variables, there is a more compact way that uses fewer binary variables. This method is sometimes referred to as the *bitwise* representation of an integer variable (assuming $z^L \geq 0$):

$$z = \sum_{k=0}^{K} 2^k y_k \tag{4.3}$$

$$z^L \leq \sum_{k=0}^{K} 2^k y_k \leq z^U \tag{4.4}$$

where $K = [\log_2(z^U - z^L)]$ with "[.]" denoting the integer part of a real number. For example, if $z \in \{0, 1, 2, 3, 4, 5\}$ then

$$K = \left[\log_2\left(z^U - z^L\right)\right] = \left[\log_2(5-0)\right] = [2.322] = 2 \text{ and}$$
$$z = 2^0 y_0 + 2^1 y_1 + 2^2 y_2 = y_0 + 2y_1 + 4y_2$$

Constraint 4.4 is needed to ensure that the original lower and upper bounds on z are not violated (e.g., for $y_1 = y_2 = y_3 = 1$ the sum above becomes equal to 7). Even though the bitwise representation requires the fewest number of binary variables, it does not always yield the fastest solving MILP representation (see also Section 4.3). In practice, both representations must be tested before arriving at the most efficient problem formulation.

4.1 MODELING WITH BINARY VARIABLES

In general, binary variables are used in applications involving a yes/no decision on the presence or absence of variables and/or constraints. In the following section, we highlight some of the most frequently occurring uses of binary variables with a particular focus on the analysis of metabolic networks.

4.1.1 Continuous Variable On/Off Switching

In the analysis of metabolic networks, binary variables are typically used as reaction on/off switches. They can be used to either impose a knockout or quantify a minimality requirement in the objective function. For example, to switch-off the flux of a reaction j a binary variable defined as follows:

$$y_j = \begin{cases} 1, & \text{if reaction } j \text{ is active} \\ 0, & \text{otherwise} \end{cases}, \quad \forall j \in J$$

is used to impose the following constraint:

$$\mathrm{LB}_j y_j \le v_j \le \mathrm{UB}_j y_j, \quad \forall j \in J \tag{4.5}$$

Here, LB_j and UB_j are lower and upper bounds on the reaction flux v_j, respectively, and J is the set of reactions in the metabolic network. The reaction flux is forced to zero when the binary variable y_j assumes a value of zero, whereas it can vary between its lower and upper bounds when y_j is equal to one.

Sometimes, the total number of switched-off (or on) reactions is under control. The following constraint ensures that the total number of inactivated reactions in a metabolic network is at most K.

$$\sum_{j \in J} \left(1 - y_j\right) \le K \tag{4.6}$$

where J denotes the set of reactions in the network. The use of binary variables allows for the straightforward enforcement of bounds (or fixed values) on the total number of reactions that are active (or inactive) in a metabolic network.

4.1.2 Condition-Dependent Variable Switching

In some cases, switching-off (or on) a continuous variable is dependent upon the value of another continuous variable. For example, consider a case where reaction j_1 must be inactive whenever reaction j_2 is inactive. The binary variables that model the switching-on/off of the two reactions are defined as before (Constraint 4.5). It often helps to tabulate all consistent combinations of binary variable values as shown in Table 4.1 before arriving at the appropriate constraint(s) to impose the conditional variable switching.

TABLE 4.1 Analyzing All Cases for a Conditional Reaction Removal

y_{j_2}	Allowed Values for y_{j_1}
1	0 or 1
0	0

By perusing the entries of Table 4.1, it becomes apparent that the following constraint is feasible if and only if a binary value combination is chosen from Table 4.1.

$$y_{j1} \leq y_{j2} \tag{4.7}$$

4.1.3 Condition-Dependent Constraint Switching

Similar to variable switching, sometimes one or more constraints must be turned on or off when a condition is true or false, respectively. Consider an inequality constraint $g(x) \leq 0$ that should be active only if a condition is true. A binary variable can be used to model whether or not the desired condition holds true:

$$y = \begin{cases} 1, & \text{if the desired condition is true} \\ 0, & \text{otherwise} \end{cases}$$

The conditional constraint activation can be captured within an optimization model as follows:

$$g(x) \leq M(1-y) \tag{4.8}$$

where M is a large positive number (*big-M representation*). Observe that if y is equal to one then we recover the original constraint; whereas if y is equal to zero the constraint becomes inactive as it is satisfied for every value of x (assuming that a large enough value is selected for M).

The conditional imposition of an equality constraint $h(x) = 0$ follows the same line of reasoning by considering that an equality constraint can be viewed as two inequalities with opposite inequality signs.

$$-M(1-y) \leq h(x) \leq M(1-y) \tag{4.9}$$

Constraint 4.9 ensures that $h(x)$ is equal to zero when y is equal to one, thereby recovering the original equality constraint. In contrast, $h(x)$ is free to assume any value (whose magnitude is less than M) when y is equal to zero, thus rendering the equality constraint inactive. It is worth noting that the selection of the "big M" value must be carefully made so that it remains larger in magnitude than $h(\boldsymbol{x})$ for every possible value of x within the feasible region.

4.1.4 Modeling AND Relations

Sometimes, we need to model a situation where multiple conditions need to be simultaneously satisfied (AND operator) for an outcome to be activated. The product of binary variables each representing one such condition captures an AND relation.

TABLE 4.2 Truth Table for the Product of Two Binary Variables $z = y_1 y_2$

y_1	y_2	$z = y_1 y_2$ (y_1 AND y_2)
1	1	1
1	0	0
0	1	0
0	0	0

For example, whenever two separate genes code for a multiunit protein complex catalyzing a metabolic reaction, an AND relation links them in the GPR map (see Chapter 3). Assuming that a binary variable is defined for each gene denoting whether a gene is present or absent, the product of these two binary variables captures the activity or inactivity of the corresponding metabolic reaction upon any combination of gene knockouts.

The product of binary variables gives rise to nonlinear terms turning an MILP problem to a mixed-integer nonlinear programming (MINLP) problem. Therefore, they need to be recast as equivalent sets of linear constraints for the optimization problem to be solvable by MILP solvers. Assume that variable z is the product of two binary variables y_1 and y_2(i.e., $z = y_1 y_2$). A truth table is useful for identifying all necessary and sufficient constraints to linearize this product (see Table 4.2):

As we can see in the truth table, we always have $0 \le z \le 1$. In addition, the following constraints hold under all four combinations:

$$z \le y_1 \tag{4.10}$$

$$z \le y_2 \tag{4.11}$$

These relations provide an upper bound on z; however, a lower bound is also needed to ensure that z assumes a value of one when both y_1 and y_2 are equal to one. This additional restriction can be imposed by adding the following constraint:

$$z \ge y_1 + y_2 - 1 \tag{4.12}$$

Note that z does not need to be declared as a binary variable as the earlier three constraints along with the non-negativity constraint ($z \ge 0$) ensure that z can take only a value of either zero or one. This exact linearization technique can be easily generalized for the product of more than two binary variables as follows:

$$z = \prod_{j=1}^{M} y_j \tag{4.13}$$

$$z \le y_j \quad \forall j \in \{1,\dots,M\}$$

$$z \ge \sum_{j=1}^{M} y_j - (M-1) \tag{4.14}$$

$$z \ge 0$$

4.1.5 Modeling OR Relations

Binary variables can also be used to model OR relations. This need arises when at least one of multiple conditions is sufficient for a process or outcome to be activated. For example, if two genes code for isozymes for a metabolic reaction these genes are related with an OR relation in the GPR map (see Chapter 3). The activity of either of the two genes is sufficient for the reaction to be present. After defining a binary variable for each gene, specific constraints can be derived that model the OR relationship. As before, a truth table is helpful for identifying all required constraints (see Table 4.3)

TABLE 4.3 Truth Table for Logical Conditions Related by OR

y_1	y_2	z (y_1 OR y_2)
1	1	1
1	0	1
0	1	1
0	0	0

It is easy to verify that the following constraints capture all the requirements imposed by the truth table:

$$z \geq y_1 \tag{4.15}$$

$$z \geq y_2 \tag{4.16}$$

$$z \leq y_1 + y_2 \tag{4.17}$$

$$z \leq 1 \tag{4.18}$$

As before, z does not need to be declared as a binary variable as these four constraints together ensure that it can assume only a value of zero or one. This technique can be readily generalized for more than two OR conditions as follows:

$$z \geq y_j, \quad \forall j \in \{1, 2, \ldots, M\} \tag{4.19}$$

$$z \leq \sum_{j=1}^{M} y_j \tag{4.20}$$

$$z \leq 1$$

Here, if any one of the M outcomes is true then z becomes equal to one. Conversely, z becomes equal to zero only when all y_j's are equal to zero.

4.1.6 Exact Linearization of the Product of a Continuous and a Binary Variable

Products between binary and continuous variables frequently arise in optimization models. Given that this product is a nonlinear term, it needs to be transformed to a set of linear constraints if MILP solvers are to be used. Let $z = xy$ be the product of continuous variable $x^L \le x \le x^U$ and binary variable y. The following constraint ensures that z assumes a value of zero when $y = 0$ and z varies between x^L and x^U when $y = 1$.

$$x^L y \le z \le x^U y \tag{4.21}$$

However, this constraint does not enforce $z = x$ when $y = 1$. This is accomplished by the following constraint:

$$x - x^U(1-y) \le z \le x - (1-y)x^L \tag{4.22}$$

It is always important to check whether Constraints 4.21 and 4.22 are mutually consistent. When Constraint 4.21 is active (i.e., $y = 0$, therefore zeroing the value of z), Constraint 4.22 allows z to freely vary between a negative $(x - x^U)$ and a positive $(x - x^L)$ number. When Constraint 4.22 is active (i.e., $y = 1$, enforcing $z = x$), Constraint 4.21 keeps z unrestricted between x^L and x^U. The exact linearization of xy is a powerful demonstration of how seemingly nonlinear terms involving binary variables can be recast as linear constraints allowing the optimization model to remain linear and enabling the use of MILP solvers such as CPLEX [1] and GUROBI [2]. Using the same ideas the exact linearization of more complex terms such as y^2x or y_1y_2x can be achieved. It is worth noting that in contrast to the product of a binary and a continuous variable, the product of two continuous variables is a nonlinear term that cannot be equivalently recast as a set of linear constraints.

4.1.7 Modeling Piecewise Linear Functions

The above described exact linearization techniques can be combined together to model more complex situations. One such example is ensuring that for a given value of x, a univariate piecewise linear function $\phi(x)$ is calculated correctly using only linear constraints. A piecewise linear function is a function that is composed of linear segments defined over different intervals for x (see Fig. 4.1). Note that even though function $\phi(x)$ is composed of linear segments, it is nonlinear (and in general nonconvex) over the entire range $[x_0, x_N]$. In fact, it could even be discontinuous. Piecewise linear functions are frequently used to approximate univariate

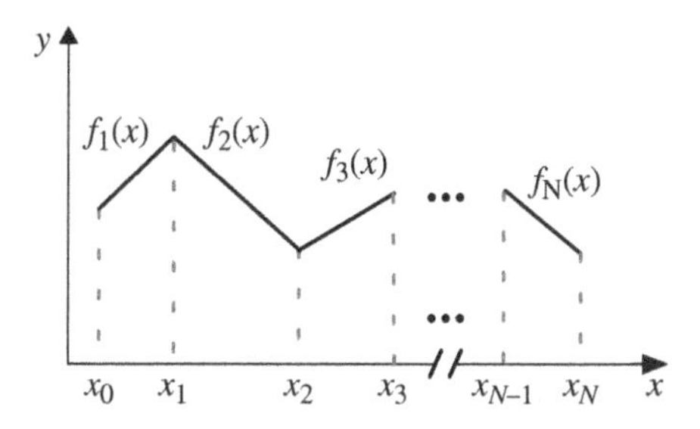

FIGURE 4.1 A graphical representation of a piecewise linear function.

nonlinear/nonconvex functions with a set of linear segments. This enables the use of a linear representation that is compatible with the use of LP or MILP solvers by using linear transformation techniques.

Let $\phi(x)$ denote a piecewise linear function where $\phi(x) = f_i(x)$ for $x_{i-1} \le x \le x_i, \forall i \in \{1, 2, \ldots, N\}$. By defining binary variables that model which linear function $f_i(x)$ to use for calculating $\phi(x)$, it is possible to recast $\phi(x)$ as a set of linear constraints. We first define binary variables to denote different sections of the overall x range as follows:

$$y_i = \begin{cases} 1, & \text{if } x_{i-1} \le x \le x_i \\ 0, & \text{otherwise} \end{cases} \quad \forall i \in \{1, 2, \ldots, N\}$$

This condition can be imposed by defining two new binary variables, y_i^1 and y_i^2 that track whether a given x is larger than x_{i-1} and smaller than x_i, respectively. The following two constraints model this requirement:

$$x - x_{i-1} + (x_{i-1} - x_0)(1 - y_i^1) \ge 0 \tag{4.23}$$

$$x_i - x + (x_N - x_i)(1 - y_i^2) \ge 0 \tag{4.24}$$

Note that $x_{i-1} \le x \le x_i$ requires that both y_i^1 and y_i^2 are equal to one. As described earlier, this AND relation can be imposed by defining a new binary variable y_i that is equal to the product of the two binary variables $(y_i = y_i^1 y_i^2)$ linearized using the following constraints (see Section 4.1.4):

$$y_i \le y_i^1 \tag{4.25}$$

$$y_i \le y_i^2 \tag{4.26}$$

$$y_i \ge y_i^1 + y_i^2 - 1 \tag{4.27}$$

$$y_i \ge 0 \tag{4.28}$$

The following constraints can then be used to select the appropriate function $f_i(\boldsymbol{x})$ based on what line segment the value of x falls within.

$$\phi(x) = \sum_{i=1}^{N} z_i \tag{4.29}$$

$$z_i = y_i f_i(x) \tag{4.30}$$

$$\sum_{i=1}^{N} y_i = 1 \tag{4.31}$$

Constraint 4.31 ensures that only one binary variable corresponding to the appropriate line segment is active at a time. The product of variable y_i and continuous variable $f_i(x)$ (Eq. 4.30) can be exactly linearized using the technique described in Section 4.1.6 as follows:

$$y_i f_i^{\mathrm{L}} \le z_i \le y_i f_i^{\mathrm{U}}, \quad \forall i \in \{1,2,\ldots,N\} \tag{4.32}$$

$$f_i(x) - f_i^{\mathrm{U}}(1-y_i) \le z_i \le f_i(x) - f_i^{\mathrm{L}}(1-y_i), \quad \forall i \in \{1,2,\ldots,N\} \tag{4.33}$$

where f_i^{L} and f_i^{U} are the lower and upper bounds on functions $f_i(x)$, respectively. By using Constraints 4.23–4.29 and 4.31–4.33, a linear description of the calculation of $\phi(x)$ is achieved that is compatible with the use of MILP solvers. This enables the integration of univariate nonlinear expressions in MILP problem formulations through piecewise linear approximation.

4.2 SOLVING MILP PROBLEMS

MILP problems are much more difficult to solve than the corresponding LP problems. Even though the feasible region of an MILP is substantially reduced, as some of the variables can assume only integer variables, a number of optimality properties that make LP problems relatively easy to solve are lost. In particular, an optimum solution (if it exists) no longer has to occur at a vertex point of the feasible region. This means that all interior MILP feasible points need to be examined yielding an NP-complete problem [3]. NP-hard problems belong to a class of equivalent problems whose worst-case solution increases in non-polynomial order with the size of the problem [4].

It is tempting to try to find the solution of an MILP problem by simply relaxing the integrality condition on the integer variables (i.e., $\boldsymbol{y} \in \mathbb{R}^M, 0 \le \boldsymbol{y} \le 1$ instead of $\boldsymbol{y} \in \{0,1\}^M$), solving the resulting linear programming problem (called *LP relaxation* of the MILP) and rounding-off the solution to the nearest integer. The following example pictorially demonstrates why this approach is untenable in practice.

Example 4.1
Find the solution to the LP relaxation of the following integer programming (IP) problem and compare the rounded-off solutions with the integer optimal solution

$$\begin{aligned}
&\text{minimize} \quad z = -y_1 - y_2 \\
&\text{subject to} \\
&-2y_1 + 2y_2 \le 1 \\
&16y_1 - 14y_2 \le 7 \\
&y_1, y_2 \ge 0 \\
&y_1, y_2 \in \mathbb{Z}
\end{aligned}$$

Solution: By solving the LP relaxation of the earlier problem (where $y_1, y_2 \in \mathbb{R}$), we obtain the following

$$\begin{aligned}
\boldsymbol{y}^* &= (7, 7.5) \\
z_{\mathrm{LP}}^* &= -14.5
\end{aligned}$$

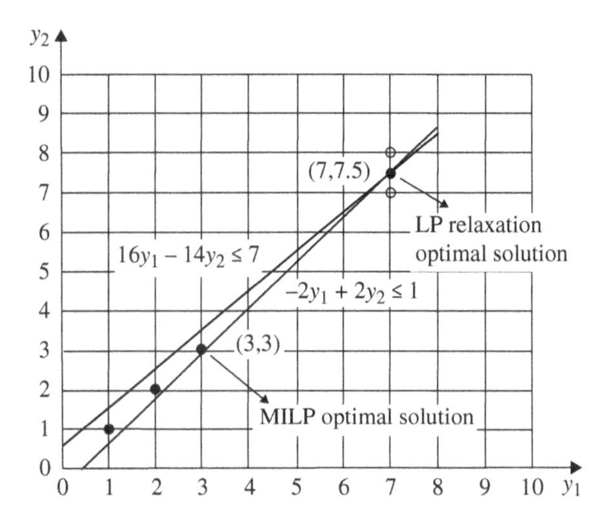

FIGURE 4.2 Pictorial illustration of the feasible region and optimal LP relaxation and integer solutions for Example 4.1.

Two possibilities emerge for the optimal solution of the original IP by rounding-off the LP solution: $\boldsymbol{y}^* = (7,7)$ and $\boldsymbol{y}^* = (7,8)$. However, as shown in Figure 4.2 both of these solutions are infeasible. The optimal solution for the earlier IP problem is in fact $\boldsymbol{y}^* = (3,3)$ (far away from the LP relaxation solution point) with an objective function value of $z^* = -6$. □

An alternative solution procedure is to exhaustively enumerate all combinations of integer variables, solve the resulting LP problems and declare the one with the best objective function value as the optimal solution. This strategy may work for small problems (i.e., only a few integer variables); however, it quickly becomes intractable for larger problems as the number of combinations grows exponentially with the number of integer or binary variables. For example, for a 0–1 MILP with 100 binary variables one needs to solve $2^{100} \sim 1.27 \times 10^{30}$ LP problems!

Despite the inherent complexity of MILP problems, a number of efficient algorithms have been proposed and successfully applied to medium- and/or large-scale problems. All these algorithms rely on the solution of a number of carefully crafted LP relaxation problems that as tightly as possible inscribe the feasible region of the MILP problem. The most widely used techniques rely on a branch-and-bound procedure.

4.2.1 Branch-and-Bound Procedure for Solving MILP Problems

The key idea of the branch-and-bound algorithm is to successively divide the original problem, which is difficult to solve directly, into smaller subproblems. It proceeds by solving the LP relaxation of these subproblems in order to eliminate (fathom) subregions of the feasible space that are guaranteed to not contain the optimal solution and updating the best so far integer solution. This iterative process terminates when only a single point (or small subregion) containing the optimum solution remains after excluding all other portions of the feasible region. We formally introduce some of the concepts used before presenting the algorithmic steps for the branch-and-bound method.

Relaxation A problem [R] is a relaxation of any general optimization problem [P] with the same objective function if

$$\mathrm{FR}(P) \subseteq \mathrm{FR}(R)$$

where FR denotes the feasible region. The following properties are a consequence of the earlier definition:

(i) If [R] is infeasible then [P] is also infeasible.
(ii) Assuming [P] is a minimization problem then $z_R \leq z_P$ (i.e., the optimal solution of [R] provides a lower bound on the optimal solution of [P]).
(iii) If the optimal solution of [R] is feasible for [P] then it is also the optimal solution for [P].

There are several ways of obtaining a relaxation problem for a given optimization problem. For example, one way is to simply remove one or more constraints from the original problem. The preferred way to obtain a relaxation for MILP problems is through an LP problem where the integrality condition for some or all variables is relaxed.

Branching for MILP Problems A set of subproblems $[P_1], [P_2], \ldots, [P_K]$ are a *branching* (also called *separation* or *partitioning*) of an original optimization problem [P] if the union of their mutually non-overlapping feasible regions recovers the feasible region of the original problem [P]:

$$\mathrm{FR}(P_1) \cup \mathrm{FR}(P_2) \cup \ldots \cup \mathrm{FR}(P_K) = \mathrm{FR}(P) \text{ and}$$
$$\mathrm{FR}(P_1) \cap \mathrm{FR}(P_2) \cap \ldots \cap FR(P_K) = \varnothing$$

It follows from the definition that a feasible solution for any of the subproblems is a feasible solution of the original problem [P]. There are several ways of "branching" an optimization problem. For MILP problems, one can choose an arbitrary integer variable (i.e., y_1) and location (e.g., between values 3 and 4) for branching leading to two subproblems as follows:

$$\begin{array}{ll}
\text{minimize } z = cx + dy \quad [P_1] & \text{minimize } z = cx + dy \quad [P_2] \\
\text{subject to} & \text{subject to} \\
\boldsymbol{Ax} + \boldsymbol{By} \geq \boldsymbol{b} & \boldsymbol{Ax} + \boldsymbol{By} \geq \boldsymbol{b} \\
\boldsymbol{x} \in \mathbb{R}^N, \boldsymbol{y} \in \mathbb{Z}^M & \boldsymbol{x} \in \mathbb{R}^N, \boldsymbol{y} \in \mathbb{Z}^M \\
y_1 \leq 3 & y_1 \geq 4
\end{array}$$

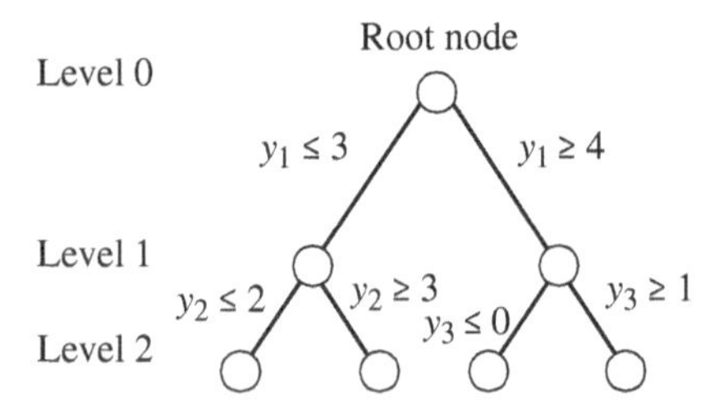

FIGURE 4.3 A schematic representation of a branch-and-bound tree.

Each one of these two subproblems can be divided into smaller subproblems by further branching on other or the same integer variable. These successive subdivisions can be graphically represented by an inverted tree, where nodes correspond to subproblems and edges denote the separation of a subproblem (*parent* node) into smaller subproblems (*children* nodes). The original problem is also referred to as the *root* node (see Fig. 4.3). Because all subproblems associated with children nodes are restrictions of the subproblem associated with the parent node, their solution provides an upper bound on the solution to the parent problem (assuming a minimization problem). This also applies for their respective LP relaxations thus providing a hierarchy of minimization problems with higher in value minima as we move deeper into the branch-and-bound tree.

Branching Precedence Order There are a number of precedence order rules employed for the selection of the next subproblem to branch further in the branch-and-bound algorithm. This choice can dramatically affect the speed of convergence to the MILP optimum solution:

(i) *Best-bound search*: Selects the subproblem with the best (i.e., the lowest for a minimization problem) objective function value for its LP relaxation. This choice usually leads to the fewest number of candidate subproblems for branching as the selected subproblem is likely to generate an integer solution with the lowest value.

(ii) *Depth-first search*: The next subproblem that is solved is the one added to the list of candidate subproblems most recently (i.e., it is one of the children nodes of the current subproblem). The advantage of this method is that the Simplex method can be restarted by re-optimizing the LP associated with the parent node upon imposing the restrictions implied by the children nodes. This method, however, may require solving an unnecessarily large number of subproblems before finding an optimal solution.

(iii) *Breadth-first search*: According to this strategy, the next subproblem to branch is an unfathomed/unexplored node at the highest possible level in the branch-and-bound tree. Based on this criterion, all nodes at the same level must be explored before moving to a lower level. The advantage of this branching strategy is that a balanced branch-and-bound tree is constructed. However, convergence to the optimal solution which is typically found lower in the branch-and-bound tree may be delayed.

After a subproblem to branch further is selected the next decision is to choose which integer variable and location to partition or branch. Branching is performed by adding constraints $y \le [y^{LP}]$ and $y \ge [y^{LP}]+1$, respectively, in the two resulting subproblems where y^{LP} is the optimal value of the integer variable y in the LP relaxation problem and $[y^{LP}]$ is the integer part of y^{LP}. Typically, the integer variable whose current LP relaxation value is farthest from an integer value is selected for branching. For example, if after solving the relaxed problem $y_1 = 0.5$ and $y_2 = 0.9$ typically y_1 is chosen for branching as y_2 is more likely to assume an integer value of one in the LP relaxation of the resulting subproblems in the children nodes.

It is worth noting that there are typically trade-offs among these branching precedence order methods and no single method works the best for all problems. One needs to empirically determine the best node selection method (or best combinations thereof) for each specific MILP problem structure. Modern MILP solvers often provide ways of controlling the node selection and branching strategy by the user in addition to default settings. For example, in CPLEX [1], the node selection and branching directions are controlled by setting parameters *VarSel* and *BrDir*, respectively.

Fathoming Fathoming is defined as the action of removing part of the feasible region of the original problem from any further consideration. This can be accomplished if (i) we ascertain that the feasible region of a subproblem cannot contain a solution better than the best integer solution found so far, or (ii) an optimal integer solution to this subproblem is fortuitously found by solving its LP relaxation.

Branch-and-Bound Algorithm Using LP Relaxations for Bounding The following steps conceptually summarize the sequence of steps carried out by modern MILP solvers:

Step 1 (*Initialization*):

1.1 Solve the LP relaxation of [MILP].

  (i) If the LP relaxation is infeasible, then problem [MILP] is infeasible. Stop.

  (ii) If the solution to the LP relaxation is integer, then we have found the optimal solution to [MILP]. Stop.

  (iii) Otherwise, add [MILP] to the list of candidate subproblems and set $z^{LB} = z_{LP}$, where z^{LB} and z_{LP} denote the so far tightest lower bound on the solution of [MILP] and the optimal objective function value of the LP relaxation of [MILP], respectively.

1.2 Let z_{incb} denote the best integer solution found so far (*incumbent*). If there is a known feasible solution to [MILP], then set z_{incb} equal to the corresponding objective function value. Otherwise set z_{incb} to $+\infty$. Note that the incumbent also corresponds to the tightest upper bound so far on the solution of [MILP] (z^{UB}).

Step 2 (*Branching/separation*):

2.1 *Node selection*: Select a candidate subproblem [CS] to branch based on a precedence order criterion discussed earlier.

2.2 *Branching*: Separate the selected [CS] by branching on one of the integer variables with a non-integer optimal value in the LP relaxation solution. Add both children subproblems to the list of candidate subproblems.

Step 3 (*Bounding*):
The following steps are performed for each child node of the selected candidate subproblem [CS]:

3.1 *LP relaxation*: Solve the LP relaxation for each of the two subproblems associated with the two children nodes and denote its optimal objective function value by z_{LP}.

3.2 *Fathoming*: Apply the following tests to each one of the subproblems.

(i) If the LP relaxation is infeasible, then fathom the node because no integer solution exists in this subregion of the feasible space.

(ii) If $z_{LP} > z_{incb}$, then fathom the node because this subregion is guaranteed to not contain the optimal solution as it cannot improve upon the incumbent solution.

(iii) If the solution to the LP relaxation is integer, then fathom the node as no further branching on this node is required.

(iv) If none of these three conditions are satisfied, then add the child node to the list of candidate subproblems to be further partitioned.

3.3 *Bounding*: Update current best upper and lower bounds

(i) If condition (iii) of Step 3.2 is satisfied, then update the incumbent integer solution (which is equivalent to the upper bound z^{UB}) as $z^{UB} = z_{incb} = \min(z_{incb}, z_{LP})$.

(ii) If condition (iv) in Step 3.2 is satisfied, then update the current best lower bound z^{LB} as follows:

$$z^{LB} = \min_{k \in UF} \left(z_{LP}^{k} \right)$$

where UF is the set of all unfathomed candidate subproblems and z_{LP}^{k} is the LP relaxation objective function value associated with subproblem k. Note that the sequence of upper and lower bounds are by design non-increasing and non-decreasing, respectively.

Step 4 (*Optimality test*):
If the list of candidate subproblems is empty and $z_{incb} = +\infty$, then terminate. The problem is infeasible. If the list is empty but $z_{incb} \neq +\infty$, terminate and report the current incumbent as the optimal solution. Otherwise, go back to Step 2.

A key factor that affects the computational efficiency of the branch-and-bound algorithm is how close the solution of the subproblems LP relaxations is to an integer solution. The distance between these two solutions is referred to as the *integrality gap*. It is common in most practical applications to set the integrality gap to a small value instead of zero in the interest of converging faster to a "near" optimal solution. In some applications, a near-optimal solution within a reasonable amount of time suffices. In such cases, the branch-and-bound iterations can be instructed to terminate if the lower and upper bounds are close enough (i.e., $|z^{UB} - z^{LB}| <$ tol, where tol is a pre-specified convergence tolerance). Term $|z^{UB} - z^{LB}|$ is sometimes referred to as the *absolute optimality gap*.

Example 4.2

Solve the following MILP using the branch-and-bound algorithm.

$$\begin{aligned}&\text{minimize} \quad z = 4y_1 - 6y_2\\&\text{subject to}\\&-y_1 + y_2 \le 1\\&y_1 - 3y_2 \le 9\\&3y_1 + y_2 \le 15\\&y_1, y_2 \ge 0\\&y_1, y_2 \in \mathbb{Z}\end{aligned}$$

Solution:

1. *Initialization*: The solution to the LP relaxation of this problem is as follows:

$$\mathbf{y}_{\text{LP}}^* = (1.5, 2.5), \quad z_{\text{LP}} = -9 = z^{\text{LB}}$$

   This solution is not integer and therefore the problem is added to the list of candidate subproblems. In addition, we set $z_{\text{incb}} = z^{\text{UB}} = +\infty$.
2. *Branching*: The original problem is the only available candidate subproblem. The problem is partitioned (branched) into two separate regions. The non-integer value of both variables is equally distant from an integer. We break the tie by choosing to branch on y_1 by adding $y_1 \le 1$ and $y_1 \ge 2$ as constraints to each one of the children subproblems, respectively.
3. *Bounding*: The solution for the LP relaxation problem of the subproblem with $y_1 \le 1$ is as follows:

$$\mathbf{y}_{\text{LP}}^* = [1, 2], \quad z_{\text{LP}} = -8$$

   Because the solution is integer (i.e., case (iii) of Step 3.2), this node in fathomed and the incumbent solution is updated:

$$z_{\text{incb}} = z^{\text{UB}} = -8$$

   The solution to the LP relaxation of the subproblem with $y_1 \ge 2$ is as follows:

$$\mathbf{y}_{\text{LP}}^* = \left[2, \frac{7}{3}\right], \quad z_{\text{LP}} = -6$$

   This node is fathomed because $z_{\text{LP}} = -6 > z_{\text{incb}} = -8$ (i.e., Case (ii) of Step 3.2).
4. *Optimality test*: The list of candidate subproblems is empty and therefore the current incumbent is the optimal solution:

$$\mathbf{y}^* = (1, 2), \quad z^* = -8$$

□

Example 4.3
Solve the following 0–1 MILP using the branch-and-bound algorithm:

$$\begin{aligned}
&\text{minimize} \quad z = x - 4y_1 - 3y_2 - 4y_3 \\
&\text{subject to} \\
&x + y_1 + y_2 + y_3 \geq 2 \\
&x + 6y_1 + 4y_2 + 5y_3 \leq 13 \\
&x \geq 0 \\
&y_1, y_2, y_3 \in \{0,1\}
\end{aligned}$$

Solution: The steps of the branch-and-bound algorithm are illustrated using an inverted tree representation in Figure 4.4. For each node in the tree, the LP relaxation solution, current lower and upper bounds, and fathoming inference case are noted. The optimal solution to this MILP is found at node 4 and is as follows:

$$\mathbf{y}^* = (1,0,1), x^* = 0, z^* = -8$$ □

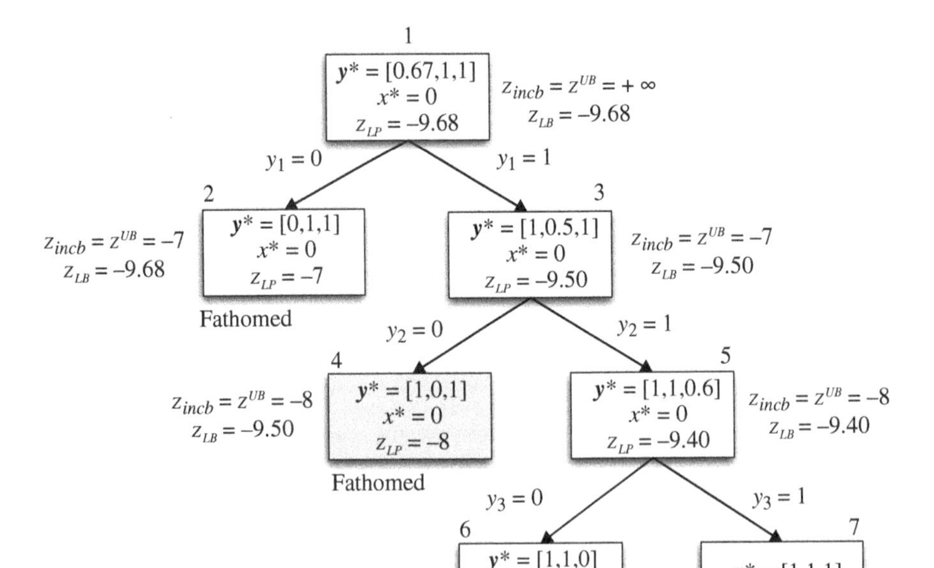

FIGURE 4.4 Branch-and-bound tree for the 0–1 MILP problem in Example 4.3. The numbers on top of the boxes denoting nodes represent the sequence of selected candidate subproblems. The optimal solution is found at node 4 after fathoming node 7.

4.2.2 Finding Alternative Optimal Integer Solutions

In addition to finding the optimum solution, it is possible to enumerate the list of all alternate optima or even rank-order suboptimal solutions. This is achieved by resolving the MILP problem upon the successive addition of new constraints referred to as *integer cuts*. An integer cut is a constraint that renders a previously identified integer solution infeasible without excluding any other feasible integer solution. This constraint is constructed in such a way that it is violated (i.e., it makes the problem infeasible) when all binary variables assume values equal to those in a previous solution. By successively appending integer cuts and resolving the MILP problem, a ranked-ordered list of alternative optimal solutions can be derived. For example, if the optimal solution for an MILP problem with three binary variables (y_1, y_2 and y_3) is $[1,0,1]$, an integer cut can be constructed as follows to exclude this solution from consideration without eliminating any other integer solutions

$$y_1 + (1 - y_2) + y_3 \leq 2$$

Observe that this constraint is violated only if y_1, y_2 and y_3 assume a value of 1, 0 and 1, respectively. This constraint can be extended as follows for a general 0–1 integer program

$$\sum_{i \in \mathrm{Ones}} y_i + \sum_{i \in \mathrm{Zeros}} (1 - y_i) \leq M - 1 \tag{4.34}$$

where M is the total number of binary variables and *Ones* and *Zeros* denote the sets of binary variables that assumed a value of one and zero, respectively, in the previous optimal solution. Multiple integer cut constraints could be appended to an MILP problem to exclude not just a single but all previously identified solutions.

4.3 EFFICIENT FORMULATION STRATEGIES FOR MILP PROBLEMS

Solution times for MILP problems can be drastically affected by the formulation of the problem. A few general rules for efficiently modeling a real-life problem with MILP are described later. Interested readers can also check more advanced reformulation techniques such as the generalized disjunctive programming [5], mixed-logic linear programming [6] or algorithms such as Bender's decomposition and Outer Approximation (see Chapter 11).

4.3.1 Using the Fewest Possible Binary Variables

The fewer the number of binary variables, the shorter the list of variables that require branching. For example, as noted in Sections 4.1.4 and 4.1.5, the introduced variable z modeling AND or OR relations does not need to be declared as a binary variable. The minimal use of binary variables in MILP equivalent problem representations generally leads to the best computational performance by MILP solvers. Nevertheless, many exceptions to this rule exist as the tightness of the LP relaxations ultimately controls the speed of convergence.

4.3.2 Fix All Binary Variables that do not Affect the Optimal Solution

Fixing the value of binary variables that are known *a priori* can significantly reduce the number of required branching operations. For example, there are often a few hundred reactions in a metabolic network that are either blocked ($y_j = 0$) or locked ($y_j = 1$) at finite values under any given examined growth condition. Therefore, it is a good strategy to fix the corresponding binary variables to exclude them from any branching operations in order to speed up convergence.

4.3.3 Group All Coupled Binary Variables

It is sometimes possible to determine the set of all binary variables that will have the same effect on the objective function value by performing a prescreening analysis such as FCF (see Chapter 3). In such a situation, only one binary variable from each coupled set needs to be considered while the remaining ones can be fixed at zero or one accordingly. For example, as noted in Chapter 3, only one reaction from each fully coupled reaction set is required to be considered for a knock-out/up/down analysis as manipulating the flux of any of these reactions will have the same effect on the flux distribution in the network as that of the others. Similarly, only one gene from a set of genes exclusively coding for a multi-unit protein complex for a metabolic reaction (i.e., when the genes are related with AND in the GPR association) needs to be considered when binary variables are used to directly impose gene deletions in the network (i.e., a single binary variable is defined for all genes).

4.3.4 Segregate Binary Variables in Constraints Rather than in the Objective Function

An MILP representation of a problem where binary variables appear only in constraints is generally faster solving than an equivalent one where binary variables appear in the objective function. It is sometimes possible to swap objective function and constraints in an optimization model to satisfy this requirement. For example, consider a strain design problem, where the objective is to maximize the flux of a target metabolite while minimizing the number of reaction eliminations needed. If binary variable y_j denotes whether or not a reaction j should be eliminated, then one may formulate the problem as follows

$$\text{minimize } z = \sum_{j \in J} y_j$$

subject to

$$v_{\text{product}} \geq v_{\text{product}}^{\text{target}}$$

... other constraints

where $v_{\text{product}}^{\text{target}}$ is a pre-specified target production level for the metabolite of interest. One can alternatively represent the same problem by swapping the objective function and the constraint on the product formation level as follows

$$\begin{aligned}
&\text{maximize } z = v_{\text{product}} \\
&\text{subject to} \\
&\sum_{j \in J} y_j \le K \\
&\text{... other constraints}
\end{aligned}$$

where K is the maximum allowable number of interventions. This problem can be solved by successively increasing the value of K and resolving until condition $v_{\text{product}} \ge v_{\text{product}}^{\text{target}}$ is met. The solution to this problem will be the same as the original formulation, however, having $\sum_{j \in J} y_j \le K$ as a constraint generally leads to tighter LP relaxations. In addition, the most important reaction eliminations are identified first for smaller values of K.

4.3.5 Use Tight Bounds for All Continuous Variables

It is a good practice to always impose tight bounds for continuous variables as they directly affect the quality of the LP relaxations. To this end, condition-specific flux variability analysis (FVA) (see Chapter 3) can be performed for metabolic networks to update the bounds on all metabolic fluxes before an MILP problem analysis is carried out. This is particularly important for maintaining the value of M in big-M problem representations as small as possible.

4.3.6 Introduce LP Relaxation Tightening Constraints

The introduction of additional constraints that do not affect the feasible region of the MILP, but help generate tighter LP relaxations can significantly speed up the solution time. These constraints are sometimes referred to as *cuts*. Finding efficient cuts for a given MILP formulation is not always a straightforward task requiring testing alternative constraints and problem reformulations. In the following, we provide a few guidelines for constructing cuts:

- Consider an MILP with the following constraints

$$\begin{aligned}
&0 \le x_i \le M y_i, \quad \forall i \in I \\
&\sum_{i \in I} y_i = 1
\end{aligned}$$

The following linear constraint would lead to a tighter LP relaxation without affecting the MILP solution

$$\sum_{i \in I} x_i \le M$$

- Another example of adding a cut is imposing constraints that guarantee that there is a connected path in a metabolic network between the uptake system and the conversion of a taken up metabolite. For example, if a reaction transporting a metabolite i from the extracellular environment into the cell is active then at least one of the intracellular reactions consuming this metabolite must be active too. This can be imposed by using the following constraint:

$$y_{\text{transport},i} \leq \sum_{j \in J^{\text{intra},i}} y_j$$

where $y_{\text{transport},i}$ is the binary variables associated with the transport reaction for metabolite i and $J^{\text{intra},i}$ is the set of intracellular reactions consuming this metabolite (i.e., $J^{\text{intra},i} = \{j | j \in J$ and j is intracellular with $S_{ij} < 0\}$). Note that all such relations in a metabolic network can be enumerated by identifying the sets of partially coupled reactions with all transport reactions (recall the concepts of "reactions affected" and "equivalent knockouts" from Chapter 3).

- Consider the following constraint where all variables y_i are binary:

$$\sum_{i \in I} a_i y_i \leq b$$

One can identify a subset S of set I such that

$$\sum_{i \in S} a_i > b$$

We can add a tightening constraint that excludes all such infeasible combinations as follows

$$\sum_{i \in S} a_i y_i \leq N_S - 1$$

where N_s is the number of binary variables in set S. For example, consider the following constraint where all variables are binary

$$3y_1 + 7y_2 + 5y_3 + 8y_4 \leq 17$$

We observe that y_2, y_3 and y_4 cannot be simultaneously equal to one as this will render the constraint infeasible. This can be prevented by introducing the following cut

$$y_2 + y_3 + y_4 \leq 2$$

This constraint will tighten the LP relaxation of the problem by eliminating the non-integer solution $\left(0, \frac{1}{2}, 1, 1\right)$.

- Consider one or more inequality constraint(s) in the form of "≤" including only nonnegative integer variables. The following procedure *may* result in a tightening cut (known as a Gomory cut):
  (i) Multiply one or more constraint(s) by a nonnegative number and add them together to form a new constraint (different values can be used for each constraint).
  (ii) Round down the coefficients on the left-hand side of the newly constructed constraint.
  (iii) Round down the constraint term on the right-hand side of the constraint.

As an example, consider the following two constraints, where variables y_1, y_2 and y_3 are binary

$$4y_1 + 5y_2 + y_3 \leq 7, \quad y_1 + y_3 \leq 1$$

Multiply the first constraint by $\frac{1}{5}$ and the second one by $\frac{1}{3}$ and add the results

$$\left(\frac{4}{5}y_1 + y_2 + \frac{1}{5}y_3\right) + \left(\frac{1}{3}y_1 + \frac{1}{3}y_3\right) \leq \frac{7}{5} + \frac{1}{3}$$

or

$$\frac{17}{15}y_1 + y_2 + \frac{8}{15}y_3 \leq \frac{26}{15}$$

We can round down the left-hand side coefficients because all variables are nonnegative and we have a "≤" constraint

$$y_1 + y_2 \leq \frac{26}{15}$$

Furthermore, we can round down the right-hand side because we have only integer variables

$$y_1 + y_2 \leq 1$$

This is a valid cut because it satisfies both of the original constraints and in addition it eliminates the non-integer solution $(0.5, 1, 0)$ which is feasible for the original set of constraints.

Observe that even though none of the three steps in this procedure has any effect on the integer feasible region, steps (ii) and (iii) can affect the tightness of the LP relaxation. In particular, step (ii) results in a looser LP relaxation problem (i.e., it allows for additional non-integer solutions) but step (iii) tightens the LP relaxation. In many (but not all) cases, the effect of step (iii) overcomes the LP relaxation loosening by step (ii). This procedure is guaranteed to result in a valid cut if constraints and multiplication coefficients are chosen in such a way that no rounding is required in step (ii). As an example, consider the following three constraints on binary variables y_1, y_2 and y_3

$$y_1 + y_2 \leq 1, \quad y_1 + y_3 \leq 1, \quad y_2 + y_3 \leq 1$$

Multiply each constraint by $\frac{1}{2}$ and add the results

$$y_1 + y_2 + y_3 \le \frac{3}{2}$$

No rounding is required for the left-hand side coefficients. By rounding down the right-hand side coefficients, we obtain the following

$$y_1 + y_2 + y_3 \le 1$$

which is a valid cut and eliminates the non-integer solution $\left(\frac{1}{2}, \frac{1}{2}, \frac{1}{2}\right)$ that is feasible for the original three constraints.

Through the judicious introduction of LP relaxation tightening constraints, MILP problems may converge significantly faster by reducing the size of the branch-and-bound tree that needs to be traversed before arriving at the optimum solution.

4.4 IDENTIFYING MINIMAL REACTION SETS SUPPORTING GROWTH

Generally, only a subset of genes is essential for cellular growth. The smallest set(s) of genes that can support growth is referred to as the minimal gene set. While many of these required genes have no (known) metabolic role, there exists a subset with distinct metabolic functions necessary for biomass formation. Identifying this subset provides insight into the minimal set of metabolic functions necessary for life. The smallest set of genes (and hence reactions) that are needed to sustain growth and ATP production for cell maintenance is referred to as the *minimal metabolic gene set* [7]. The task of identifying this minimal set can be formulated as an MILP problem that satisfies a target biomass production yield, which is a fraction c of the theoretical maximum for the wild-type strain ($0 \le c \le 1$). The formulation and solution of this MILP problem draws from many of the techniques discussed in Section 4.3.

A binary variable is defined for each reaction in the network as follows

$$y_j = \begin{cases} 1, & \text{if reaction } j \text{ is active} \\ 0, & \text{otherwise} \end{cases}$$

The following MILP formulation identifies the minimal reaction set for the metabolic model of interest:

$$\text{maximize} \sum_{j \in J} y_j$$

subject to

$$\sum_{j \in J} S_{ij} v_j = 0, \quad \forall i \in I \tag{4.35}$$

$$\text{LB}_j y_j \le v_j \le \text{UB}_j y_j, \quad \forall j \in J \tag{4.36}$$

$$v_{\text{biomass}} \geq f v_{\text{biomass}}^{\text{max,WT}} \qquad (4.37)$$
$$y_j \in \{0,1\}, \quad \forall j \in J$$

Here the objective function minimizes the number of reactions that must be active in the network in order to achieve the target biomass yield (Constraint 4.37). Solving this optimization problem is computationally intensive as hundreds of binary variables need to be activated out of thousands of possibilities captured in a genome-scale model. The problem becomes particularly challenging as the biomass yield target is lowered [7] as more reaction combinations become viable solutions. A number of reformulation and pre-processing steps described in the previous section can be employed here to alleviate some of the computational burden:

(a) Set $y_j = 1$ for all essential reactions and $y_j = 0$ for all blocked reactions. Caution must be exercised here as the set of essential and blocked reactions depend both on the uptake conditions and the imposed biomass target yield.
(b) Add constraint $y_{j_1} + y_{j_2} \geq 1$ for all synthetic lethal reaction pairs (j^1, j^2) since at least one of these reactions must be present (see Chapter 3). Note that similar to essential reactions, synthetic lethality also depends on the uptake condition and the target biomass yield.
(c) Tighten lower and upper bounds of each reaction flux using flux variability analysis (FVA) under the examined condition (see Chapter 3).
(d) Switch the objective function with the performance constraint (i.e., maximize v_{biomass} as the objective function and replace Constraint 4.37 with $\sum_{j \in J} y_j \leq K$). This segregates all binary variables within a constraint. The problem is then solved by successively increasing the value of K from an initial value equal to the number of essential reactions and iteratively solving the resulting MILP until the target biomass yield is achieved (see Fig. 4.5a). A side benefit of this approach is that the minimal gene set is identified as a function of the imposed biomass yield. Furthermore, this iterative framework can inspire heuristic rules whenever the true optimum is computationally too expensive to reach. For example, a fraction of the genes found necessary for a given value of K can be retained when the value of K is increased. This provides a way of reducing problem complexity by fixing the value of many binary variables.

It is important to note that the minimal set of reactions is context specific. A rich growth medium will impose less of a demand on internal metabolism for growth, thus requiring fewer metabolic functions than growth on a minimal medium. The imposed target of biomass yield is another factor that significantly affects the total number of reactions required for growth. As the imposed yield increases towards the theoretical maximum, more carbon recovery and redox functions are required to recover the maximum amount of energy and carbon flux thus leading to additional reactions in the minimal set (see Fig. 4.5b). Using integer cuts, one can identify multiple reaction sets with the

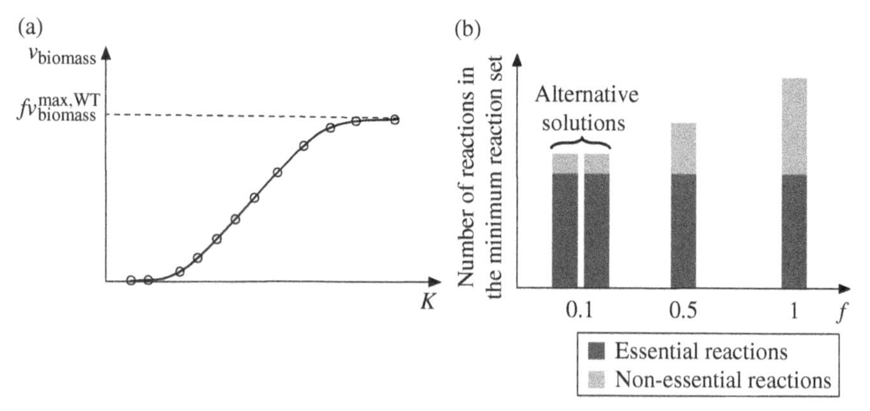

FIGURE 4.5 (a) Iteratively solving the MILP for identifying the minimal reaction sets while increasing the allowable number of active reactions (K) until the target biomass yield is met. (b) The minimum reaction sets are highly dependent on the imposed biomass target (f is the required fraction of the maximum theoretical biomass). Lower biomass targets require a smaller set of genes to maintain cellular growth. Essential reactions remain unchanged in the minimal reaction sets across different biomass targets.

same number of reactions but different combinations that satisfy the imposed biomass requirement. This quantitatively confirms that there is no single minimal reaction set but rather we have an ensemble of possible minimal sets with some of the reactions always present while the remaining may vary between members of the set. Minimal reaction sets that support growth are important to infer as they shed light on requirements imposed on metabolism under different uptake and biomass yield scenarios. However, it is important to stress that an organism with a minimal (or near minimal) reaction set will be poorly prepared to respond to genetic or environmental perturbations. Reaction redundancy is needed to respond to these perturbations and to recycle carbon and redox energy equivalents.

EXERCISES

4.1 Consider the *i*AF1260 *Escherichia coli* metabolic model [8].

- **(a)** Identify the minimum number (or as close as you can) of enzymatic reactions required for the production of biomass (use 1% of the maximum theoretical yield as the threshold) and non-growth associated ATP maintenance under aerobic minimal glucose conditions.
- **(b)** Use the GPR associations to identify the corresponding minimal gene set for growth.
- **(c)** Repeat your simulations to identify the minimum set of reactions required that ensure biomass production at a level of 10, 20, 30, …, 100% of its theoretical maximum. Interpret your results.

4.2 Develop an efficient and general MILP formulation for solving Sudoku puzzles (see www.sudoku.com). Solve the example shown in the following figure (adopted from www.sudoku.com) and confirm that there is only one solution.

8								
		3	6					
	7			9		2		
	5				7			
				4	5	7		
			1				3	
		1					6	8
		8	5				1	
	9					4		

4.3 The following figure shows the solution to a game of connecting numbered dots in a 5×5 grid. The rules of the game are as follows:

(i) No cell in the grid should be left empty.

(ii) No line can overlap with a line of another number.

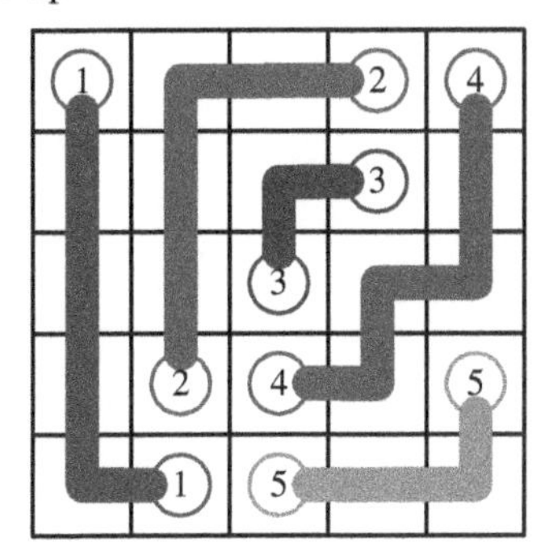

Using binary variables to represent whether a cell is occupied (or not) by a particular number, develop a general MILP formulation that solves the puzzle shown in the following figure. Confirm there is only one solution.

						1
	4	3			2	
5			5	1	6	
			6			
	3				7	
4	2		8		8	7

REFERENCES

1. IBM: IBM ILOG CPLEX Optimization Studio v12.5.1 documentation; 2015. Available at http://www-03.ibm.com/software/products/en/ibmilogcpleoptistud/ (accessed September 1, 2015).
2. Gurobi Optimization, Inc.: Gurobi Optimizer Reference Manual; 2015. Available at http://www.gurobi.com (accessed September 1, 2015).
3. Vavasis SA: *Nonlinear optimization: complexity issues*. New York: Oxford University Press; 1991.
4. Garey MR, Johnson DS: *Computers and intractability: a guide to the theory of NP-completeness*. New York: W.H. Freeman; 1991.
5. Raman R, Grossmann I: Modeling and computational techniques for logic-based integer programming. *Comput Chem Eng* 1994, **18**(7):563–578.
6. Hooker J, Osorio M: Mixed logical-linear programming. *Discrete Appl Mathemat* 1999, **97**:395–442.
7. Burgard AP, Vaidyaraman S, Maranas CD: Minimal reaction sets for *Escherichia coli* metabolism under different growth requirements and uptake environments. *Biotechnol Prog* 2001, **17**(5):791–797.
8. Feist AM, Henry CS, Reed JL, Krummenacker M, Joyce AR, Karp PD, Broadbelt LJ, Hatzimanikatis V, Palsson B: A genome-scale metabolic reconstruction for *Escherichia coli* K-12 MG1655 that accounts for 1260 ORFs and thermodynamic information. *Mol Syst Biol* 2007, **3**:121.

5

THERMODYNAMIC ANALYSIS OF METABOLIC NETWORKS

Thermodynamic analysis of metabolic pathways can shed light onto the directionality and activity limits of metabolic reactions in FBA. In this chapter, we focus on using thermodynamics to systematically assess the allowable reaction directionalities and the elimination of thermodynamically infeasible cycles (TICs) in metabolic network reconstructions using systematic optimization based methods.

5.1 THERMODYNAMIC ASSESSMENT OF REACTION DIRECTIONALITY

The directionality of a stand-alone metabolic reaction is determined by the sign of its Gibbs free energy change. A negative value implies that the forward direction will proceed with a higher flux than the backward direction resulting in a net forward flux. Variations in the concentration of reactants and products may lead to a change of the sign of the Gibbs free energy change and thus a reversal of its net direction. The following expression defines the Gibbs free energy change of a reaction j:

$$\Delta_{\mathrm{r}} G_j = \Delta_{\mathrm{r}} G_j^{\circ} + RT \ln\left(\prod_{i \in \{i | i \in I, S_{ij} \neq 0\}} x_i^{S_{ij}} \right) \tag{5.1}$$

or equivalently

$$\Delta_{\mathrm{r}} G_j = \Delta_{\mathrm{r}} G_j^{\circ} + RT \left(\sum_{i \in \{i | i \in I, S_{ij} \neq 0\}} S_{ij} \ln(x_i) \right) \tag{5.2}$$

Optimization Methods in Metabolic Networks, First Edition. Costas D. Maranas and Ali R. Zomorrodi.

where I is the set of metabolites, S_{ij} is the stoichiometric coefficient of metabolite i in reaction j, $\Delta_r G_j$ is the Gibbs free energy change of reaction j, $\Delta_r G_j^\circ$ is the standard Gibbs free energy change of reaction j (i.e., at $T = 25°\text{C}$, 1 M concentration of all aqueous reactants and products and 1 atm for the partial pressure of any gases involved in the reaction), R is the universal gas constant, T is temperature and x_i is the activity of metabolite i.

The standard Gibbs free energy change of reaction j is equal to

$$\Delta_r G_j^\circ = \sum_{i \in \{i | i \in I,\, S_{ij} \neq 0\}} S_{ij} \Delta_f G_i^\circ \tag{5.3}$$

where $\Delta_f G_i^\circ$ is the standard Gibbs free energy of formation of metabolite i determined experimentally or approximated using group contribution methods [1–3]. The activities of most metabolites are unknown; however, the mean activity in the cell is in the order of 1 mM [4]. Therefore, typically the Gibbs free energy change of metabolic reactions is calculated using 1 mM as the reference state (i.e., $x_i = 1\,\text{mM}, \forall i \in I$). Generally, a range is assessed for $\Delta_r G_j$ instead of a single value to compensate for the uncertainty in the estimated $\Delta_r G_j^\circ$ (i.e., $U_{r,est,j}$), the departure of metabolite activities from 1 mM concentrations and the contribution of the electrochemical potential and pH gradient across the cell membrane for transport reactions [5]. Minimum and maximum estimates for $\Delta_r G_j$ can be calculated by incorporating these uncertainties and deviations into Equation 5.2 as follows [5]:

$$\begin{aligned} \Delta_r G_j^{\max} &= \Delta_r G_j^\circ + \Delta G_{\text{transport}} \\ &+ RT\left(\sum_{i \in \{i | i \in I,\, S_{ij} > 0\}} S_{ij} \ln\left(x_i^{\max}\right) + \sum_{i \in \{i | i \in I,\, S_{ij} < 0\}} S_{ij} \ln\left(x_i^{\min}\right) \right) \\ &+ U_{\text{r,est,j}}, \quad \forall j \in J \end{aligned} \tag{5.4}$$

$$\begin{aligned} \Delta_r G_j^{\min} &= \Delta_r G_j^\circ + \Delta G_{\text{transport}} \\ &+ RT\left(\sum_{i \in \{i | i \in I,\, S_{ij} > 0\}} S_{ij} \ln\left(x_i^{\min}\right) + \sum_{i \in \{i | i \in I,\, S_{ij} < 0\}} S_{ij} \ln\left(x_i^{\max}\right) \right) \\ &- U_{\text{r,est,j}}, \quad \forall j \in J \end{aligned} \tag{5.5}$$

Here, $\Delta G_{\text{transport}}$ is the free energy contribution of the electrochemical potential and proton gradient across the cytoplasmic membrane, $x_i^{\min}$ and $x_i^{\max}$ are the minimum and maximum activity of metabolite i conservatively assumed to vary between 10^{-5} and 0.02 M, respectively, in the absence of any more precise data. Given that the physiological range of activities for dissolved gases such as H_2, O_2 and CO_2 is lower than that of other metabolites, their minimum activity $x_i^{\min}$ is set at 10^{-8} M whereas their maximum activity $x_i^{\max}$ is set to their respective saturation concentration in water at 298 K and 1 atm (i.e., 0.000034, 0.000055 and 0.0014 M for H_2, O_2 and CO_2, respectively) [5].

The widest possible bounds are used to include all thermodynamically feasible flux distributions and to systematically assign reversibility and directionality of reactions in a metabolic network as follows:

(i) If $\Delta_r G_j^{max} < 0$ then the reaction is irreversible in the forward direction ($v_j > 0$).
(ii) If $\Delta_r G_j^{min} > 0$ then the reaction is irreversible in the backward direction ($v_j < 0$).
(iii) If $\Delta_r G_j^{min} < 0$ and $\Delta_r G_j^{max} > 0$ then the reaction is treated as reversible ($v_j \in \mathbb{R}$).

It is important to note that experimental data may be used to further restrict the above identified directionalities. Alternatively, reaction directionality assignment has been proposed based on a probability metric representing the likelihood that a reaction's Gibbs free energy change is negative [6].

Sometimes, reactions with a positive standard Gibbs free energy change are present in pathways that carry significant flux. This is accomplished by coupling these reactions with others having a larger in magnitude but negative Gibbs free energy change in such a way that the entire reaction set becomes thermodynamically feasible. For example, even though glyceraldehyde-3-phosphate dehydrogenase (GAPD) is endergonic under standard conditions ($\Delta_r G_j^{\circ} = +6.7$ kJ/mol) when coupled with phosphoglycerate kinase (PGK) with $\Delta_r G_j^{\circ} = -18.8$ kJ/mol the overall reaction producing nicotinamide adenine dinucleotide hydride (NADH) and adenosine triphosphate (ATP) becomes thermodynamically favorable [7]:

$$\begin{array}{lll} \text{GAPD}: & \text{g3p} + \text{pi} + \text{nad} \rightarrow \text{13dpg} + \text{nadh} & \Delta_r G_j^{\circ} = +6.7 \text{ kJ/mol} \\ \text{PGK}: & \text{13dpg} + \text{adp} \rightarrow \text{3pg} + \text{atp} & \Delta_r G_j^{\circ} = -18.8 \text{ kJ/mol} \\ \text{Overall}: & \text{g3p} + \text{pi} + \text{nad} + \text{adp} \rightarrow \text{3pg} + \text{nadh} + \text{atp} & \Delta_r G_j^{\circ} = -12.1 \text{ kJ/mol} \end{array}$$

where g3p, 13dpg and 3pg denote glyceraldehyde-3-phosphate, 1,3-biphosphoglycerate and 3-phosphoglycerate, respectively. It must be noted that for a pathway operating at steady-state all overall reaction steps must have a non-positive free energy change $(\Delta_r G_j)$ under the relevant physiological conditions [7]. This requirement can be met for thermodynamically unfavorable steps by creating a large enough pool of the reactant metabolites while quickly draining the product pools.

5.2 ELIMINATING THERMODYNAMICALLY INFEASIBLE CYCLES (TICs)

5.2.1 Cycles in Cellular Metabolism

Conversion cycles are ubiquitous in metabolic networks (e.g., the tricarboxylic acid or TCA cycle and urea cycle) serving to partition an overall chemical conversion into a number of reaction steps that generally involve lower activation barriers through a sequence of molecular activation/deactivation steps (e.g., phosphorylation and addition of CoA moiety). Figure 5.1a shows the TCA cycle that uses eight separate reaction steps to convert an acetyl group into two molecules of carbon dioxide while generating three molecules of NADH and one molecule of $FADH_2$. The concept of decomposing an overall conversion chemistry into a number of cyclic steps has been exploited in metabolic engineering for many carbon and energy efficient bioconversions. For example, in the recently proposed non-oxidative glycolysis (NOG) [8] a

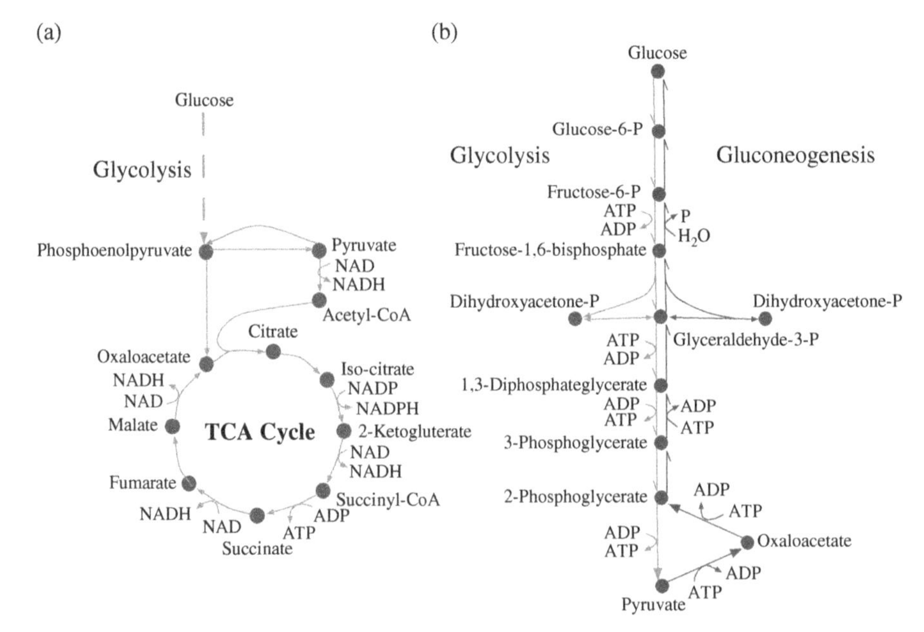

FIGURE 5.1 Two important cycles in cellular metabolism: (a) the TCA cycle and (b) the futile cycle between glycolysis and gluconeogenesis.

six-carbon sugar molecule is fully converted into three two-carbon acetyl-phosphate molecules. In the methanol condensation cycle (MCC) [9] methanol is converted into higher-chain alcohols without carbon loss or ATP expenditure.

In contrast to conversion cycles, *futile cycles* do not carry out an overall conversion but rather couple two metabolic pathways operating in opposing directions with no overall impact other than the dissipation of energy (e.g., ATP hydrolysis) in the form of heat. A well-known example is the futile cycle formed by combining glycolysis and gluconeogenesis (see Fig. 5.1b) which consumes one ATP per cycle leading to the dissipation of chemical energy into heat. Photorespiration is also a futile cycle that dissipates excess light energy into heat. Futile cycles are generally undesirable in bio-production as they reduce energy efficiency and therefore product yield; however, on occasion, they do serve important physiological roles in regulating metabolism or priming a pathway to respond quickly to a perturbation. They provide a mechanism for metabolism to assume an alert posture primed to quickly supply flux in either direction by simply blocking the flow in the opposite direction as opposed to having to assemble catalytic resources upon demand. It is important to distinguish between futile cycles, which are thermodynamically feasible cycles from thermodynamically infeasible ones that are frequently encountered in FBA.

5.2.2 Thermodynamically Infeasible Cycles

In contrast to naturally occurring cycles, TICs are caused by the permissive inclusion of reactions or reaction directionalities in a metabolic model. These cycles consist of a set of reactions forming loops such as $A \rightarrow B \rightarrow C \rightarrow A$ with no other metabolites entering or leaving. The source of thermodynamic infeasibility in these cycles is not elemental or charge imbalance. According to the second law of thermodynamics, the overall thermodynamic driving force is zero implying that no net flux can flow around this cycle [10]. FBA calculations are constrained only by stoichiometry, therefore an unbounded amount of flux through TICs with no external input of energy thus forming a perpetual machine of the second-kind which is thermodynamically impossible. The identification and elimination of TICs is necessary to prevent unbounded metabolic flows and thermodynamically infeasible flux distributions.

5.2.3 Identifying Reactions Participating in TICs

Identifying all TICs in a GSM network is a challenging task as there could be a confounding assembly of overlapping loop-forming TICs. Instead, it is much more tractable to identify all reactions that participate in one or more TICs using flux variability analysis (FVA). Infeasible loops are manifested in a metabolic model as a set of reactions able to carry an unbounded metabolic flux under finite (or even zero) substrate inputs. This implies that by applying FVA, a number of fluxes will hit their arbitrarily large (in magnitude) upper and/or lower bounds (i.e., M and/or −M, see also Chapter 3). In the subsequent sections, a number of methods are introduced to safeguard against TICs.

5.2.4 Thermodynamics-Based Metabolic Flux Analysis

Henry et al. [11] introduced a systematic method termed thermodynamic-based metabolic flux analysis (TMFA), which uses bounds on $\Delta_r G_j^{\circ}$ to impose thermodynamic consistency in FBA calculations. This is accomplished by the addition of thermodynamic constraints to FBA that restrict allowable reaction directionalities, thereby eliminating thermodynamically infeasible solutions (including cycles) from identified flux distributions.

TMFA first decomposes all reversible reactions into a forward and backward direction to track whether a reaction (in the decomposed network) can carry a nonzero flux. In addition, reactions for which the standard Gibbs free energy change $\left(\Delta_r G_j^{\circ}\right)$ data are not known and cannot be reliably estimated using group contribution methods are lumped into a new set of reactions for which the standard Gibbs free energy change can be estimated (J^{lumped}). It is important to note that reactions with unknown standard Gibbs free energy change are retained in the model and the set of lumped reactions are introduced only to impose the thermodynamic feasibility constraints on such reactions. A parameter α_{kj} is defined to trace the set of reactions comprising a lumped reaction k:

$$\alpha_{kj} = \begin{cases} 1 & \text{if reaction } j \text{ is in the set composing a lumped reaction } k \\ 0 & \text{otherwise} \end{cases}, \quad \forall k \in J^{\text{lumped}}, \; j \in J^{\text{model}}$$

where J^{model} is the set of reactions in the metabolic model (with decomposed reversible reactions containing both reactions with known or unknown standard Gibbs free energies). Two binary variable sets are introduced to capture whether reactions in the model with known $\Delta_r G_j^{\circ}$ as well as lumped reactions are thermodynamically feasible:

$$z_j = \begin{cases} 1 & \text{if reaction } j \text{ is thermodynamically feasible} \\ 0 & \text{otherwise} \end{cases}, \qquad \forall j \in J^{\text{model}}$$

$$y_k = \begin{cases} 1 & \text{if lumped reaction } k \text{ is thermodynamically feasible} \\ 0 & \text{otherwise} \end{cases}, \qquad \forall k \in J^{\text{lumped}}$$

Thermodynamic feasibility for a reaction j in the model implies that it can carry a nonzero flux only if $\Delta_r G_j$ is negative. Similarly, the constitutive reactions of a lumped reaction k can carry flux only if $\Delta_r G_k$ is negative. These conditions are enforced by using the following constraints [11]:

$$\Delta_r G_j \le (1 - z_j) M, \quad \forall j \in \left\{ j \middle| j \in J^{\text{model}} \text{ and } \Delta_r G_j^{\circ} \text{ is known} \right\} \tag{5.6}$$

$$0 \le v_j \le \text{UB}_j z_j, \quad \forall j \in J^{\text{model}} \tag{5.7}$$

$$\Delta_r G_k \le (1 - y_k) M, \quad \forall k \in J^{\text{lumped}} \tag{5.8}$$

$$\sum_{j \in J^{\text{model}}} \alpha_{kj} z_j \le \left(\sum_{j \in J^{\text{model}}} \alpha_{kj} \right) - (1 - y_k), \quad \forall k \in J^{\text{lumped}} \tag{5.9}$$

$$\Delta_r G_j = \Delta_r G_j^{\circ} + RT \left(\sum_{i \in \{ i | i \in I,\, S_{ij} \neq 0 \}} S_{ij} \ln(x_i) \right), \tag{5.10}$$

$$\forall j \in \left\{ j \middle| j \in J^{\text{model}} \text{ and } \Delta_r G_j^{\circ} \text{ is known} \right\} \cup J^{\text{lumped}}$$

Constraints 5.6 and 5.7 allow nonzero flux through a reaction j (with known $\Delta_r G_j^{\circ}$) only if $\Delta_r G_j$ is negative (observe that if $z_j = 1$ then $\Delta_r G_j \le 0$ and $0 \le v_j \le \text{UB}_j$). Similarly, Constraints 5.7, 5.8 and 5.9 allow nonzero flux through the reactions comprising lumped reaction k only if $\Delta_r G_k$ is negative. Observe that if $y_k = 0$, then $\sum_{j \in J^{\text{model}}} \alpha_{kj} z_j \le \sum_{j \in J^{\text{model}}} \alpha_{kj} - 1$ implying that at least one of the reactions forming the lumped reaction k is inactive. Equation 5.10 determines the Gibbs free energy change of a reaction j with known $\Delta_r G_j^{\circ}$ in the model or a lumped reaction k. Uncertainty in the estimation of $\Delta_r G_j^{\circ}$ can readily be incorporated into the equations [11].

These constraints can then be incorporated into FBA calculations to prevent the selection of infeasible reaction directions that can lead to the formation of TICs. Addition of these constraints to FBA results in the following MILP:

$$
\begin{aligned}
&\text{maximize } v_{\text{biomass}} \\
&\text{subject to} \\
&\sum_{j \in J^{\text{model}}} S_{ij} v_j = 0, \quad \forall i \in I \\
&0 \le v_j \le \text{UB}_j, \quad \forall j \in J^{\text{model}} \\
&\Delta_r G_j \le (1 - z_j) M, \quad \forall j \in \{ j | j \in J^{\text{model}} \ \text{ and } \ \Delta_r G_j^{\circ} \text{ is known} \} \\
&0 \le v_j \le \text{UB}_j z_j, \quad \forall j \in J^{\text{model}} \\
&\Delta_r G_k \le (1 - y_k) M, \quad \forall k \in J^{\text{lumped}} \\
&\sum_{j \in J^{\text{model}}} \alpha_{kj} z_j \le \left(\sum_{j \in J^{\text{model}}} \alpha_{kj} \right) - (1 - y_k), \quad \forall k \in J^{\text{lumped}} \\
&\Delta_r G_j = \Delta_r G_j^{\circ} + RT \left(\sum_{i \in \{ i | i \in I, S_{ij} \neq 0 \}} S_{ij} \ln(x_i) \right), \\
&\quad \forall j \in \{ j | j \in J^{\text{model}} \ \text{ and } \ \Delta_r G_j^{\circ} \ \text{ is known} \} \cup J^{\text{lumped}} \\
&z_j \in \{0,1\}, \quad \forall j \in J^{model} \\
&y_k \in \{0,1\}, \quad \forall k \in J^{\text{lumped}}
\end{aligned}
$$

Note that all continuous variables including v_j, $\Delta_r G_j$ and $X_i = \ln(x_i)$ appear linearly in the model. As mentioned before, lumped reactions participate in the thermodynamic feasibility constraints but not in the steady-state mass balances. Lower and upper bounds on variables X_i can be inferred from metabolite concentration data whenever available. However, given that such measurements are typically absent, a wide range from 10^{-5} M to 0.02 M is typically used [5] leading to inferred reaction directionalities that may on occasion be too permissive.

5.2.5 Elimination of the TICs by Applying the Loop Law

TMFA and follow-up methods [12–15] require *a priori* knowledge of the standard Gibbs free energy of formation of metabolites. These values can be found in databases (e.g., the NIST Chemical Kinetics Database) or estimated using group contribution methods [1–3]. However, the need for reliable free energy data can limit their use. Loop-less constraint-based reconstruction and analysis (ll-COBRA) is an alternative method that can eliminate TICs without making use of Gibbs free energy of formation values [10]. Infeasible cycle elimination is based on the observation that irrespective of reaction free energy change values, the flux in a TIC must be zero as required by the second law of thermodynamics [10]. This is analogous to Kirchhoff's second law for electrical circuits. Therefore, ll-COBRA cannot capture the effect of metabolite concentrations on feasible reaction directions. It only removes all TICs from the FBA solutions, and thus generally identifies only a subset of the infeasible reaction directions that can be revealed by TMFA.

Schellenberger et al. [10] introduced a continuous variable G_j as a "proxy" for the Gibbs free energy change for each reaction. G_j has the same sign as $\Delta_r G_j$, but its numerical value is arbitrary. It was shown that a given metabolic flux distribution will not contain any infeasible loops if the following conditions hold [16]:

$$-M \le G_j < -\varepsilon, \quad \forall j \in \{ j | j \in J, v_j > 0 \} \tag{5.11}$$

$$\varepsilon < G_j < M, \quad \forall j \in \{ j | j \in J, v_j < 0 \} \tag{5.12}$$

$$G_j \in \mathbb{R}, \quad \forall j \in \{ j | j \in J, v_j = 0 \} \tag{5.13}$$

$$\boldsymbol{N}_{\text{int}} \boldsymbol{G} = \boldsymbol{0} \tag{5.14}$$

where ε and M are small and big positive scalars, respectively and $\boldsymbol{G}$ is the vector of the proxy free energy changes of reactions. $\boldsymbol{N}_{\text{int}}$ is the null space of $\boldsymbol{S}_{\text{int}}$, the stoichiometric matrix of internal reactions only (obtained by removing the columns associated with transport/exchange reactions from S, the stoichiometric matrix of the entire network):

$$\boldsymbol{N}_{\text{int}} = \text{null}(\boldsymbol{S}_{\text{int}}) = \{ \boldsymbol{v}_{\text{int}} | \boldsymbol{S}_{\text{int}} \boldsymbol{v}_{\text{int}} = 0 \}$$

Observe that Equations 5.11 and 5.12 require that G_j be strictly negative or strictly positive if v_j is positive or negative, respectively. This is to avoid the possibility of having $G_j = 0$, as it provides no information about the direction of flux through the reaction j. Equation 5.13 emphasizes that G_j is free to assume any positive or negative value when v_j is equal to zero. Imposing these conditions within FBA requires the definition of the following binary variable:

$$y_j = \begin{cases} 1 & \text{if } v_j \ge 0 \\ 0 & \text{if } v_j \le 0 \end{cases}, \quad \forall j \in J$$

The loop-less condition is then incorporated into the optimization formulation of a constraint-based analysis method such as FBA by appending the following constraints:

$$-M(1 - y_j) \le v_j \le My_j, \quad \forall j \in J^{\text{internal}} \tag{5.15}$$

$$-My_j + \varepsilon(1 - y_j) \le G_j \le -\varepsilon y_j + M(1 - y_j), \quad \forall j \in J^{\text{internal}} \tag{5.16}$$

$$\begin{aligned} &\boldsymbol{N}_{\text{int}} \boldsymbol{G} = \boldsymbol{0} \\ &y_j \in \{0,1\}, \quad \forall j \in J^{\text{internal}} \\ &G_j \in \mathbb{R}, \quad \forall j \in J^{\text{internal}} \end{aligned} \tag{5.17}$$

where J^{internal} denotes the set of internal reactions. These constraints render FBA calculations into MILP problems denoted as ll-FBA:

$$
\begin{aligned}
&\text{maximize} \quad v_{\text{biomass}} \\
&\text{subject to} \\
&\sum_{j \in J} S_{ij} v_j = 0, \quad \forall i \in I \\
&LB_j \leq v_j \leq UB_j, \quad \forall j \in J, \\
&-M\left(1-y_j\right) \leq v_j \leq M y_j, \quad \forall j \in J^{\text{internal}} \\
&\boldsymbol{N}_{\text{int}} \boldsymbol{G} = 0, \\
&y_j \in \{0,1\}, \quad \forall j \in J^{\text{internal}} \\
&G_j \in \mathbb{R}, \quad \forall j \in J^{\text{internal}} \\
&v_j \in \mathbb{R}, \quad \forall j \in J
\end{aligned}
$$

5.2.6 Elimination of the TICs by Modifying the Metabolic Model

Both TMFA and ll-COBRA do not make modifications to the model. Instead, they use additional constraints to safeguard against tracing thermodynamically infeasible loops. Alternatively, a proactive way of addressing such infeasible loops is to iteratively use FVA and make modifications to the model so as to eliminate all unbounded reactions, thereby removing thermodynamic inconsistencies. In general, thermodynamically infeasible loops may arise due to the incorrect incorporation of reactions in the model, overly permissive reaction directionalities or missing regulatory constraints. One can thus remedy these problems through successive model modifications that eliminate all occurrences of reactions with unbounded flux. The nature of TICs and corresponding model changes can be broadly grouped into three categories (see Fig. 5.2) as follows:

(i) Thermodynamically infeasible reaction directions may lead to the emergence of cycles. Directly restricting the directionality of often a single reaction (using the Gibbs free energy change information as a guide) can eliminate one or more overlapping TICs and unbounded fluxes (Fig. 5.2a).

(ii) The presence of linearly dependent reversible reactions in a model leads to the formation of a TIC. This linear dependence may arise when both the lumped and the de-lumped form of a reversible reaction sequence is included. The elimination of the cycle requires that a linearly dependent and thus dispensable reaction is removed (Fig. 5.2b).

(iii) The presence of multiple compartments can give rise to TICs whenever interconversion of two metabolites is possible in two separate compartments along with a cost-free (passive) mode of transport between them (Fig. 5.2c). Elimination of

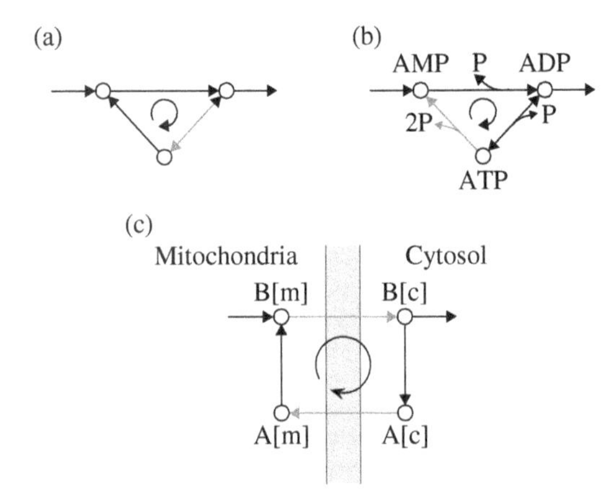

FIGURE 5.2 Three possible ways to fix TICs by modifying the metabolic model. Reactions shown in gray are the ones that need to be modified using these strategies. (a) Restrict the directionality of reactions; (b) eliminate one reaction from a linearly dependent set; and (c) add transport costs.

these cycles can be achieved by identifying and including in the transport reaction stoichiometry the associated cost (e.g., proton or ATP use). Alternatively, preventing one direction of transport (if no verified corresponding transporter is available) will also resolve the thermodynamic inconsistency.

Note that all these model modifications must be carefully corroborated using data and literature information as in most cases there exist multiple ways of "breaking" a cycle. In addition, the potential for cycle formation is dependent on substrate availability and regulatory restrictions. In Chapter 6, we discuss how to "fill in" gaps in metabolic networks using optimization based methods. Gap-bridging strategies must be carefully scrutinized to ensure that no TICs are created as a consequence of the reaction additions into the model.

EXERCISES

5.1 Revisit the *i*AF1260 metabolic model of *Escherichia coli* [17].

- **(a)** How many reactions participate in TICs under aerobic conditions with glucose as the carbon source?
- **(b)** From the list of reactions identified in (i), find three TICs and suggest ways of eliminating them.
- **(c)** How does the set of reactions with unbounded fluxes change for a rich growth medium where the uptake of all carbon substrates is allowed?
- **(d)** Does the maximum biomass flux change if the ll-COBRA algorithm is implemented for the *i*AF1260 model?

(e) Use the flux variability analysis (FVA) (see Chapter 3) in conjunction with ll-COBRA constraints and biomass flux fixed at its maximum obtained under (d) to confirm that no reaction has an unbounded flux.

REFERENCES

1. Jankowski MD, Henry CS, Broadbelt LJ, Hatzimanikatis V: Group contribution method for thermodynamic analysis of complex metabolic networks. *Biophys J* 2008, **95**(3): 1487–1499.
2. Mavrovouniotis ML: Group contributions for estimating standard gibbs energies of formation of biochemical compounds in aqueous solution. *Biotechnol Bioeng* 1990, **36**(10):1070–1082.
3. Alberty RA: Calculation of standard transformed Gibbs energies and standard transformed enthalpies of biochemical reactants. *Arch Biochem Biophys* 1998, **353**(1):116–130.
4. Albe KR, Butler MH, Wright BE: Cellular concentrations of enzymes and their substrates. *J Theor Biol* 1990, **143**(2):163–195.
5. Henry CS, Jankowski MD, Broadbelt LJ, Hatzimanikatis V: Genome-scale thermodynamic analysis of Escherichia coli metabolism. *Biophys J* 2006, **90**(4):1453–1461.
6. Fleming RM, Thiele I, Nasheuer HP: Quantitative assignment of reaction directionality in constraint-based models of metabolism: application to Escherichia coli. *Biophys Chem* 2009, **145**(2–3):47–56.
7. Voet D, Voet JG: *Biochemistry*, 4th edn. Hoboken, NJ: John Wiley & Sons; 2011.
8. Bogorad IW, Lin TS, Liao JC: Synthetic non-oxidative glycolysis enables complete carbon conservation. *Nature* 2013, **502**(7473):693–697.
9. Bogorad IW, Chen CT, Theisen MK, Wu TY, Schlenz AR, Lam AT, Liao JC: Building carbon-carbon bonds using a biocatalytic methanol condensation cycle. *Proc Natl Acad Sci U S A* 2014, **111**(45):15928–15933.
10. Schellenberger J, Lewis NE, Palsson B: Elimination of thermodynamically infeasible loops in steady-state metabolic models. *Biophys J* 2011, **100**(3):544–553.
11. Henry CS, Broadbelt LJ, Hatzimanikatis V: Thermodynamics-based metabolic flux analysis. *Biophys J* 2007, **92**(5):1792–1805.
12. Jol SJ, Kümmel A, Terzer M, Stelling J, Heinemann M: System-level insights into yeast metabolism by thermodynamic analysis of elementary flux modes. *PLoS Comput Biol* 2012, **8**(3):e1002415.
13. Zamboni N, Kümmel A, Heinemann M: anNET: a tool for network-embedded thermodynamic analysis of quantitative metabolome data. *BMC Bioinformat* 2008, **9**:199.
14. Martínez VS, Quek LE, Nielsen LK: Network thermodynamic curation of human and yeast genome-scale metabolic models. *Biophys J* 2014, **107**(2):493–503.
15. Zhu Y, Song J, Xu Z, Sun J, Zhang Y, Li Y, Ma Y: Development of thermodynamic optimum searching (TOS) to improve the prediction accuracy of flux balance analysis. *Biotechnol Bioeng* 2013, **110**(3):914–923.
16. Noor E, Lewis NE, Milo R: A proof for loop-law constraints in stoichiometric metabolic networks. *BMC Syst Biol* 2012, **6**:140.
17. Feist AM, Henry CS, Reed JL, Krummenacker M, Joyce AR, Karp PD, Broadbelt LJ, Hatzimanikatis V, Palsson B: A genome-scale metabolic reconstruction for Escherichia coli K-12 MG1655 that accounts for 1260 ORFs and thermodynamic information. *Mol Syst Biol* 2007, **3**:121.

6

RESOLVING NETWORK GAPS AND GROWTH PREDICTION INCONSISTENCIES IN METABOLIC NETWORKS

In this chapter, we outline basic concepts and methods based on LP and MILP for pinpointing and bridging gaps in metabolic models as well as for reconciling model growth prediction inconsistencies with experiments.

6.1 FINDING AND FILLING NETWORK GAPS IN METABOLIC MODELS

All metabolic models are inherently incomplete containing network gaps. These gaps are usually in the form of missing reactions in the network arising as a result of missed gene annotations due to the lack of experimental or homology evidence. Identifying and bridging these gaps are key tasks in the reconstruction of high-quality genome-scale metabolic (GSM) models. The presence of gaps in a model may lead to erroneous predictions about growth or product yields and incorrect genetic intervention strategies for the overproduction of a target metabolite.

6.1.1 Categorization of Gaps in a Metabolic Model

Gaps in metabolic reconstructions are manifested as dead-end metabolites and blocked reactions (see Chapter 3 for the definition of blocked reactions). In this chapter, we primarily focus on dead-end metabolites. Resolution of these problem

Optimization Methods in Metabolic Networks, First Edition. Costas D. Maranas and Ali R. Zomorrodi.

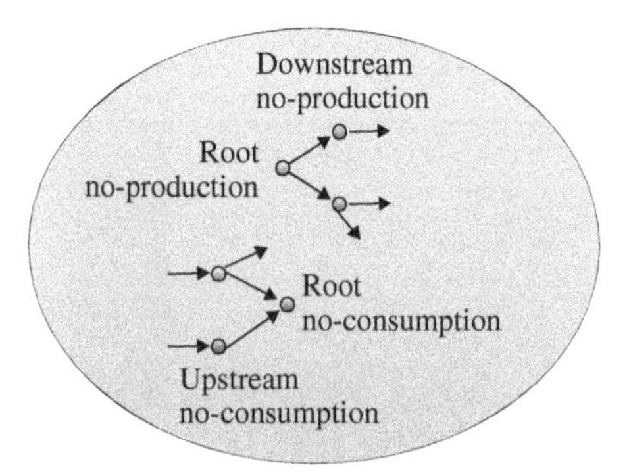

FIGURE 6.1 Graphical illustration of different types of problem metabolites in a metabolic network [1].

metabolites automatically restores metabolic flow in blocked reactions. Two major types of dead-end metabolites can be identified in a metabolic network [1] (see Fig. 6.1):

1. *Root no-production metabolites*: Metabolites with neither an intracellular producing reaction nor an uptake reaction.
2. *Root no-consumption metabolites*: Metabolites with neither an intracellular consuming reaction nor an export reaction.

Imposing the steady-state condition under flux balance analysis (FBA) implies that there will be no flux through reactions consuming or producing these dead-end metabolites under any growth condition. The effect of root no-production and no-consumption metabolites is propagated downstream and upstream in the network, respectively, giving rise to new sets of problem metabolites that cannot be produced (i.e., *downstream no-production* metabolites) or consumed (i.e., *upstream no-consumption* metabolites) even though they have both producing and consuming reactions or transport systems associated with them (see Fig. 6.1). By restoring the connectivity of root problem metabolites, both downstream and upstream problem metabolites are also reconnected. The reverse does not necessarily hold true.

6.1.2 Gap Finding

Identification of root problem metabolites in a metabolic network is straightforward. Root no-production metabolites can be identified by scanning all columns of the stoichiometric matrix of the network for each metabolite i and checking whether there exists at least one positive coefficient for irreversible reactions or at least one nonzero coefficient for reversible reactions. If no such entries are found, then metabolite i is a root no-production metabolite. Root no-consumption metabolites can be identified in a similar manner.

Identification of downstream no-production metabolites cannot be achieved simply by inspecting the stoichiometric matrix. They can be identified by solving a sequence of LP problems (one per metabolite) that maximize the flow through that particular metabolite. If the maximum flow is zero, then the model cannot produce that metabolite. An alternative way of finding all no-production metabolites relies on the solution of a single MILP problem (i.e., GapFind [1]). Formulation of the GapFind procedure requires the definition of the following two binary variables:

$$x_i = \begin{cases} 1 & \text{if metabolite } i \text{ can be produced in the network} \\ 0 & \text{otherwise} \end{cases}$$

$$w_{ij} = \begin{cases} 1 & \text{if reaction } j \text{ producing metabolite } i \text{ in the network is active} \\ 0 & \text{otherwise} \end{cases}$$

for $i \in I$ and $j \in J$ with I and J being the set of metabolites and reactions in the network, respectively. The following MILP can then identify all root and downstream no-production metabolites:

$$\text{maximize } z = \sum_{i \in I} x_i \quad [\text{GapFind}]$$

subject to

$$\varepsilon - M(1 - w_{ij}) \leq S_{ij} v_j \leq M w_{ij}, \quad \forall i \in I, j \in \{j | j \in J \text{ and } S_{ij} \neq 0\} \tag{6.1}$$

$$\sum_{j' \in J'} w_{ij'} \geq x_i, \quad \forall i \in I \tag{6.2}$$

$$\text{where } J' = \{j | j \in J^{\text{irrev}} \text{ and } S_{ij} > 0\} \cup \{j | j \in J^{\text{rev}} \text{ and } S_{ij} \neq 0\}$$

$$LB_j \leq v_j \leq UB_j, \quad \forall j \in J \tag{6.3}$$

$$\sum_{j \in J} S_{ij} v_j \geq 0, \quad \forall i \in I^{\text{cytosol}} \tag{6.4}$$

$$\sum_{j \in J} S_{ij} v_j = 0, \quad \forall i \notin I^{\text{cytosol}} \tag{6.5}$$

$$x_i \in \{0,1\}, \quad \forall i \in I$$

$$w_{ij} \in \{0,1\}, \quad \forall i \in I, j \in J$$

Here, J^{irrev} and J^{rev} denote the set of irreversible and reversible reactions, respectively, I^{cytosol} denotes the set of metabolites present in the cytosol, M denotes wide reaction bounds, and ε is a scalar defining a minimum threshold of reaction activity. The objective function maximizes the number of metabolites with nonzero metabolic flux through them. Constraint 6.1 imposes the definition of binary variable w_{ij} ensuring that if an irreversible reaction producing metabolite i or a reversible reaction in which metabolite i participates is active, then its corresponding binary variable w_{ij} assumes a value of one. Constraint 6.3 imposes the definition of binary variable x_i, which assumes a value of zero only when none of the reactions producing metabolite i are active. Constraint 6.3 restricts the reaction fluxes to vary between their pre-specified lower and upper bounds (see Table 3.1). Note that here the lower bound for all exchange reactions is set to $-M$ (i.e., all metabolites for which there exists an exchange reaction in the model are allowed to be taken up). This assumes that elucidation of blocked metabolites is performed for a rich growth medium containing all metabolites for which an uptake pathway exists. Similar to blocked reactions, the set of blocked metabolites is a context-dependent network property that is a function of the composition of the growth medium. Constraint 6.4 states that a default consumption or export pathway exists for all cytosolic metabolites due to (i) the diluting effect of the cell division acting as a metabolic sink, (ii) the consumption of metabolites by non-metabolic reactions, (iii) the participation of metabolites in macromolecular production pathways and (iv) the passive diffusion through the cell membrane [1]. Finally, Constraint 6.5 indicates that the steady-state mass balance should hold for all non-cytosolic metabolites (i.e., for those present in the extracellular compartment or in the internal compartments such as the mitochondria [in multicompartment models]). This optimization problem simultaneously identifies all root and downstream no-production metabolites in the network. Any metabolite i for which the optimal value of x_i is zero is a (root or downstream) no-production metabolite. The list of downstream no-production metabolites can be recovered by subtracting the set of root no-production metabolites (identified using the method described earlier) from the list of all metabolites for which $x_i = 0$. This procedure can be adjusted in a straightforward manner to identify non-cytosolic no-consumption metabolites (i.e., those present in internal compartments).

This MILP representation of GapFind provides a foundation for introducing the GapFill procedure that provides reconnection strategies for as many network gaps as possible. Alternatively, an LP-based identification of gaps is also possible whereby no-production metabolites are identified *one at a time* by solving a sequence of LP problems for metabolite i (instead of a single MILP) as follows:

$$\text{maximize} \sum_{j \in \{j | j \in J^{\text{irrev}}, S_{ij} > 0\} \cup \{j | j \in J^{\text{rev}}, S_{ij} \neq 0\}} S_{ij} v_j$$

subject to

$$LB_j \leq v_j \leq UB_j, \quad \forall j \in J \tag{6.6}$$

$$\sum_{j \in J} S_{ij} v_j \geq 0, \quad \forall i \in I^{\text{cytosol}} \tag{6.7}$$

$$\sum_{j \in J} S_{ij} v_j = 0, \quad \forall i \notin I^{\text{cytosol}} \tag{6.8}$$

Metabolite i is a no-production metabolite if the maximum value of the objective function is zero. The total flux through a metabolite needs to be carefully assessed so as to exclude flux due to a thermodynamically infeasible cycles (TICs) (see Chapter 5).

6.1.3 Gap Filling

Once the problem metabolites are identified, the next step is to attempt to restore their connectivity with the rest of the network. Three basic strategies are available to restore flow for no-production metabolites (see Fig. 6.2):

(a) Relaxing the irreversibility constraints on existing reactions in the network (i.e., converting irreversible reactions to reversible).
(b) Adding new reactions to the model from external reaction databases such as MetaCyc [2], KEGG [3], Model SEED [4], and MetRxn [5].
(c) Adding transport reactions between the cytosol and extracellular environment or between internal compartments and the cytosol.

GapFill is an MILP-based method that systematically attempts to restore flow for no-production metabolites through the minimal use of the three aforementioned mechanisms [1]. All added reactions to the model are treated as reversible unless their directionality is pre-specified. GapFill requires the definition of three sets of binary variables:

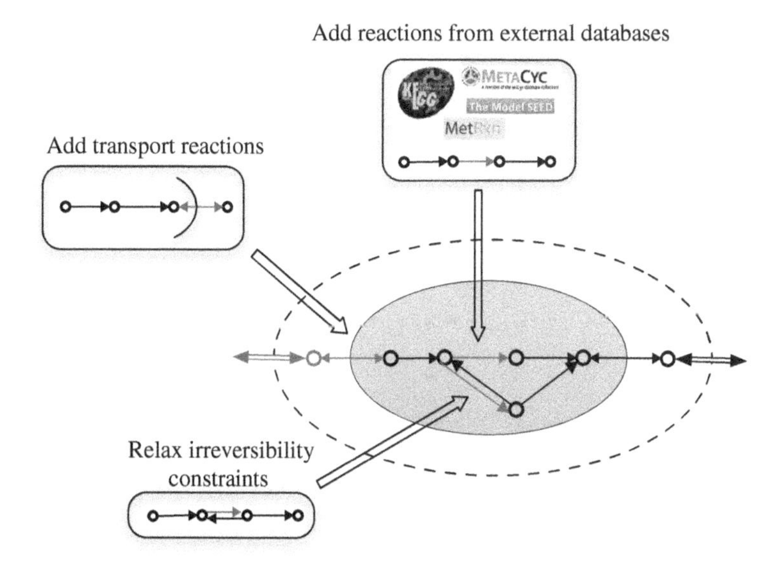

FIGURE 6.2 Mechanisms used in GapFill [1] to fix no-production metabolites.

$$x_j = \begin{cases} 1 & \text{if the reversibility of a reaction } j \text{ is relaxed} \\ 0 & \text{otherwise} \end{cases} \quad \forall j \in J^{\text{model,irrev}}$$

$$y_j = \begin{cases} 1 & \text{if reaction } j \text{ is added to the model} \\ 0 & \text{otherwise} \end{cases} \quad \forall j \in J^{\text{database}} \bigcup J^{\text{transport}}$$

$$w_{i^*j} = \begin{cases} 1 & \text{if reaction } j \text{ producing } i^* \text{ is active} \\ 0 & \text{otherwise} \end{cases} \quad \forall j \in J^{\text{model}} \bigcup J^{\text{database}} \bigcup J^{\text{transport}}$$

where J^{model}, $J^{\text{model,irrev}}$, J^{database} and $J^{\text{transport}}$ denote the set of reactions in the original model, irreversible reactions in the original model, reactions in database, and candidate transport reactions, respectively. The task of fixing a no-production metabolite i^* while minimally perturbing the original model is formulated as the following MILP:

$$\text{minimize} \quad z = \sum_{j \in J^{\text{model,irrev}}} x_j + \sum_{j \in J^{\text{database}} \bigcup J^{\text{transport}}} y_j \quad [\text{GapFill}]$$

subject to

$$\sum_{j \in J^{\text{model}} \bigcup J^{\text{database}} \bigcup J^{\text{transport}}} S_{ij} v_j \geq 0, \quad \forall i \in I^{\text{cytoslic}} \tag{6.9}$$

$$\sum_{j \in J^{\text{model}} \bigcup J^{\text{database}} \bigcup J^{\text{transport}}} S_{ij} v_j = 0, \quad \forall i \notin I^{\text{cytoslic}} \tag{6.10}$$

$$\varepsilon - M\left(1 - w_{i^*j}\right) \leq S_{ij} v_j \leq M w_{i^*j}, \quad \forall j \in \left\{ j \mid j \in J, S_{i^*j} \neq 0 \right\} \tag{6.11}$$

$$\sum_{j \in J} w_{i^*j} \geq 1 \tag{6.12}$$

$$LB_j \leq v_j \leq UB_j, \quad \forall j \notin j^{\text{model, rev}} \tag{6.13}$$

$$-Mx_j \leq v_j \leq UB_j, \quad \forall j \in J^{\text{model,irrev}} \tag{6.14}$$

$$-My_j \leq v_j \leq My_j, \quad \forall j \in J^{\text{database}} \bigcup J^{\text{transport}} \tag{6.15}$$

$$y_j \in \{0,1\}, \quad \forall j \in J^{\text{database}} \bigcup J^{\text{transport}}$$

$$x_j \in \{0,1\}, \quad \forall j \in J^{\text{model, irrev}}$$

$$w_{i^*j} \in \{0,1\}, \quad \forall j \in J^{\text{model}} \bigcup J^{\text{database}} \bigcup J^{\text{transport}}$$

The objective function involves the minimization of the number of modifications to the original model. Constraints 6.9 and 6.10 play the same role as Constraints 6.4 and 6.5 in the GapFind formulation. Constraint 6.11 defines binary variable w_{i^*j}. Constraint 6.12 ensures that at least one reaction producing i^* is active in response to the model modifications. Equation 6.13 sets the bounds for reversible reactions (i.e., J^{rev}) in the model. Finally, Constraints 6.14 and 6.15 control the irreversibility relaxation and reaction/transporter addition into the model, respectively. This optimization problem is solved separately for each no-production metabolite i^*. If the problem is infeasible, then the no-production metabolite i^* cannot be fixed using any of the three mechanisms previously described. Integer cuts (see Chapter 4) are typically used to identify alternate gap filling strategies. This is necessary as GapFill only suggests flow restoration hypotheses. The validity of model modifications must be ascertained on a case-by-case basis. This optimization problem can be adjusted to fix non-cytosolic no-consumption metabolites in multicompartment metabolic models or to fill gaps in a specified (i.e., rich or minimal) growth media. Gap filling is generally performed first for biomass precursors that cannot be produced by the network. Once biomass production is restored in the network, the next step is to bridge the remaining gaps in the model. The following stepwise procedure codifies current practices in gap filling:

Step 1: Identification of all gaps in the model. Using GapFind described in Section 6.1.2 or other tools, the root and downstream no-production metabolites are identified. Biomass components that cannot be produced are then identified from the list of no-production metabolites.

Step 2: Generation of alternative gap filling hypotheses. The GapFill or other tools are used to identify flow restoration hypotheses (i.e., from three to five alternatives using integer cuts) for blocked biomass components.

Step 3: Exclusion of hypotheses that create TICs. Hypotheses from Step 2 are screened for the introduction of TICs (see Chapter 5).

Step 4: Selection of gap filling hypothesis. Gap filling hypotheses not forming TICs are subsequently examined for evidence of validity (see Section 6.3 for details). Generally, model modifications are ranked based on parsimony and whether the model modification fixes multiple gaps.

Step 5: Generation of gap filling hypotheses for non-biomass metabolites. Non-biomass no-production metabolites identified in Step 1 are next examined and gap-filled in a similar manner as those for biomass constituents (see Steps 2,3, and 4).

There exist a number of other methods of automated gap filling procedures [6–10] that build upon the basic GapFill workflow. For example, MetabolIc Reconstruction via functionAl GEnomics (MIRAGE) identifies missing reactions in the network by integrating metabolic flux analysis and functional genomics (e.g., phylogenetic profiles and gene expression) data [9]. In a recent study [6], genomic and sequence homology information is used in tandem to calculate a likelihood score for genes and reactions. The gene score quantifies the likelihood that a gene carries a specific annotated function, while the reaction score assigns a probability score that a specific reaction should be in the metabolic model. Reactions are then added to the model to fill gaps according to the computed likelihood scores.

Note that in some cases, gaps in the model do not imply omitted functionalities. Sometimes, gaps are generated because a reaction was erroneously added to the model and there is no way to connect it with the rest of metabolism. Therefore, it is important to explore whether the corresponding blocked reaction should indeed be part of the model for persistent gaps. Application of an automated gap filling to 130 organisms showed that, on average, 56 reactions were added to each model to restore growth [4]. Over 50% of the added reactions were involved in cofactor or cell wall biosynthesis [4].

6.2 RESOLVING GROWTH PREDICTION INCONSISTENCIES

The gold standard for testing the accuracy of metabolic reconstructions is to contrast model biomass yield predictions of single or multiple gene knockouts with *in vivo* growth data for different growth media [11]. This comparison for single gene deletion experiments leads to four possible outcomes: GG, NGNG, GNG and NGG (see Fig. 6.3a). In cases GG and NGNG both model and experimental data agree by either denoting growth (G) or no growth (NG), respectively. In case GNG, the model predicts that biomass can be formed but experiments show no growth. In contrast, in case NGG the model predicts that biomass cannot be formed, however, *in vivo* growth is observed for the mutant strain. Generally, a threshold for biomass yield is prespecified below which a mutant strain is classified as not growing. This threshold varies among studies [12–15] between 1% and 33% of the theoretical maximum biomass flux for the wild-type strain. Similarly, a viability threshold is also chosen

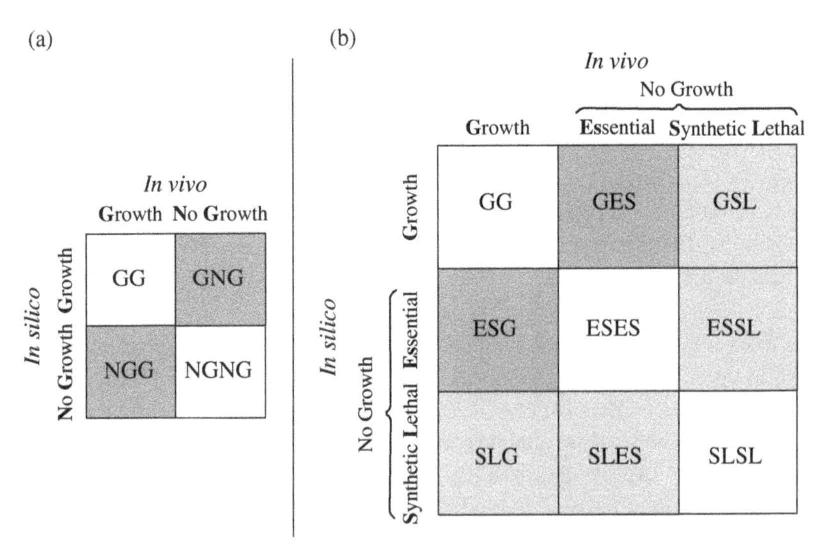

FIGURE 6.3 Classification of inconsistencies between model predictions and experimental observations: (a) single gene deletion experiments and (b) double gene deletion experiments.

(e.g., 33% from the wild-type [16]) for the classification of experimental growth data to account for measurement errors.

6.2.1 Quality Metrics for Quantifying the Accuracy of Metabolic Models

Once the consistencies and inconsistencies between the model and experiments are identified, a number of quality metrics that integrate this information into a single numeric value (usually a percentage) are used to quantify the accuracy of the model:

$$\text{Sensitivity} = \text{TVR}\left(\text{true viable rate}\right) = \frac{\text{number of GGs}}{\text{number of GGs} + \text{number of NGGs}} \times 100$$

$$\text{Specificity} = \text{TIR}\left(\text{true inviable rate}\right) = \frac{\text{number of NGNGs}}{\text{number of NGNGs} + \text{number of GNGs}} \times 100$$

$$\text{FVR}\left(\text{false viable rate}\right) = \frac{\text{number of GNGs}}{\text{number of NGNGs} + \text{number of GNGs}} \times 100$$

$$\text{FIR}\left(\text{false inviable rate}\right) = \frac{\text{number of NGGs}}{\text{number of GGs} + \text{number of NGGs}} \times 100$$

Specificity and sensitivity are the most widely used metrics [13, 14, 17]. These metrics can be extended for multiple gene deletions [13].

6.2.2 Automated Reconciliation of Growth Prediction Inconsistencies Using GrowMatch

A number of procedures have been proposed for reconciling growth prediction inconsistencies [12, 13, 18–20]. In this chapter, we describe GrowMatch, which is based on separate MILP formulations for resolving NGG and GNG inconsistencies.

Resolution of NGG Inconsistencies NGG inconsistencies arise when the GSM model underestimates the metabolic capabilities of the organism due to the absence of relevant reactions/pathways. These inconsistencies can thus be fixed by identifying and adding these missing functionalities to the model. GrowMatch employs the same three mechanisms used in GapFill (see Section 6.1.3) to restore *in silico* growth for the mutant strains thereby converting NGGs to GGs one at a time. This can be formulated as the following MILP to resolve a particular NGG inconsistency corresponding to the deletion of gene g^*:

$$\text{minimize} \quad z = \sum_{j \in J^{\text{database}} \bigcup J^{\text{transport}}} y_j + \sum_{j \in J^{\text{model,irrev}}} x_j \quad \left[\text{GrowMatch - NGG}\right]$$

subject to

$$\sum_{j \in J^{\text{model}} \bigcup J^{\text{database}} \bigcup J^{\text{transport}}} S_{ij} v_j = 0, \quad \forall i \in I \tag{6.16}$$

$$v_j = 0, \quad \forall j \in J^{g^*} \tag{6.17}$$

$$v_{\text{biomass}} \geq f\, v_{\text{biomass}}^{\text{max,WT}} \tag{6.18}$$

$$LB_j \leq v_j \leq UB_j, \quad \forall j \in j^{\text{model, rev}} \tag{6.19}$$

$$-Mx_j \leq v_j \leq UB_j, \quad \forall j \in J^{\text{model, irrev}} \tag{6.20}$$

$$-My_j \leq v_j \leq My_j, \quad \forall j \in J^{\text{database}} \bigcup J^{\text{transport}} \tag{6.21}$$
$$y_j \in \{0,1\}, \quad \forall j \in J^{\text{database}} \bigcup J^{\text{transport}}$$

where J^{g^*} denotes the set of reactions that must be eliminated due to the deletion of gene g^* (based on the gene-reaction map enforced by Constraint 6.17), f is a pre-specified viability threshold ($0 \leq f \leq 1$) and $v_{\text{biomass}}^{\text{max,WT}}$ is the maximum theoretical biomass flux computed for the wild-type network. Binary variables y_j and x_j are defined as in GapFill (see Section 6.1.3). Constraint 6.18 requires the biomass flux to meet the minimum level imposed by the viability threshold. Constraints 6.19, 6.20 and 6.21 are identical to 6.13, 6.14 and 6.15, respectively. This optimization problem is solved for one NGG inconsistency at a time and alternative correction strategies can be found by using integer cuts. It is worth noting that, as discussed in Chapter 4, a generally more efficient MILP representation of the same problem involves maximizing the biomass flux as the objective function and replacing Constraint 6.18 with

$$\sum_{j \in J^{\text{database}} \bigcup J^{\text{transport}}} y_j + \sum_{j \in J^{\text{model,irrev}}} x_j \leq K \tag{6.22}$$

where K is a pre-specified number determining the maximum allowable number of model perturbations. One may start with $K = 1$, solve the problem and continue increasing the value of K until $v_{\text{biomass}} \geq f\, v_{\text{biomass}}^{\text{max,WT}}$. The procedure is repeated for all instances of NGG inconsistencies. It is important to stress that the resolution strategy of one NGG inconsistency may propagate in the model and lead to the resolution of additional NGGs and/or the invalidation of existing NGNG cases by converting them into GNGs. Therefore, the global effect of all generated NGG resolution strategies must be assessed carefully.

Resolution of GNG Inconsistencies GNG inconsistencies arise when the GSM model overestimates the metabolic capabilities of an organism alluding to the presence of reactions/transport systems that are erroneously present in the GSM model or missing regulatory restrictions under the examined condition. The one-by-one reconciliation of these mismatches using GrowMatch involves identifying minimal reaction suppressions lowering the maximum biomass flux below a pre-specified viability threshold in the perturbed network, thereby converting the GNG inconsistency to a NGNG consistency. Binary variables are used to determine whether a reaction needs to be suppressed/eliminated:

$$y_j = \begin{cases} 0 & \text{if a reaction } j \text{ is suppressed/eliminated} \\ 1 & \text{otherwise} \end{cases}$$

Resolution of a GNG inconsistency associated with the deletion of a gene g^* can then be accomplished using the following bilevel optimization problem (the solution procedure for bilevel formulations will be discussed in detail in Chapter 8).

$$\text{minimize } v_{\text{biomass}} \quad [\text{GrowMatch-GNG}]$$
$$\text{subject to}$$
$$\begin{bmatrix} \text{maximize } v_{\text{biomass}} & \\ \text{subject to} & \\ \sum_{j \in J} S_{ij} v_j = 0, & \forall i \in I \\ LB_j y_j \le v_j \le UB_j y_j, & \forall j \in J \end{bmatrix} \quad \begin{matrix} \\ \\ (6.23) \\ (6.24) \end{matrix}$$

$$y_j = 0, \quad \forall j \in J^{g^*} \tag{6.25}$$

$$\sum_{j \in J} (1 - y_j) \le K \tag{6.26}$$

$$y_j \in \{0,1\}, \quad \forall j \in J$$

where K is a pre-specified value determining the maximum allowable number of reaction suppressions. This is a nested optimization problem referred to as a *bilevel problem*, where one of the constraints is another optimization problem (see Chapter 8 for a detailed description of bilevel optimization problems and their solution methods). The inner optimization problem maximizes the biomass formation of the perturbed network subject to the network stoichiometry (Constraint 6.23) and reaction suppressions imposed by the outer problem (Constraint 6.24). Constraint 6.25 ensures the elimination of reactions that must not carry any flux due to the deletion of gene g^*. The outer problem minimizes the maximum biomass formation by identifying K reactions to suppress (Constraint 6.26). In essence, this minmax bilevel structure finds the K reaction eliminations that can drive the maximum biomass formation potential of the network at its lowest level even when fluxes in the network are re-apportioned to maximize biomass formation. If this biomass level is below the pre-defined viability threshold then the K eliminations become candidates for removal from the model. If not, the value of K is increased by one and the problem is resolved. This optimization problem is solved for each GNG occurrence separately. Alternative GSM model correction strategies can be obtained by using integer cuts (see Chapter 4).

It is important to note that removal of essential reactions should be avoided by fixing their binary variable values to one. Similarly, synthetic lethal reactions must not be simultaneously eliminated (e.g., by adding constraint $y_{j_1} + y_{j_2} \geq 1$, if j_1 and j_2 form a synthetic lethal pair). Other reactions that must not be suppressed/eliminated include the biomass reaction, non-growth-associated maintenance ATP (i.e., ATPM or NGAM) and exchange reactions. As noted in Chapter 4, the computational performance of the algorithm can be improved by fixing the binary variables associated with blocked reactions and by considering only one reaction from each fully coupled reaction set for suppression/elimination.

Note that growth prediction inconsistencies are growth medium composition dependent. Implementation of GrowMatch thus requires a careful delineation of every compound that can be taken up. The set of nutrients available in a minimal (e.g., M9) medium is a subset of those available in a rich (e.g., LB) medium, therefore any NGG inconsistency in the rich medium will also apply for the minimal medium under the same limiting carbon source and aeration conditions. For the same reason, all GNG inconsistencies derived for a minimal growth medium will also apply for the rich growth medium. Consequently, it is advisable to first resolve NGG inconsistencies under a rich medium before finding possibly additional ones under a minimal medium. The reverse order of resolution applies for GNG inconsistencies.

Global and Conditional Modifications Overall, the resolution of NGG and GNG inconsistencies is expected to increase the sensitivity and specificity of the model, respectively. However, this may not always be the case as the suggested correction strategies by GrowMatch for resolving a GNG or NGG inconsistency may conflict with one or more correct model predictions (i.e., GG or NGNG). In particular, the resolution of an NGG may convert one or more NGNG consistencies to new GNG inconsistencies. Similarly, fixing a GNG may convert one or more GG consistencies to new NGG inconsistencies. Such modifications are thus flagged as *conditional modifications* and are generally rejected from consideration. This is because while the resolution of NGGs (GNGs) using conditional modifications improves the sensitivity (specificity), it worsens the specificity (sensitivity) of the model due to the newly emerged inconsistencies. Modifications that do not invalidate any correct model predictions are referred to as *global modifications* and undergo further examination (see Section 6.3) before acceptance.

Instead of relying on the concept of conditional and global modifications, Henry et al. [20] suggested two new optimization-based procedures termed *gap filling reconciliation* and *gap generation reconciliation* to improve the quality of the entire model after incorporating all suggested modifications. The gap filling reconciliation step maximizes the correction of NGGs, while minimizing the number of modifications to the model as well as the number of newly formed GNGs. Alternatively, the gap generation reconciliation step maximizes the correction of GNGs with the minimum number of modifications to the model, while also minimizing the newly emerged NGGs.

6.2.3 Resolution of Higher-Order Gene Deletion Inconsistencies

The original version of GrowMatch uses only gene essentiality data to identify and fix model/experiment inconsistencies. This procedure was later extended to make use of higher-order gene deletion (i.e., synthetic lethality) data providing additional layers of model correction (Extended GrowMatch) [13]. Contrasting model predictions for double deletions [21] with available experimental double gene deletion data reveals a number of additional ways that model and experiment may disagree (see Fig. 6.3b). Here, no growth may arise due to either gene essentiality (ES) or synthetic lethality (SL). Cases ESG and GES are identical to NGG and GNG for single gene deletions. A GSL inconsistency refers to a mismatch where the *in silico* deletion of a gene pair allows for growth but experiment reveals that they form a synthetic lethal (SL) pair. Similarly, ESSLs are inconsistencies where one or both genes are essential (ES) *in silico,* however, they form a synthetic lethal pair (SL) *in vivo*. Finally, SLG and SLES denote mismatches where the model implies that only the simultaneous deletion of both genes is lethal (SL), however, experimental data show either growth when simultaneously deleting both genes (G) or essentiality (ES) of one or both genes, respectively. These types of inconsistencies can be generalized to higher-order gene knockouts in a straightforward manner.

Extended GrowMatch can also correct for another type of inconsistency referred to as *auxotrophy mismatches*. They correspond to inconsistencies where the essentiality of a single gene deletion or the synthetic lethality of double (or higher order) gene knockout is correctly predicted by the model, however, model predictions for supplementation rescue (i.e., auxotrophy) scenarios are inconsistent with experimental observations [13]. For example, experiment may imply that growth can be restored for a gene mutant strain if additional compounds are added to the growth medium in contrast to model predictions showing no growth even in the presence of these compounds. These inconsistencies can be treated exactly the same way as NGG or SLG mismatches.

Extended GrowMatch "extends" the optimization problems [GrowMatch - NGG] and [GrowMatch - GNG] to account for higher-order gene deletions inconsistencies described earlier [13]. In addition, these optimization problems were modified to directly identify gene (rather than reaction) suppressions. This was achieved by defining a binary variable for each gene g in the model as follows:

$$w_g = \begin{cases} 0 & \text{if a gene } g \text{ is suppressed/deleted} \\ 1 & \text{otherwise} \end{cases}$$

The impact of gene suppressions/deletions on the network is then captured by the addition of new constraints mathematically describing the gene–protein–reaction (GPR) map for each reaction. These constraints relate reaction fluxes v_j with binary variables w_g for genes appearing in the GPR of the reaction j. Interested readers are encouraged to refer to Ref. [13] for a detailed description.

A number of online tools to perform gap filling and to reconcile growth inconsistencies using the methods described above (and beyond) are currently available. Examples of these tools include the Model SEED [4], the Department of Energy (DOE) Systems Biology Knowledgebase (KBase) (http://kbase.us) and Pathway Tools [22].

6.3 VERIFICATION OF MODEL CORRECTION STRATEGIES

It is important to note that GapFill, GrowMatch, Extended GrowMatch and other similar procedures are hypothesis generation tools, not an unsupervised way of fixing gaps and model/experiment inconsistencies. The use of integer cuts is warranted to trace alternative solutions, which should then be manually scrutinized for their biological relevance before they can be incorporated into the model. In the following, we briefly describe various methods to test the veracity of each type of model modification suggested by these procedures.

- Relaxation of the irreversibility constraint on existing reactions in the model can be tested using three independent methods: (i) Check the reversibility of this reaction in well-curated metabolic models of other organisms, particularly those that are phylogenetically close to the organism of interest. (ii) Query the reversibility of this reaction in global databases of biochemical reactions such as MetaCyc [2] or ModelSeed [4]. (iii) Verify the reversibility of a reaction by computing the range of its Gibbs free energy change (ΔG) [23] (see Chapter 5). If the maximum of ΔG is negative or its minimum is positive, then the reaction can reliably be treated as irreversible. For all other cases, the reversibility of the reaction cannot be excluded.
- Addition of new reactions to the model from external databases can be verified by mining the literature and/or by performing a bidirectional BLAST analysis between the enzymes catalyzing these new reactions and those present in the genome of the organism of interest. Generally, a p-value of 10^{-5} or lower [6, 13, 24] is needed before reliably assigning the function to the model. Ancillary pieces of data such as a gene loci within an operon coding for a present pathway [25], synteny in the genomic organization between two organisms [26], genomic proximity [25], etc. have also been used.
- Addition of transport reactions to the model can be verified by using a variety of checks: (i) Mine the literature to find potential clues about the presence of desired transport mechanisms in the organism of interest. (ii) Query the TransportDB database [27] for secretion and export pathways. (iii) Examine whether metabolites with similar structures have known transport reactions in the metabolic network. (iv) Query databases such as MetaCyc [2] to check for the presence of the desired transport mechanisms in other related species. For example, in the case of multicompartment models, one can check the presence of similar transport mechanisms in other multicompartment organisms or in multicellular species.
- Reaction suppressions from a model can be verified by searching the literature to find potential regulatory constraints under the examined condition or by analyzing gene expression data to check if the expression of the desired gene is below a pre-specified threshold under the examined condition. In addition, sometimes genes/reactions are added to the model based on weak homology evidence or an incorrect annotation and thus can be removed from the model.

Generally, identifying reactions that should not be part of the model is a much harder task than adding reactions required in the model to replicate an observed phenotype. Therefore, it is a good practice to only add reactions with a strong evidence supporting their existence in the model and leave reactions for which there is little of conflicting evidence out of the model.

Care must be exercised when adding new reactions to the model for debottlenecking flux through one or more metabolites as these additions often lead to the generation of TICs (see Chapter 5 for more details). This is particularly problematic for compartmentalized models. The use of integer cuts is an important tool for generating alternate resolution strategies avoiding the ones that create TICs. Furthermore, when adding new reactions to the model, one needs to ensure consistency in metabolite naming and elemental balance across all reactions and proper cofactor balancing when adding reactions to a model to fill the network gaps.

Finally, it is important to stress that not all network gaps or growth inconsistencies are fixable given our incomplete understanding of an organism's physiology. For example, in some cases an organism may be living symbiotically with another organism thus providing (mutual) backup for resolving network gaps in their respective metabolic networks. In other cases, some blocked reactions could be evolutionary relics of a pathway that is in the process of being eliminated. In general, even after careful scrutiny only, at best, approximately half of the gaps in a reconstructed network are fixed with corrections that have high confidence.

EXERCISE

6.1 Consider the metabolic model for archaeon *M. barkeri* [28].

- **(a)** Identify all blocked reactions. How many are they?
- **(b)** Use GapFill to suggest reconnection strategies for at least 10 of them. You can use reactions from the *E. coli* model *i*AF1260 [29] as candidates for reconnecting blocked reactions in the *M. barkeri* model.

REFERENCES

1. Satish Kumar V, Dasika MS, Maranas CD: Optimization based automated curation of metabolic reconstructions. *BMC Bioinformat* 2007, **8**:212.
2. Caspi R, Altman T, Billington R, Dreher K, Foerster H, Fulcher CA, Holland TA, Keseler IM, Kothari A, Kubo A *et al.*: The MetaCyc database of metabolic pathways and enzymes and the BioCyc collection of Pathway/Genome Databases. *Nucleic Acids Res* 2014, **42**(Database issue):D459–471.
3. Kanehisa M, Goto S, Sato Y, Kawashima M, Furumichi M, Tanabe M: Data, information, knowledge and principle: back to metabolism in KEGG. *Nucleic Acids Res* 2014, **42**(Database issue):D199–205.

4. Henry CS, DeJongh M, Best AA, Frybarger PM, Linsay B, Stevens RL: High-throughput generation, optimization and analysis of genome-scale metabolic models. *Nat Biotechnol* 2010, **28**(9):977–982.
5. Kumar A, Suthers PF, Maranas CD: MetRxn: a knowledgebase of metabolites and reactions spanning metabolic models and databases. *BMC Bioinformat* 2012, **13**:6.
6. Benedict MN, Mundy MB, Henry CS, Chia N, Price ND: Likelihood-based gene annotations for gap filling and quality assessment in genome-scale metabolic models. *PLoS Comput Biol* 2014, **10**(10):e1003882.
7. Agren R, Liu L, Shoaie S, Vongsangnak W, Nookaew I, Nielsen J: The RAVEN toolbox and its use for generating a genome-scale metabolic model for Penicillium chrysogenum. *PLoS Comput Biol* 2013, **9**(3):e1002980.
8. Orth JD, Palsson B: Systematizing the generation of missing metabolic knowledge. *Biotechnol Bioeng* 2010, **107**(3):403–412.
9. Vitkin E, Shlomi T: MIRAGE: a functional genomics-based approach for metabolic network model reconstruction and its application to cyanobacteria networks. *Genome Biol* 2012, **13**(11):R111.
10. Thiele I, Vlassis N, Fleming RM: fastGapFill: efficient gap filling in metabolic networks. *Bioinformatics* 2014, **30**(17):2529–2531.
11. Thiele I, Palsson B: A protocol for generating a high-quality genome-scale metabolic reconstruction. *Nat Protoc* 2010, **5**(1):93–121.
12. Kumar VS, Maranas CD: GrowMatch: an automated method for reconciling in silico/in vivo growth predictions. *PLoS Comput Biol* 2009, **5**(3):e1000308.
13. Zomorrodi AR, Maranas CD: Improving the iMM904 S. cerevisiae metabolic model using essentiality and synthetic lethality data. *BMC Syst Biol* 2010, **4**:178.
14. Mo ML, Palsson BO, Herrgård MJ: Connecting extracellular metabolomic measurements to intracellular flux states in yeast. *BMC Syst Biol* 2009, **3**:37.
15. Shlomi T, Berkman O, Ruppin E: Regulatory on/off minimization of metabolic flux changes after genetic perturbations. *Proc Natl Acad Sci U S A* 2005, **102**(21):7695–7700.
16. Joyce AR, Reed JL, White A, Edwards R, Osterman A, Baba T, Mori H, Lesely SA, Palsson B, Agarwalla S: Experimental and computational assessment of conditionally essential genes in Escherichia coli. *J Bacteriol* 2006, **188**(23):8259–8271.
17. Suthers PF, Dasika MS, Kumar VS, Denisov G, Glass JI, Maranas CD: A genome-scale metabolic reconstruction of Mycoplasma genitalium, iPS189. *PLoS Comput Biol* 2009, **5**(2):e1000285.
18. Herrgård MJ, Fong SS, Palsson B: Identification of genome-scale metabolic network models using experimentally measured flux profiles. *PLoS Comput Biol* 2006, **2**(7):e72.
19. Reed JL, Patel TR, Chen KH, Joyce AR, Applebee MK, Herring CD, Bui OT, Knight EM, Fong SS, Palsson BO: Systems approach to refining genome annotation. *Proc Natl Acad Sci U S A* 2006, **103**(46):17480–17484.
20. Henry CS, Zinner JF, Cohoon MP, Stevens RL: iBsu1103: a new genome-scale metabolic model of Bacillus subtilis based on SEED annotations. *Genome Biol* 2009, **10**(6):R69.
21. Suthers PF, Zomorrodi A, Maranas CD: Genome-scale gene/reaction essentiality and synthetic lethality analysis. *Mol Syst Biol* 2009, **5**:301.
22. Karp PD, Paley S, Romero P: The Pathway Tools software. *Bioinformatics* 2002, **18** Suppl 1:S225–232.

23. Henry CS, Jankowski MD, Broadbelt LJ, Hatzimanikatis V: Genome-scale thermodynamic analysis of Escherichia coli metabolism. *Biophys J* 2006, **90**(4):1453–1461.
24. Saha R, Suthers PF, Maranas CD: Zea mays iRS1563: a comprehensive genome-scale metabolic reconstruction of maize metabolism. *PLoS One* 2011, **6**(7):e21784.
25. Green ML, Karp PD: Using genome-context data to identify specific types of functional associations in pathway/genome databases. *Bioinformatics* 2007, **23**(13):i205–211.
26. Zheng XH, Lu F, Wang ZY, Zhong F, Hoover J, Mural R: Using shared genomic synteny and shared protein functions to enhance the identification of orthologous gene pairs. *Bioinformatics* 2005, **21**(6):703–710.
27. Ren Q, Chen K, Paulsen IT: TransportDB: a comprehensive database resource for cytoplasmic membrane transport systems and outer membrane channels. *Nucleic Acids Res* 2007, **35**(Database issue):D274–279.
28. Feist AM, Scholten JC, Palsson B, Brockman FJ, Ideker T: Modeling methanogenesis with a genome-scale metabolic reconstruction of Methanosarcina barkeri. *Mol Syst Biol* 2006, **2**:2006.0004.
29. Feist AM, Henry CS, Reed JL, Krummenacker M, Joyce AR, Karp PD, Broadbelt LJ, Hatzimanikatis V, Palsson B: A genome-scale metabolic reconstruction for Escherichia coli K-12 MG1655 that accounts for 1260 ORFs and thermodynamic information. *Mol Syst Biol* 2007, **3**:121.

7

IDENTIFICATION OF CONNECTED PATHS TO TARGET METABOLITES

Genome-scale metabolic (GSM) models and reaction databases [1–4] are often used to identify promising pathways that produce a metabolite of interest given a carbon substrate. Computationally, this task can be abstracted as the identification of all shortest metabolic routes connecting the substrate and the target metabolite(s). These routes can then subsequently be evaluated with respect to carbon and energy efficiency, thermodynamic feasibility, enzyme kinetics, etc. The identified routes also need to be linked with the native metabolism of the production host as they may involve one or more heterologous steps. In this chapter, we focus on mixed-integer linear programming (MILP)-based approaches for searching through a reaction list and/or GSM model for pathways toward a target product.

7.1 USING MILP TO IDENTIFY SHORTEST PATHS IN METABOLIC GRAPHS

A metabolic network can be represented as a directed graph (see Fig. 7.1) where nodes denote metabolites and directed edges represent reactions. A variety of graph-based methods have been proposed to identify all possible metabolic routes between a source and a target metabolite using shortest-path algorithmic implementations. Examples include MetaRoute [6], PathMiner [7], Pathway Tools [8, 9], PathFinder [10], PathComp [11], PathProd [12] and UM-BBD Pathway Prediction System [13]. A notable departure from this collection of methods is the Biochemical Network Integrated Computational Explorer (BNICE) algorithm [14], which uses graph-based representations of biochemical and enzyme reaction rules to generate novel biosynthesis pathways by successively applying reaction chemistry operators on metabolites represented as molecular graphs.

Optimization Methods in Metabolic Networks, First Edition. Costas D. Maranas and Ali R. Zomorrodi.

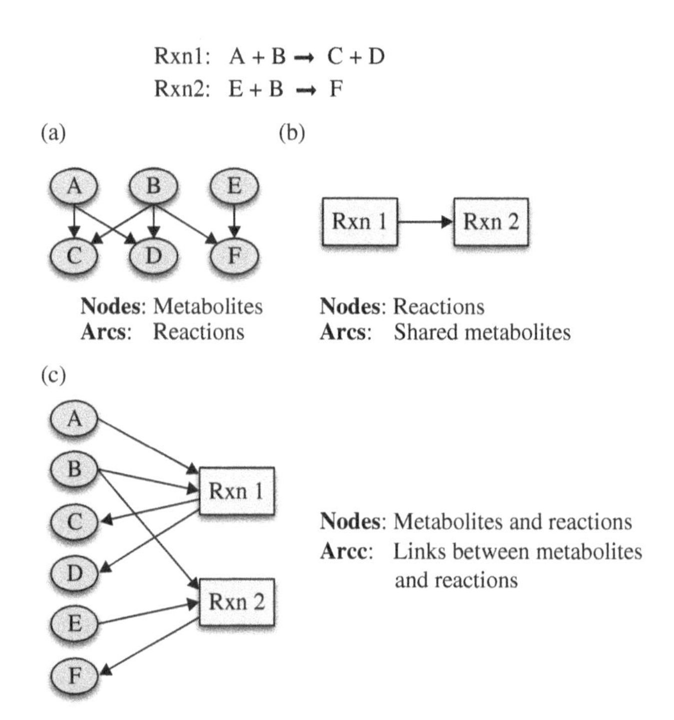

FIGURE 7.1 Different ways of representing a metabolic network using a directed graph illustrated for a simple network of two reactions. (a) Metabolite graph, (b) reaction graph and (c) bipartite graph. Source: Adapted from Ref. [5].

One of the shortcomings of the graph-based pathway identification methods is that the reaction stoichiometry is generally not captured directly. Consequently, some identified routes may be carbon or energy inefficient or even infeasible lacking a steady-state FBA solution. Therefore, identified pathways are generally assessed *a posteriori* through FBA analysis. An alternative to this two-step procedure is to integrate stoichiometric information with path-finding algorithms. The concepts of *flux path* and *effective carbon exchange* proposed by Pey et al. [15] provide an elegant solution to this challenge. A flux path is a path (as each node is visited only once) between a source and a target node that can operate under steady-state conditions. An effective carbon exchange between reactant A and product B in a chemical reaction of the form (A + C → B + D) captures if any carbon atoms from reactant A end up in product B. This concept allows for automatically tracking the main carbon exchange path implied by a reaction pathway avoiding connections through cofactor sharing or small molecules (e.g., water). The effective carbon exchange is encoded within a binary parameter (see also Fig. 7.2) for each reaction j capturing if there is an effective carbon exchange between an input metabolite i (with $S_{ij} < 0$) and an output metabolite i' (with $S_{ij} > 0$):

$$d_{ii'j} = \begin{cases} 1 & \text{if there exists a carbon exchange between metabolites } i \text{ and } i' \text{ in reaction } j \\ 0 & \text{otherwise} \end{cases}$$

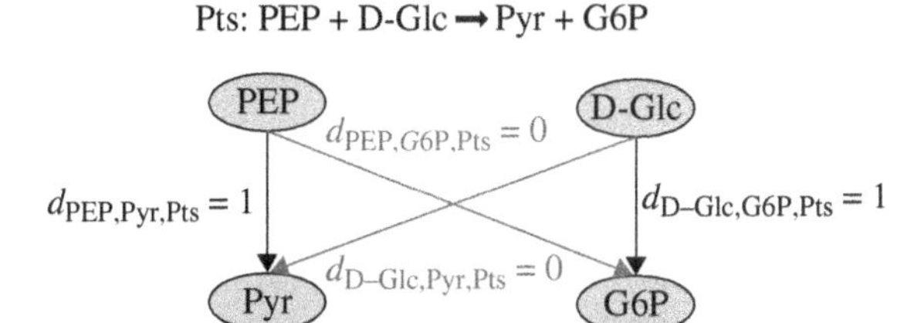

FIGURE 7.2 An example of the effective carbon exchange concept. Here, effective carbon exchange occurs only between D-Glc and G6P and between PEP and Pyr. PEP, D-Glc, Pyr, and G6P denote phosphoenolpyruvate, D-glucose, pyruvate, and glucose-6-phosphote, respectively. Source: Adapted from Ref. [15] with permission of BioMed Central.

Identifying flux paths involving an effective carbon exchange requires the integration of traditional path finding approaches with FBA constraints. In this section, we introduce an MILP model proposed by Pey et al. [15] to identify the shortest flux paths between a pair of metabolites in a metabolic network.

The goal here is to identify all shortest flux paths between a start metabolite s and an end metabolite e. All reversible reactions in the network are decomposed into their irreversible forward and backward counterparts such that all reaction fluxes in the network are nonnegative. The metabolic network is represented using a metabolite graph representation (see Fig. 7.1a). In the following subsections, we describe the constraints needed to define a valid flux path between metabolites s and e.

Path Finding Constraints Identifying flux paths is equivalent to determining the directed edges connecting metabolite s and e. A binary variable is defined to capture whether a directed edge is present in the flux path:

$$y_{ii'} = \begin{cases} 1 & \text{if the edge linking metabolite } i \text{ to metabolite } i' \text{ is active in the flux path} \\ 0 & \text{otherwise} \end{cases}$$

The following constraints are needed to impose the conditions of a valid path between metabolites s and e:

(a) Only one directed edge can leave node s and enter node e:

$$\sum_{i \in I} y_{si} = 1 \quad \text{and} \tag{7.1}$$

$$\sum_{i \in I} y_{ie} = 1 \tag{7.2}$$

where I is the set of metabolites in the network.

(b) No edge should enter s or leave e:

$$\sum_{i \in I} y_{is} = 0 \text{ and} \tag{7.3}$$

$$\sum_{i \in I} y_{ei} = 0 \tag{7.4}$$

(c) The number of edges entering any metabolite i should be equal to the number of edges leaving (thus ensuring a linear path):

$$\sum_{i' \in I - \{s,e\}} y_{ii'} = \sum_{i' \in I - \{s,e\}} y_{i'i}, \quad \forall i \in I \tag{7.5}$$

(d) No metabolite can be revisited:

$$\sum_{i' \in I} y_{i'i} \leq 1, \quad \forall i \in I \tag{7.6}$$

Stoichiometric Constraints Steady-state component balances are imposed as in FBA (see Chapter 3 for details):

$$\sum_{j \in J} S_{ij} v_j = 0, \quad \forall i \in I \tag{7.7}$$

where J denotes the set of reactions in the network. Another binary variable is introduced to capture whether reaction j carries a nonzero flux under the steady-state condition.

$$z_j = \begin{cases} 1 & \text{if reaction } j \text{ carries a non-zero flux} \\ 0 & \text{otherwise} \end{cases}$$

This definition is imposed by using the following constraint:

$$\varepsilon z_j \leq v_j \leq z_j UB_j, \quad \forall j \in J \tag{7.8}$$

where ε is a small number and UB_j is an upper bound on reaction j. Note that in the original treatment in Ref. [15] fluxes are scaled to assume values between zero and one; however, this scaling does not affect the generality of the procedure.

The simultaneous activity of both forward and backward directions of a reversible reaction is prevented as follows:

$$z_{j_f} + z_{j_b} \leq 1, \quad \forall \ j_f, j_b \in \left\{ j \middle| j \in J, S_{ij_f} = -S_{ij_b}, \ \forall i \in I \right\} \tag{7.9}$$

where j_f and j_b denote the forward and backward reactions associated with a reversible reaction j.

Linking Path Finding and Stoichiometric Constraints Any path defined by Constraints 7.1 to 7.6 must conform to the steady-state condition stated by Constraint 7.7. This implies that if an edge $i \rightarrow i'$ is included in the path, then at least one reaction j involving an effective carbon exchange between i and i' is active. Flux paths involving an effective carbon exchange are also referred to as *carbon flux paths* [15]. Notably, $y_{ii'}$ can *a priori* be fixed at zero if $\sum_{j \in J} d_{ii'j} = 0$ since in that case there is no possible carbon exchange reaction involving metabolites i and i'. An edge between i and i' can be included in the carbon flux path only if there exists at least one reaction j with $d_{ii'j} = 1$. This requirement can be enforced using the following constraint:

$$y_{ii'} \le \sum_{j \in \{j | j \in J, d_{ii'j} = 1\}} z_j, \quad \forall i, i' \in I \text{ and } i \ne i' \tag{7.10}$$

This constraint implies that if the edge ii' is included in the flux path at least one reaction with $d_{ii'j} = 1$ must be active assuming a non-zero flux (see Eq. 7.8), while satisfying the steady-state mass balance condition (Eq. 7.7). Note that all fluxes in the network have to satisfy the steady-state condition irrespective of whether they are included in the carbon flux path.

Objective function Constraints 7.1 to 7.10 define the necessary conditions for any feasible carbon flux path between metabolites s and e. The shortest path can be identified by using the following objective function:

$$\text{minimize} \sum_{i \in I} \sum_{i' \in I, i' \ne i} y_{ii'}$$

Optimization Model The shortest carbon flux path between a source metabolite s and an end metabolite e can thus be formulated as the following MILP problem:

$$\text{minimize} \sum_{i \in I} \sum_{i' \in I, i' \ne i} y_{ii'}$$

subject to

$$\sum_{i \in I} y_{si} = 1$$

$$\sum_{i \in I} y_{ie} = 1$$

$$\sum_{i \in I} y_{is} = 0$$

$$\sum_{i \in I} y_{ei} = 0$$

$$\sum_{i' \in I - \{s,e\}} y_{ii'} = \sum_{i' \in I - \{s,e\}} y_{i'i}, \quad \forall i \in I$$

$$\sum_{i' \in I} y_{i'i} \le 1, \quad \forall i \in I$$

$$\sum_{j \in J} S_{ij} v_j = 0, \quad \forall i \in I$$

$$\varepsilon z_j \le v_j \le z_j UB_j, \quad \forall j \in J$$

$$z_{j_f} + z_{j_b} \le 1, \quad \forall j_f, j_b \in \{ j \mid j \in J, S_{ij_f} = -S_{ij_b}, \ \forall i \in I \}$$

$$y_{ii'} \le \sum_{j \in \{j | j \in J, d_{ii'j} = 1\}} z_j, \quad \forall i, i' \in I, \quad i \ne i'$$

$$y_{ii'} \in \{0,1\}, \quad \forall i, i' \in I$$

$$z_j \in \{0,1\}, \quad \forall j \in J$$

$$v_j \ge 0, \quad \forall j \in J$$

Alternatively, one can specify the length of the pathway

$$\sum_{i \in I} \sum_{i' \in I, i \neq i'} y_{ii'} = L \tag{7.11}$$

and identify the one that maximizes the yield for target metabolite e

$$\text{maximize} \sum_{j \in J} S_{ej} v_j$$

Alternative shortest carbon flux paths can be found by using the following integer cut

$$\sum_{i \in I} \sum_{i' \in I, i' \neq i} y_{ii'}^{\text{opt},k} y_{ii'} \leq \sum_{i \in I} \sum_{i' \in I, i \neq i'} y_{ii'}^{\text{opt},k} - 1, \quad \forall\ k \in K = \{1,2,3,\ldots\} \tag{7.12}$$

where K indicates the set of previously found solutions and $y_{ii'}^{\text{opt},k}$ is the optimal value of binary variable $y_{ii'}$ in the previous solution $k \in K$.

This analysis introduces a way of integrating a stoichiometric description of metabolism embedded within FBA with a graph-based description of reactions. It can be used as is for tracing pathways from a source to a sink metabolite or it can become part of a more complex optimization formulation that uses free energy, DNA microarray data, flux measurements, metabolomics data, etc. to screen metabolic paths.

7.2 USING MILP TO IDENTIFY NON-NATIVE REACTIONS FOR THE PRODUCTION OF A TARGET METABOLITE

An alternative to the use of graphs is the integration of GSM models with reaction compilations found in databases for prospecting pathways linking source and target metabolites. The GSM model of an organism can be expanded by allowing for the addition of non-native reactions from databases of biochemical reactions such as MetaCyc [1], KEGG [2], Model SEED [3] and MetRxn [4]. This enables assessing all possibilities for extending the metabolism of a microbial production host to new products that could not be natively produced (see Fig. 7.3). Examples of such investigations include identifying new metabolic routes leading to the production of 1-butanol [16, 17] and 1,4-butanediol [18] in *Escherichia coli.*

The identification of candidate non-native pathways can be formulated as an MILP problem. Generally the minimum number of non-native reactions is sought that upon appending to the GSM model, the production of the non-native metabolite i^* with a desired target yield of $b \dfrac{\text{mmol}}{\text{gDW} \cdot \text{h}}$ per mole of limiting carbon source (e.g., glucose) uptake is achieved. In addition, a minimum level of biomass formation is often imposed along with product formation. This biomass production levels is generally expressed as a fraction of the theoretical maximum biomass yield (i.e., $100c\%$ where c is a scalar between zero and one).

Let J^{model} and J^{database} denote the set of reactions in the GSM model (i.e., native reactions) and external database of reactions (i.e., candidate non-native reactions),

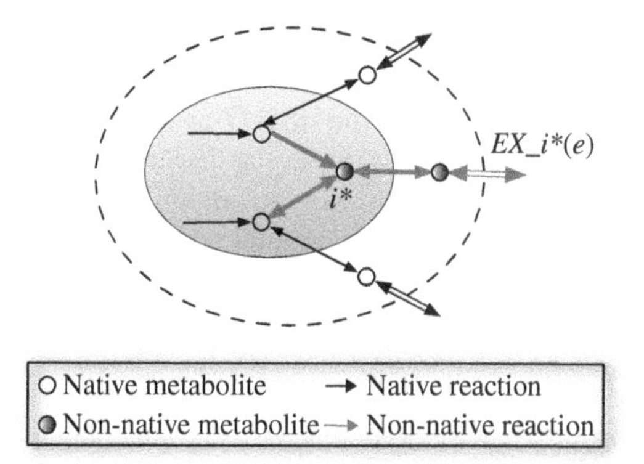

FIGURE 7.3 Addition of non-native reactions to a GSM model leading to the production of a non-native metabolite i^*.

respectively. A binary variable y_j is defined as follows to indicate if a non-native reaction is appended to the model:

$$y_j = \begin{cases} 1, & \text{if non-native reaction } j \text{ is added to the model} \\ 0, & \text{otherwise} \end{cases}$$

The production of the non-native metabolite i^* is modeled by the addition of a transport reaction (transferring i^* between cytosol [c] and extracellular space [e]) and the corresponding exchange reaction to the model. The following MILP problem can then be formulated to identify reactions that maximize the production of metabolite i^* using at most K non-native reactions:

$$\text{maximize} \quad v_{EX\_i^*(e)}$$

$$\text{subject to}$$

$$\sum_{j \in J^{\text{model}} \bigcup J^{\text{database}}} S_{ij} v_j = 0, \quad \forall i \in I \bigcup I^{\text{database}} \tag{7.13}$$

$$LB_j \leq v_j < UB_j, \quad \forall j \in J^{\text{model}} \tag{7.14}$$

$$LB_j y_j \leq v_j \leq UB_j y_j, \quad \forall j \in J^{\text{database}} \tag{7.15}$$

$$v_{\text{biomass}} \geq f \, v_{\text{biomass}}^{\text{max,WT}} \tag{7.16}$$

$$\sum_{j \in J^{\text{database}}} y_j \leq K \tag{7.17}$$

$$y_j \in \{0,1\}, \quad \forall j \in J^{\text{database}}$$

where I^{database} is the set of metabolites participating in reactions in J^{database}, $\text{EX}\_i^*(\text{e})$ is the exchange reaction corresponding to metabolite i^* and $v_{\text{biomass}}^{\text{max,WT}}$ is the maximum theoretical biomass flux for the wild-type strain. Constraints 7.13 and 7.14 are standard FBA constraints (see Chapter 3), Constraint 7.15 enforces the definition of the binary variable y_j, Constraint 7.16 imposes the minimum required level of biomass formation and Constraint 7.17 restricts the total number of added reactions to the model to a pre-determined value K. Generally, this MILP problem is iteratively solved for increasing values of K until the target yield for i^* is met $\left(\text{i.e., } \frac{v_{\text{EX}\_i^*(e)}}{v_{\text{EX}\_\text{glucose(e)}}} \geq b\right)$. More complex procedures that assess product yield at maximum biomass yield based on bilevel optimization formulations are discussed in Chapter 8. In a recent study, Campodonico et al. [19] proposed a computational workflow termed GEM-Path (Genome-scale Model Pathway Predictor), which integrates a variety of methods including retrosynthetic path prediction algorithms based on biochemical reaction operators, filtering procedures using GSM models and reaction promiscuity analysis to identify all non-native production pathways for a microbial host such as *E. coli*.

When adding new reactions to the model, care must be exercised to ensure that (i) the native and non-native reactions and metabolites use the same naming conventions, (ii) there is no overlap between native and non-native reaction sets (i.e., $J^{\text{model}} \cap J^{\text{database}} = \varnothing$), (iii) all reactions are cofactor and elemental balanced and (iv) no thermodynamically infeasible cycles (TICs) are created.

7.3 DESIGNING OVERALL STOICHIOMETRIC CONVERSIONS

A limitation of many graph or MILP-based path-finding approaches is that they aim to trace pathways connecting a single substrate to a single product even though most metabolic engineering tasks recruit many more metabolites and reactions than the ones along the main carbon conversion path. Cofactor usage, co-reactant choices and stoichiometric ratios are important pathway design decision variables that have to be taken into account. This motivated the development of a two-stage computational procedure for overall stoichiometry and pathway design [20]. This procedure first optimizes the overall stoichiometry of conversion while taking into account co-reactants/co-products and maintaining thermodynamic feasibility. This is followed by identifying intervening reactions from a database that link the chosen reactants and products in the optimized stoichiometric ratios. These two stages are described in the following sections.

7.3.1 Determining the Stoichiometry of Overall Conversion

The goal of this initial scoping phase termed optStoic [20] is to identify a thermodynamically feasible overall conversion that optimizes a desired performance criterion such as maximally utilizing a limiting carbon resource toward a target product, co-utilization of other carbon substrates (e.g., gaseous reactants), co-production of valuable by-products or cogeneration of redox resources for biomass. For the sake of simplicity of presentation, optStoic is derived here for the following bimolecular reaction:

$$a\text{A} + c\text{C} \rightarrow b\text{B} + d\text{D}$$

where A and B are the main reactant and product and C and D denote an alternative limiting resource and a by-product, respectively. optStoic is then formulated as follows [20]:

$$\underset{a,b,c,d}{\text{maximize}}/\underset{a,b,c,d}{\text{minimize}}\, f(a,b,c,d) \quad [\text{optStoic}]$$

subject to

$$n_{\text{B}q}b + n_{\text{D}q}d - n_{\text{A}q}a - n_{\text{C}q}c = 0, \quad \forall q \in Q \tag{7.18}$$

$$e_{\text{B}}b + e_{D}d - e_{A}a - e_{C}c = 0 \tag{7.19}$$

$$\begin{aligned}&\Delta G_{\text{B}}^{\text{f},0}b + \Delta G_{\text{D}}^{\text{f},0}d - \Delta G_{\text{A}}^{\text{f},0}a - \Delta G_{\text{C}}^{\text{f},0}c \\ &\quad + RT\left(b \ln x_{\text{B}} + d \ln x_{\text{D}} - a \ln x_{\text{A}} - c \ln x_{\text{C}}\right) \leq \Delta G^{\text{target}}\end{aligned} \tag{7.20}$$

$$h(a,b,c,d) = 1 \tag{7.21}$$

$$a, b, c, d \in \mathbb{R}^{+}$$

where Q is the set of elements found in metabolites A, B, C and D (e.g., C, O, N, P, S, H and Fe), n_{iq} is the number of atoms of element q in metabolite i, e_i is the charge of metabolite i, $\Delta G_i^{\text{f},0}$ is the Gibbs free energy of formation of metabolite i under the standard thermodynamic conditions, x_i is the activity of metabolites i, where $i = \{\text{A,B,C,D}\}$, and $f(a,b,c,d)$ is an arbitrary function of the overall stoichiometric coefficients quantifying a desired performance criterion. Examples include maximum yield for primary product B (maximize b while a is fixed), or the maximum co-utilization of free reactant C for a given product yield (maximize c for a fixed b) or combinations thereof.

optStoic is a linear programming problem that identifies the optimal stoichiometric coefficients a, b, c and d for the pre-specified performance objective $f(a,b,c,d)$. Constraints 7.18 and 7.19 impose elemental and charge balances on the overall conversion, respectively while Constraint 7.20 imposes overall thermodynamic feasibility by requiring the Gibbs free energy change of the overall conversion (see Chapter 5) to be less than or equal to a pre-specified target ΔG^{target} (where $\Delta G^{\text{target}} < 0$). Whenever activity coefficients x_{A}, x_{B}, x_{C} and x_{D} are known, they can be provided as input so as to quantify departures from standard conditions. Constraint 7.21 describes the scaling with respect to a *basis* metabolite by setting its stoichiometric coefficient to one. The basis metabolite is usually either the limiting carbon source (i.e., $a = 1$) or the primary product (i.e., $b = 1$). Function $h(a,b,c,d)$ in Constraint 7.21 generalizes the description of this scaling decision. For example if $a = 1$, then

$h(a,b,c,d) = a = 1$. All stoichiometric coefficients are declared as positive real variables. However, if a metabolite could serve as a reactant or product in the overall conversion reaction, then its stoichiometric coefficient is declared as a real variable. Consequently, optStoic can be used in a prospective fashion to design an optimal overall stoichiometric conversion by allowing a number of nonzero stoichiometric coefficients for metabolites selected from an exhaustive list (see [20] for details).

7.3.2 Identifying Reactions Steps Conforming to the Identified Overall Stoichiometry

Once the stoichiometry of the overall conversion is selected, the next step is to identify the smallest network of external reactions from a database that must be added to the host network in order to achieve the optimized stoichiometric ratios determined by optStoic. To this end, exchange reactions for the metabolites present in the overall stoichiometry must first be added to the metabolic network. Identification of the minimal set of reactions matching the desired overall stoichiometry is formulated as the following MILP problem [20] similar to the earlier path-finding procedures:

$$\underset{y_j}{\text{minimize}} \sum_{j \in J^{\text{database}}} y_j \quad [\text{minRxn}]$$

subject to

$$\sum_{j \in J^{\text{model}} \cup J^{\text{database}}} S_{ij} v_j = 0, \quad \forall i \in I^{\text{host}} \cup I^{\text{database}} \tag{7.22}$$

$$v_{\text{EX\_A(e)}} = -a,\ v_{\text{EX\_C(e)}} = -c,\ v_{\text{EX\_B(e)}} = b,\ v_{\text{EX\_D(e)}} = d \tag{7.23}$$

$$v_j = 0, \quad \forall j \in J^{\text{host,exch.}} - \{\text{EX\_A}(\text{e}), \text{EX\_B}(\text{e}), \text{EX\_C}(\text{e}), \text{EX\_D}(\text{e})\} \tag{7.24}$$

$$LB_j \le v_j \le UB_j, \quad \forall j \in J^{\text{host}} \tag{7.25}$$

$$LB_j y_j \le v_j \le UB_j y_j, \quad \forall j \in J^{\text{database}} \tag{7.26}$$

$$v_j \in \mathbb{R}, \quad \forall j \in J^{\text{host}} \cup J^{\text{database}}$$

$$y_j \in \{0,1\}, \quad \forall j \in J^{\text{database}}$$

Here I^{host}, J^{host} and $J^{\text{host,exch.}}$ are the set of metabolites, reactions and exchange reactions, respectively in the metabolic model of the host, while I^{database} and J^{database} are the set of metabolites and reactions in the external database. $\text{EX\_A}(\text{e})$, $\text{EX\_C}(\text{e})$, $\text{EX\_B}(\text{e})$ and $\text{EX\_D}(\text{e})$ denote the exchange reactions for A, C, B and D, respectively. Binary variables y_j control whether or not a reaction j from J^{database} should be included in the host network. The objective function of minRxn minimizes the number of reactions in J^{database} that are added to the host network.

Constraint 7.22 ensures a steady-state mass balance for all metabolites in the host network and database. Constraints 7.23 and 7.24 enforce the optimal stoichiometry of the overall conversion and the flux of exchange reactions for A, B, C and D equal to their respective stoichiometric coefficients (designed by optStoic). The flux of all other exchange reactions in the network are set to zero. By disallowing the uptake and export of any other metabolites not specified in the overall stoichiometry, all flux from the reactant metabolite(s) is routed toward the target metabolite(s) at their optimal stoichiometric ratios. Constraint 7.25 defines the lower and upper bounds of the reaction fluxes in the host while Constraint 7.26 enforces the definition of binary variables y_j.

The minRxn MILP formulation can be computationally challenging to solve as the number of binary variables, which is equal to the number of reactions in J^{database} can be quite large (several thousands). A related LP formulation, termed minFlux that is significantly less computationally taxing was suggested as a surrogate of min-Rxn [20]. Instead of directly minimizing the total number of reactions in the designed network, minFlux minimizes the total metabolic flux of the chosen reactions in J^{database}. This modified objective function could be viewed as an approximation of the total enzyme-load imposed on the host organism [21]. minFlux is formulated as follows [20]:

$$\text{minimize} \sum_{j \in J^{\text{database}}} w_j \quad [\text{minFlux}]$$

subject to

$$\sum_{j \in J^{\text{model}} \bigcup J^{\text{database}}} S_{ij} v_j = 0, \quad \forall i \in I^{\text{host}} \bigcup I^{\text{database}} \tag{7.22}$$

$$v_{\text{EX\_A(e)}} = -a,\ v_{\text{EX\_C(e)}} = -c,\ v_{\text{EX\_B(e)}} = b, v_{\text{EX\_D(e)}} = d \tag{7.23}$$

$$v_j = 0, \quad \forall j \in J^{\text{host,exch.}} - \{\text{EX\_A(e)}, \text{EX\_B(e)}, \text{EX\_C(e)}, \text{EX\_D(e)}\} \tag{7.24}$$

$$LB_j \le v_j \le UB_j, \quad \forall j \in J^{\text{host}} \bigcup J^{\text{databse}} \tag{7.25}$$

$$w_j \ge v_j, \quad \forall j \in J^{\text{database}} \tag{7.27}$$

$$w_j \ge -v_j, \quad \forall j \in J^{\text{database}} \tag{7.28}$$

$$v_j \in \mathbb{R}, \quad \forall j \in J^{\text{host}} \bigcup J^{\text{database}}$$

$$w_j > 0, \quad \forall j \in J^{\text{database}}$$

minFlux minimizes the sum of the absolute values of reaction fluxes to be added from J^{database}. Absolute values are linearized by introducing a new variable w_j for reaction j using Constraints 7.27 and 7.28 (see Chapter 2). Additional constraints can be added to both minRxn and minFlux to restrict all reactions to a single production host and maintain the total number of reactions with a positive standard Gibbs free energy change at a minimum (see Ref. [20] for details).

Example 7.1

Design an optimal overall stoichiometry and pathway for the conversion of glucose to succinate in *E. coli*. (a) Use optStoic to identify an overall stoichiometry that maximizes the molar yield of succinate production from one mole of glucose as the primary carbon substrate. Allow for the uptake or export of one additional carbon containing co-metabolite (carbon monoxide or carbon dioxide) as well as water and protons in the design of the overall stoichiometry. Required information about these metabolites is given in Table 7.1. Ensure that the free energy change of the overall conversion is less than −5 kcal at $T = 298$ K. (b) Use the minRxn formulation to check whether the overall stoichiometry of conversion is achievable in *E. coli* under the aerobic condition using the *i*AF1260 metabolic model [22]. If not, find the minimum number of non-native reactions chosen from a database of curated reactions given in Ref. [20] that must be added to the *i*AF1260 network to enable the overall stoichiometry. Use $1 \frac{\text{mmol}}{\text{gDW} \cdot \text{h}}$ of glucose uptake as the basis.

Solution: (a) Glucose and succinate have dedicated roles as reactant and product, respectively in the overall conversion, however the designation of CO, CO_2, H_2O and H^+ is uncertain. Therefore, the stoichiometric coefficients corresponding to glucose and succinate (i.e., $S_{C_6H_{12}O_6}$ and $S_{C_4H_4O_4^{2-}}$) are declared as positive variables while the ones for CO, CO_2, H_2O and H^+ (i.e., S_{CO}, S_{CO_2}, S_{H_2O} and S_{H^+}) are denoted as real. The optStoic formulation for this problem is as follows:

$$\text{maximize} \quad S_{C_4H_4O_4^{2-}} \quad [\text{optStoic}]$$

subject to

$$4S_{C_4H_4O_4^{2-}} - 6S_{C_6H_{12}O_6} + S_{CO} + S_{CO_2} = 0$$

$$4S_{C_4H_4O_4^{2-}} - 6S_{C_6H_{12}O_6} + S_{CO} + 2S_{CO_2} + S_{H_2O} = 0$$

$$4S_{C_4H_4O_4^{2-}} - 12S_{C_6H_{12}O_6} + 2S_{H_2O} + S_{H^+} = 0$$

$$-2S_{C_4H_4O_4^{2-}} + S_{H^+} = 0$$

$$S_{C_6H_{12}O_6} = 1$$

$$217.7S_{C_6H_{12}O_6} - 162.27\,S_{C_4H_4O_4^{2-}} - 32.82S_{CO} - 94.35S_{CO_2} - 57.1S_{H_2O} - 9.53S_{H^+} + (0.001989)\,298$$

$$\left(\left(-S_{C_6H_{12}O_6}\right)\ln 0.001 + S_{C_4H_4O_4^{2-}} \ln 0.001 + S_{CO} \ln 0.00095 + S_{CO_2} \ln 0.001\right) \le -5$$

$$S_{C_4H_4O_4^{2-}}, S_{C_6H_{12}O_6} \in \mathbb{R}^+$$

$$S_{CO}, S_{CO_2}, S_{H_2O}, S_{H^+} \in \mathbb{R}$$

TABLE 7.1 Required Information for Metabolites Participating in the Overall Conversion of Glucose to Succinate (See Example 7.1)

Metabolite	Formula	Charge	$\Delta G^{f,0}\left(\frac{\text{kcal}}{\text{mol}}\right)$	Activity (x_i)
Glucose	$C_6H_{12}O_6$	0	−217.70	0.001
Succinate	$C_4H_4O_4^{2-}$	−2	−162.27	0.001
Carbon monoxide	CO	0	−32.82	0.00095
Carbon dioxide	CO_2	0	−94.35	0.001
Water	H_2O	0	−57.10	1
Proton	H^+	+1	−9.53	1

The optimal values of the stoichiometric coefficients are identified by solving this LP problem:

$$S^*_{CO} = -2, S^*_{CO_2} = 0, S^*_{H_2O} = 0, S^*_{H^+} = 4 \text{ and } S^*_{C_4H_4O_4^{2-}} = 2$$

Therefore, the optimum overall stoichiometry that maximizes succinate yield from glucose is as follows:

$$C_6H_{12}O_6 + 2CO \rightarrow 2C_4H_4O_4^{2-} + 4H^+, \quad \Delta G^{\text{overall}} = -75.16 \text{ kcal}$$

Note that using just the standard Gibbs free energy of formation of the participating metabolites without the activities correction would have recovered the same overall stoichiometry in this example. Not surprisingly, CO was chosen as the preferred co-reactant over CO_2 due to its more reduced state providing a higher succinate yield $\left(2\frac{\text{mol succinate}}{\text{mol glucose}} \text{ for CO vs } 1.71\frac{\text{mol succinate}}{\text{mol glucose}} \text{ for } CO_2\right)$. Here, the elemental balance and not the thermodynamic feasibility constraint is the limiting factor in further improving succinate yield as the total number of available hydrogen atoms in glucose determines the maximum number of CO molecules that can be reduced to succinate. Existing efforts aimed at succinate overproduction under aerobic conditions have only used CO_2 as the co-metabolite with much lower experimentally achieved production yields $\left(0.85\frac{\text{mol succinate}}{\text{mol glucose}} [23]\right)$. Microbial CO utilization efforts have been limited to acetate, ethanol and butyrate production [24].

(b) Exchange and transport reactions for carbon monoxide (CO) are first added to the *i*AF1260 model to allow for the uptake/production of this metabolite (the model contains exchange and transport reactions for the rest of metabolites already). An FBA problem was solved for the *i*AF1260 model by the addition of Constraints 7.23 and 7.24 to determine whether the metabolic network of *E. coli* can support this stoichiometry. This problem was found to be infeasible implying that the *i*AF1260 model is incapable of finding a feasible pathway from glucose to succinate with the imposed overall stoichiometry identified in (a). As a result, minFlux/minRxn are solved next to

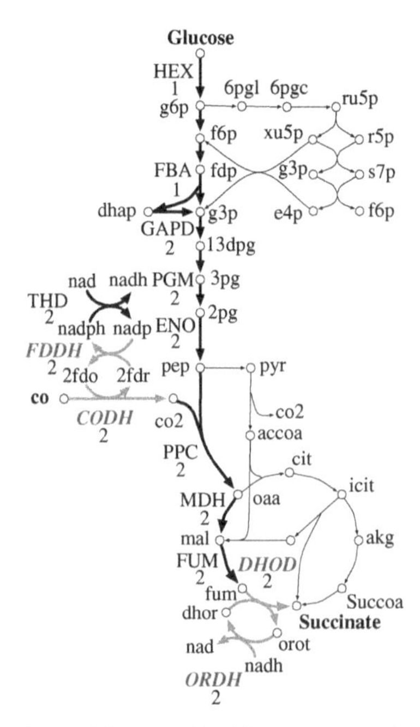

FIGURE 7.4 Optimal pathway (shown with thicker arrows along with corresponding flux values) for maximizing succinate yield in *E. coli* from glucose and carbon monoxide (see Example 7.1). Non-native reactions are shown in gray with italicized reaction names. All native reaction and metabolite names follow those in *i*AF1260 mode [22].

identify the minimal number of non-native reactions that should be added to *i*AF1260 from the reaction database in Ref. [20]. Figure 7.4 shows the pathway that achieves the overall stoichiometry by the addition of four non-*E. coli* reactions (shown in gray) to the *i*AF1260 model. These reactions include carbon monoxide: ferredoxin oxidoreductase (CODH), ferredoxin: NADP oxidoreductase (FDDH), dihydroorotate: fumarate dehydrogenase (DHOD) and orotate: nadh dehydrogenase (ORDH). In this pathway design, one mole of glucose is first oxidized to two moles of phosphoenolpyruvate (pep) through glycolysis. Pep is carboxylated to oxaloacetate (oaa) by pep carboxylase (PPC) that also fixes the additional carbon supplied by CO. Oaa is finally converted to succinate by the reversal of the non-oxidative branch of the TCA cycle. Non-native reactions DHOD and ORDH are needed to compensate for the repression of the native fumarate reductase (FRD) activity under aerobic conditions. Interestingly, earlier studies reported that both DHOD and FDDH are present and active in *E. coli* [25, 26] (though absent in *i*AF1260). ORDH is present in *S. cerevisiae* [27] and CODH in proteobacteria such as *C. carboxydovorans* or *P. thermocarboxydovorans* [28]. Note that all the enzymes for the non-native reactions are active under aerobic conditions. In an earlier effort, FRD [29] in *E. coli* was engineered to remain active under aerobic conditions. minFlux suggests an alternate combination of reactions (i.e., ORDH and DHOD) to match the overall stoichiometry. This example, demonstrates how optStoic can identify a co-reactant that offers the potential for a higher carbon yield. A GAMS implementation of this example is available on the book's website. □

EXERCISES

7.1 Use the formulation described in Section 7.1 on the *i*AF1260 model of *E. coli* [22] to check whether a carbon flux path exists from acetyl-CoA to glucose-6-phosphate when the glyoxylate shunt reactions (i.e., ICL and MALS) (i) are present and (ii) when they are absent/inactive.

7.2 Consider the production of non-native metabolite triacetic acid lactone (TAL) in *E. coli.*

(a) Find the minimum number of heterologous reactions from an external database of biochemical reactions [1–4] enabling a nonzero production of TAL in the *i*AF1260 model of *E. coli* [22].

(b) Starting from a single non-native reaction, check how the maximum production flux of TAL varies with each additional reaction up to a maximum of five reactions.

Consider aerobic glucose minimal conditions in all cases with a glucose uptake of $10\frac{\text{mmol}}{\text{g dW}}$.

REFERENCES

1. Caspi R, Altman T, Billington R, Dreher K, Foerster H, Fulcher CA, Holland TA, Keseler IM, Kothari A, Kubo A *et al.*: The MetaCyc database of metabolic pathways and enzymes and the BioCyc collection of Pathway/Genome Databases. *Nucleic Acids Res* 2014, **42**(Database issue):D459–471.
2. Kanehisa M, Goto S, Sato Y, Kawashima M, Furumichi M, Tanabe M: Data, information, knowledge and principle: back to metabolism in KEGG. *Nucleic Acids Res* 2014, **42**(Database issue):D199–205.
3. Henry CS, DeJongh M, Best AA, Frybarger PM, Linsay B, Stevens RL: High-throughput generation, optimization and analysis of genome-scale metabolic models. *Nat Biotechnol* 2010, **28**(9):977–982.
4. Kumar A, Suthers PF, Maranas CD: MetRxn: a knowledgebase of metabolites and reactions spanning metabolic models and databases. *BMC Bioinformat* 2012, **13**:6.
5. Deville Y, Gilbert D, van Helden J, Wodak SJ: An overview of data models for the analysis of biochemical pathways. *Brief Bioinform* 2003, **4**(3):246–259.
6. Blum T, Kohlbacher O: MetaRoute: fast search for relevant metabolic routes for interactive network navigation and visualization. *Bioinformatics* 2008, **24**(18):2108–2109.
7. McShan DC, Rao S, Shah I: PathMiner: predicting metabolic pathways by heuristic search. *Bioinformatics* 2003, **19**(13):1692–1698.
8. Karp PD, Paley S, Romero P: The Pathway Tools software. *Bioinformatics* 2002, **18** Suppl 1:S225–232.
9. Karp PD, Paley SM, Krummenacker M, Latendresse M, Dale JM, Lee TJ, Kaipa P, Gilham F, Spaulding A, Popescu L *et al.*: Pathway Tools version 13.0: integrated software for pathway/genome informatics and systems biology. *Brief Bioinform* 2010, **11**(1):40–79.

10. Goesmann A, Haubrock M, Meyer F, Kalinowski J, Giegerich R: PathFinder: reconstruction and dynamic visualization of metabolic pathways. *Bioinformatics* 2002, **18**(1):124–129.
11. Kanehisa M, Goto S, Hattori M, Aoki-Kinoshita KF, Itoh M, Kawashima S, Katayama T, Araki M, Hirakawa M: *From genomics to chemical genomics: new developments in KEGG. Nucleic Acids Res* 2006, **34**(Database issue):D354–357.
12. Moriya Y, Shigemizu D, Hattori M, Tokimatsu T, Kotera M, Goto S, Kanehisa M: PathPred: an enzyme-catalyzed metabolic pathway prediction server. *Nucleic Acids Res* 2010, **38**(Web Server issue):W138–143.
13. Ellis LB, Roe D, Wackett LP: The University of Minnesota Biocatalysis/Biodegradation Database: the first decade. *Nucleic Acids Res* 2006, **34**(Database issue):D517–521.
14. Hatzimanikatis V, Li C, Ionita JA, Henry CS, Jankowski MD, Broadbelt LJ: Exploring the diversity of complex metabolic networks. *Bioinformatics* 2005, **21**(8):1603–1609.
15 Pey J, Prada J, Beasley JE, Planes FJ: Path finding methods accounting for stoichiometry in metabolic networks. *Genome Biol* 2011, **12**(5):R49.
16. Atsumi S, Cann AF, Connor MR, Shen CR, Smith KM, Brynildsen MP, Chou KJ, Hanai T, Liao JC: Metabolic engineering of Escherichia coli for 1-butanol production. *Metab Eng* 2008, **10**(6):305–311.
17. Shen CR, Lan EI, Dekishima Y, Baez A, Cho KM, Liao JC: Driving forces enable high-titer anaerobic 1-butanol synthesis in Escherichia coli. *Appl Environ Microbiol* 2011, **77**(9):2905–2915.
18. Yim H, Haselbeck R, Niu W, Pujol-Baxley C, Burgard A, Boldt J, Khandurina J, Trawick JD, Osterhout RE, Stephen R *et al.*: Metabolic engineering of Escherichia coli for direct production of 1,4-butanediol. *Nat Chem Biol* 2011, **7**(7):445–452.
19. Campodonico MA, Andrews BA, Asenjo JA, Palsson BO, Feist AM: Generation of an atlas for commodity chemical production in Escherichia coli and a novel pathway prediction algorithm, GEM-Path. *Metabolic Eng* 2014, **25**:140–158.
20 Chowdhury A, Maranas C: Designing overall stoichiometric conversions and intervening metabolic reactions. *Sci Rep* 2015, **5**:16009, doi: 10.1038/srep16009.
21. Lewis NE, Hixson KK, Conrad TM, Lerman JA, Charusanti P, Polpitiya AD, Adkins JN, Schramm G, Purvine SO, Lopez-Ferrer D *et al.*: Omic data from evolved E. coli are consistent with computed optimal growth from genome-scale models. *Mol Syst Biol* 2010, **6**:390.
22. Feist AM, Henry CS, Reed JL, Krummenacker M, Joyce AR, Karp PD, Broadbelt LJ, Hatzimanikatis V, Palsson BO: A genome-scale metabolic reconstruction for Escherichia coli K-12 MG1655 that accounts for 1260 ORFs and thermodynamic information. *Mol Syst Biol* 2007, **3**:121.
23. Thakker C, Martínez I, San KY, Bennett GN: Succinate production in *Escherichia coli. Biotechnol J* 2012, **7**(2):213–224.
24. Köpke M, Mihalcea C, Bromley JC, Simpson SD: Fermentative production of ethanol from carbon monoxide. *Curr Opin Biotechnol* 2011, **22**(3):320–325.
25. Jensen KF, Larsen S: Dihydroorotate dehydrogenase of *Escherichia coli. Methods Mol Biol* 2003, **228**:11–21.
26. Wan JT, Jarrett JT: Electron acceptor specificity of ferredoxin (flavodoxin):NADP+ oxidoreductase from *Escherichia coli. Arch Biochem Biophys* 2002, **406**(1):116–126.
27. Jouhten P, Wiebe M, Penttilä M: Dynamic flux balance analysis of the metabolism of Saccharomyces cerevisiae during the shift from fully respirative or respirofermentative metabolic states to anaerobiosis. *FEBS J* 2012, **279**(18):3338–3354.

28. King GM, Weber CF: Distribution, diversity and ecology of aerobic CO-oxidizing bacteria. *Nat Rev Microbiol* 2007, **5**(2):107–118.
29. Iuchi S, Kuritzkes DR, Lin EC: Three classes of Escherichia coli mutants selected for aerobic expression of fumarate reductase. *J Bacteriol* 1986, **168**(3):1415–1421.
30. Deville Y, Gilbert D, van Helden J, Wodak SJ: An overview of data models for the analysis of biochemical pathways. *Briefings Bioinformat* 2003, **4**(3):246–259.

8

COMPUTATIONAL STRAIN DESIGN

The need for the sustainable production of fuels and chemicals has spearheaded the use of microbial production hosts as chemical factories to produce a wide range of biofuel candidates and bio-renewables. Microbial metabolism is generally primed for fast growth, which may often be in direct competition with overproduction goals. As a result, the product yield of wild-type strains is usually far below the maximum theoretical yield. Therefore, the microbial production of a desired chemical often requires engineering of the gene content and circuitry of the microbial host (through gene knock-outs/ins/ups/downs) in order to redirect the metabolic flow toward the product of interest. The design of carbon efficient and stable engineered microbial strains is an overarching challenge in biotechnology spanning biofuels, bio-renewables, secondary metabolism products and drugs. The design of microbial hosts with product yields close to the theoretical maximum generally requires multiple, often nonintuitive metabolic interventions. For example, the overexpression of genes located in the terminal pathway of a target metabolite to achieve a satisfactory product yield target is often insufficient. Additional manipulations upstream may be needed to route carbon flux toward the production pathway, block the production of competing by-products, and generate favorable redox ratios. Identifying such interventions and quantifying their impact on product yield require a system-wide understanding of cellular metabolism. This task is complicated by the fact that organisms have evolved to counteract any externally imposed genetic or environmental perturbations. This makes the design of stable engineered microbial hosts particularly challenging.

In response to these challenges, a growing list of computational strain design procedures has emerged [1–16] benefiting from rapid advancements in the development and reconstruction of genome-scale metabolic (GSM) models. Most of these approaches rely upon a mathematical (or evolutionary) optimization framework that

Optimization Methods in Metabolic Networks, First Edition. Costas D. Maranas and Ali R. Zomorrodi.

search over a set of potential genetic interventions. In this chapter, we introduce some of the core computational strain design protocols based on bilevel and mixed-integer linear programming (MILP) formulations.

8.1 EARLY COMPUTATIONAL TREATMENT OF STRAIN DESIGN

Gene knockouts are modeled, as in earlier chapters, with a binary variable that decide whether or not a reaction j in the network must be eliminated:

$$y_j = \begin{cases} 0 & \text{if reaction } j \text{ is eliminated} \\ 1 & \text{otherwise} \end{cases}$$

One may suggest that it suffices to solve an MILP problem to identify reaction eliminations leading to the overproduction of the metabolite of interest by extending the maximum theoretical yield calculation problem introduced in Chapter 3. This MILP problem for the overproduction of product P in a microbial strain under the aerobic minimal medium with glucose as the carbon source can be formally stated as follows (see Fig. 8.1):

$$\text{maximize } z = v_{\mathrm{EX\_P(e)}}$$

$$\text{subject to}$$

$$\sum_{j \in J} S_{ij} v_j = 0, \quad \forall i \in I \tag{8.1}$$

$$v_{\mathrm{EX\_glc(e)}} \geq -v_{\mathrm{glc}}^{\mathrm{uptake}} \tag{8.2}$$

$$v_{\mathrm{EX\_O_2(e)}} \geq -v_{\mathrm{O_2}}^{\mathrm{uptake}} \tag{8.3}$$

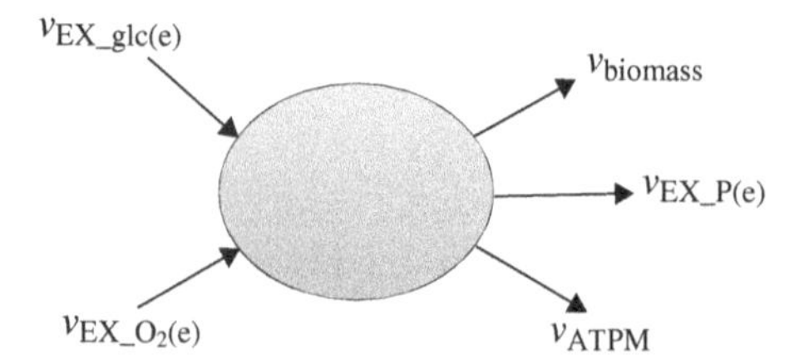

FIGURE 8.1 Metabolic sources and sinks in FBA-based strain design methods. $v_{\mathrm{EX\_glc(e)}}$, $v_{\mathrm{EX\_O_2(e)}}$ and $v_{\mathrm{EX\_P(e)}}$ denote the flux of exchange reactions for glucose, oxygen and the product of interest P, respectively and v_{ATPM} and v_{biomass} represent the flux of the non-growth associated ATP maintenance and biomass reactions.

$$v_{\text{ATPM}} = v_{\text{ATP}}^{\text{maint.}} \tag{8.4}$$

$$v_{\text{biomass}} \geq f v_{\text{biomass}}^{\text{max, WT}}, \quad (0 \leq f \leq 1) \tag{8.5}$$

$$LB_j y_j \leq v_j \leq UB_j y_j, \quad \forall j \in J \tag{8.6}$$

$$\sum_{j \in J} (1 - y_j) \leq K \tag{8.7}$$

$$y_j \in \{0,1\}, \quad v_j \in \mathbb{R}, \quad \forall j \in J$$

where I is the set of metabolites in the network, v_j is the flux of a reaction j, S_{ij} is the stoichiometric coefficient of metabolite i in reaction j, LB_j and UB_j are the lower and upper bounds on the flux of reaction j, $v_{\text{EX\_P(e)}}$, $v_{\text{EX\_glc(e)}}$ and $v_{\text{EX\_O}_2\text{(e)}}$ denote the flux of the exchange reactions for the product of interest P, glucose and oxygen, respectively, and v_{ATPM} is the flux of the non-growth-associated maintenance ATP reaction (ATPM or NGAM). In addition, v_{biomass} is the flux of the biomass reaction and $v_{\text{biomass}}^{\text{max,WT}}$ is the maximum theoretical biomass flux (computed for the wild-type strain using FBA). Constraint 8.1 represents the steady-state component balance. Constraints 8.2 and 8.3 set the maximum allowable uptake for glucose and oxygen to $v_{\text{glc}}^{\text{uptake}}$ and $v_{\text{O}_2}^{\text{uptake}}$, respectively (recall from Chapter 3 that a negative value for exchange reaction fluxes imply uptake). Constraint 8.4 sets the required demand for ATPM flux to the prespecified value of $v_{\text{ATP}}^{\text{maint.}}$. The values for $v_{\text{glc}}^{\text{uptake}}$, $v_{\text{O}_2}^{\text{uptake}}$ and $v_{\text{ATP}}^{\text{maint.}}$ for *Escherichia coli* as the microbial host are typically set to 10, 20 and 8.39 $\frac{\text{mmol}}{\text{gDW} \cdot \text{h}}$, respectively (see Chapter 3). Constraint 8.5 requires the biomass formation in the network to attain a minimum threshold, which is expressed as a fraction (typically $f = 0.1$) of the maximum theoretical biomass in the wild-type strain. Constraint 8.6 enforces the definition of binary variable y_j and Constraint 8.7 restricts the total number of knockouts to a user-specified value K.

The limitation of this formulation is that imposing one or more gene knockouts does not account for the fact that the organism will likely re-apportion fluxes in the metabolic network so as to restore growth rather than overproduce the desired product. In response to this, the objective function could be augmented as follows:

$$\text{maximize} \ v_{\text{EX\_P(e)}} + c v_{\text{biomass}}$$

where c is a positive coefficient. However, even if weight c is carefully selected, the presence of a single aggregate objective function fails to recognize that there exist two separate and often conflicting drivers of metabolic fluxes in the metabolic network with different "dials" at their disposal. On one hand, the organism recruits the regulatory machinery to restore/sustain growth in response to a genetic perturbation. On the other, the metabolic engineer prunes away reactions (through gene knockouts) from the metabolic network so as to maximize flux towards the target product.

8.2 OPTKNOCK

OptKnock recognizes that organisms tend to counteract any externally imposed genetic perturbations through the redirection of metabolic flux to restore cellular growth. It aims to design reaction eliminations that reshape network connectivity in such a way that the production of the target metabolite is maximized while the organism still maintains maximum biomass production yield under the constraints imposed by gene knockouts. Often, this renders the targeted overproduction an obligatory by-product of biomass formation [1]. This is achieved by using a bilevel optimization framework which explicitly accounts for two competing objective functions (instead of using a single aggregate objective function). Bilevel problems are nested optimization problems where one (or more) constraints are implicitly expressed as an optimization problem. The main optimization problem is usually referred to as the outer-level problem, whereas the inner ones are called inner-level problems. The outer problem in OptKnock captures the metabolic engineering objective (e.g., to maximize the production flux of the target metabolite), while the inner problem optimizes a cellular fitness objective (e.g., maximization of the biomass production flux). OptKnock is posed as a bilevel optimization problem as follows for the overproduction of a target metabolite under the aerobic minimal medium with glucose as the carbon source:

$$\text{maximize} \quad z = v_{\text{EX\_P(e)}} \quad [\text{OptKnock}]$$

subject to

$$\left[\begin{array}{lll} \text{maximize } v_{\text{biomass}} & & \\ \text{subject to} & & \\ \sum_{j \in J} S_{ij} v_j = 0, & \forall i \in I & (8.1) \\ v_{\text{EX\_glc(e)}} \geq -v_{\text{glc}}^{\text{uptake}} & & (8.2) \\ v_{\text{EX\_O}_2\text{(e)}} \geq -v_{\text{O}_2}^{\text{uptake}} & & (8.3) \\ v_{\text{ATPM}} = v_{\text{ATP}}^{\text{maint.}} & & (8.4) \\ v_{\text{biomass}} \geq f v_{\text{biomass}}^{\text{max, WT}}, & (0 \leq f \leq 1) & (8.5) \\ LB_j y_j \leq v_j \leq UB_j y_j, & \forall j \in J & (8.6) \end{array}\right]$$

$$\sum_{j \in J} (1 - y_j) \leq K \tag{8.7}$$

$$y_j \in \{0,1\}, \quad v_j \in \mathbb{R}, \quad \forall j \in J$$

where binary variable y_j and all constraints are the same as those defined in Section 8.1. The outer problem in OptKnock identifies reaction candidates for elimination that maximize the production of the target metabolite P, while the inner problem redistributes metabolic fluxes so as to maximize the biomass formation in the perturbed network subject to the reaction eliminations imposed by the outer problem (Constraint 8.6). OptKnock suggests knockouts that attempt to couple growth production with the production of the maximum possible amount of the target product. Note that binaries y_j are variables for the outer problem but are parameters for the inner problem. Any alternative (convex) objective function such as the minimization of metabolic adjustment (MOMA) [17] (see Chapter 10), maximization of ATP exchange or the minimization of total flux through all reactions can be used in the inner problem if they are deemed to be better surrogates for cellular fitness.

8.2.1 Solution Procedure for OptKnock

Even though all constraints present in OptKnock are linear, the bilevel optimization problem is nonlinear because one of the constraints is implicitly expressed as an optimization problem. Therefore, one needs to convert this bilevel structure to a standard single-level optimization problem. This can be achieved by aggregating the constraints of the inner problem with those of its dual while imposing the strong duality condition (see Chapter 2). Dual variables associated with primal constraints (i.e., the inner problem) are as follows:

$$
\begin{array}{lll}
\text{maximize} & z_p = v_{\text{biomass}} & \\
\text{subject to} & & \\
\sum_{j \in J} S_{ij} v_j = 0, & \forall i \in I & \leftarrow \ \lambda_i \in \mathbb{R} \\
-v_{\text{EX\_glc(e)}} \le v_{\text{glc}}^{\text{uptake}} & & \leftarrow \ \mu_{\text{EX\_glc(e)}}^{\text{LB}} \ge 0 \\
-v_{\text{EX\_O}_2\text{(e)}} \le v_{\text{O}_2}^{\text{uptake}} & & \leftarrow \ \mu_{\text{EX\_O}_2\text{(e)}}^{\text{LB}} \ge 0 \\
v_{\text{ATPM}} \le v_{\text{ATP}}^{\text{maint.}} & & \leftarrow \ \mu_{\text{ATPM}}^{\text{UB}} \ge 0 \\
-v_{\text{ATPM}} \le -v_{\text{ATP}}^{\text{maint.}} & & \leftarrow \ \mu_{\text{ATPM}}^{\text{LB}} \ge 0 \\
-v_{\text{biomass}} \le -f v_{\text{biomass}}^{\text{max,WT}} & & \leftarrow \ \mu_{\text{biomass}}^{\text{LB}} \ge 0 \\
v_j \le UB_j y_j, & \forall j \in J - \{\text{ATPM}\} & \leftarrow \ \mu_j^{\text{UB}} \ge 0 \\
-v_j \le -LB_j y_j, & \forall j \in J - \{\text{EX\_glc(e), EX\_O}_2\text{(e), ATPM, biomass}\} & \leftarrow \ \mu_j^{\text{LB}} \ge 0
\end{array}
$$

The dual problem can then be written with constraints for the ATPM and biomass requirements as well as the limiting nutrient dual variables incorporated in Constraints 8.10 and 8.11 as follows (see Example 2.4 for details):

$$
\text{minimize } z_d = \sum_{i \in I} 0 \lambda_i + \sum_{j \in J} UB_j y_j \mu_j^{\text{UB}} + \sum_{j \in J} \left(-LB_j y_j\right) \mu_j^{\text{LB}}
$$

subject to

$$
\sum_{i \in I} S_{ij} \lambda_i + \mu_j^{\text{UB}} - \mu_j^{\text{LB}} = 0, \quad \forall j \in J - \{\text{biomass}\} \tag{8.8}
$$

$$\sum_{i \in I} S_{i,\text{biomass}} \lambda_i + \mu_{\text{biomass}}^{\text{UB}} - \mu_{\text{biomass}}^{\text{LB}} = 1 \tag{8.9}$$

$$0 \le \mu_j^{\text{LB}} \le \mu_j^{\text{LB,max}}, \quad \forall j \in J \tag{8.10}$$

$$\begin{aligned} &0 \le \mu_j^{\text{UB}} \le \mu_j^{\text{UB,max}}, \quad \forall j \in J \\ &\lambda_i \in \mathbb{R}, \quad \forall i \in I \end{aligned} \tag{8.11}$$

where $\mu_j^{\text{LB,max}}$ and $\mu_j^{\text{UB,max}}$ are the maximum values (or upper bounds) of μ_j^{LB} and μ_j^{UB}, respectively. Constraints for $v_{EX\_glc(e)}$, $v_{EX\_O2(e)}$, v_{ATPM} and $v_{biomass}$ are subsumed within Constraints 8.10 and 8.11 for brevity of presentation. The bilevel problem [OptKnock] can now be recast as the following single-level mixed-integer optimization problem:

$$\text{maximize} \quad z = v_{\text{EX\_P(e)}}$$

subject to

Primal constraints

$$\sum_{j \in J} S_{ij} v_j = 0, \quad \forall i \in I \tag{8.1}$$

$$LB_j y_j \le v_j \le UB_j y_j, \quad \forall j \in J \tag{8.6}$$

Dual constraints

$$\sum_{i \in I} S_{ij} \lambda_i + \mu_j^{\text{UB}} - \mu_j^{\text{LB}} = 0, \quad \forall j \in J - \{\text{biomass}\} \tag{8.8}$$

$$\sum_{i \in I} S_{i,\text{biomass}} \lambda_i + \mu_{\text{biomass}}^{\text{UB}} - \mu_{\text{biomass}}^{\text{LB}} = 1 \tag{8.9}$$

$$0 \le \mu_j^{\text{LB}} \le \mu_j^{\text{LB,max}}, \quad \forall j \in J \tag{8.10}$$

$$0 \le \mu_j^{\text{UB}} \le \mu_j^{\text{UB,max}}, \quad \forall j \in J \tag{8.11}$$

Strong duality

$$v_{\text{biomass}} = \sum_{j \in J} UB_j y_j \mu_j^{\text{UB}} - \sum_{j \in J} LB_j y_j \mu_j^{\text{LB}} \tag{8.12}$$

Allowable number of knockouts

$$\sum_{j \in J} (1 - y_j) \le K \tag{8.7}$$

$$y_j \in \{0,1\}, \quad v_j \in \mathbb{R}, \quad \forall j \in J$$

$$\lambda_i \in \mathbb{R}, \quad \forall i \in I$$

Here, Equation 8.12 imposes the strong duality theorem condition by setting the objective function of the primal equal to that for the dual. This automatically renders all nonoptimal values for the maximization of v_{biomass} of the metabolic fluxes (and dual variables) infeasible. In other words, the newly formatted single-level optimization problem can only be feasible when the inner problem is optimized. Caution must be exercised here in interpreting an optimized inner-level problem as it is quite possible for the optimal biomass flux to become zero in the absence of Constraint 8.5. Whenever this happens, the inner problem in essence disappears collapsing the OptKnock optimization problem into a simple maximum theoretical yield of product calculation. To avoid this, one needs to impose a minimum biomass production level through specifying a nonzero lower bound for the biomass flux $\left(\text{LB}_{\text{biomass}} = f\, v_{\text{biomass}}^{\text{max,WT}}\right)$ as described earlier (see Constraint 8.5).

This dualization of the inner problem forms the basis for replacing the inner-level optimization problem with a set of explicit constraints. All bilinear terms in Constraint 8.12 can be systematically linearized using the techniques described in Chapter 4 to convert the problem to an MILP. In addition, one can analyze the terms appearing in this constraint in order to simplify and *a priori* remove terms guaranteed to be equal to zero. The value of the first term, $UB_j y_j \mu_j^{\text{UB}}$ can be resolved depending on whether the reaction is removed (i.e., $y_j = 0$) or present (i.e., $y_j = 1$). If the reaction is eliminated $(y_j = 0)$ then $UB_j y_j \mu_j^{\text{UB}} = 0$ and the term disappears from the equation. Alternatively, if the reaction is not removed (i.e., $y_j = 1$) then the value of $UB_j y_j \mu_j^{\text{UB}}$ depends on the upper bound and corresponding dual variable values.

(i) If $UB_j = M$ then the constraint is inactive (i.e., $v_j < UB_j$) and $\mu_j^{\text{UB}} = 0$ according to the complementary slackness (CS) condition (see Chapter 2) implying that $UB_j y_j \mu_j^{\text{UB}} = 0$.

(ii) If $UB_j < M$ then the constraint can become active ($v_j = UB_j$) with $\mu_j^{\text{UB}} \neq 0$ based on the CS condition and $UB_j y_j \mu_j^{\text{UB}}$ is simplified to $UB_j \mu_j^{\text{UB}}$.

By inspecting the first term on the right-hand side of Constraint 8.12 (dual objective function), the only surviving term is $\left(v_{\text{ATP}}^{\text{maint.}}\right)\mu_{\text{ATPM}}^{\text{UB}}$. Similarly, analysis of the second term $LB_j y_j \mu_j^{\text{LB}}$ gives rise to the following possibilities. If the reaction is eliminated (i.e., $y_j = 0$), then term $LB_j y_j \mu_j^{\text{LB}}$ can be eliminated. However, if the reaction is not removed (i.e., $y_j = 1$), the term must be further scrutinized based on the value of the lower bound on reaction fluxes:

(iii) If $LB_j = -M$ then the constraint is inactive ($v_j > -M$) and the CS condition requires that $\mu_j^{\text{LB}} = 0$ implying that term $LB_j y_j \mu_j^{\text{LB}}$ is eliminated. This is true for reversible reactions with non-limiting bounds and exchange reactions corresponding to metabolites present in the growth medium in excess (e.g., water).

(iv) If $LB_j = 0$ then $LB_j y_j \mu_j^{\text{LB}} = 0$. This is true for irreversible reactions and exchange reactions corresponding to metabolites not present in the growth medium.

(v) If $LB_j \neq 0$ or $-M$ then constraint $v_j \geq LB_j$ can become active and μ_j^{LB} can be nonzero according to the CS condition. In this case, $LB_j y_j \mu_j^{\mathrm{LB}}$ is simplified to $LB_j \mu_j^{\mathrm{LB}}$. This typically holds for exchange reactions corresponding to limiting resources in the medium (e.g., carbon source and limiting nitrogen source).

Overall, the only terms that survive in the dual objective function are the ones associated with reactions that impose performance requirements on biomass and ATPM and reactions that set upper bounds on limiting resources which are oxygen and glucose. Note that the list of limiting resources is organism, growth medium and product specific. For example, for a photosynthetic organism CO_2, photon flux or even nitrate could become the limiting resource. Therefore, the remaining right-hand side terms for Constraint 8.12 are $\left(f v_{\mathrm{biomass}}^{\mathrm{max,WT}}\right)\mu_{\mathrm{biomass}}^{\mathrm{LB}}$, $\left(v_{\mathrm{ATP}}^{\mathrm{maint.}}\right)\mu_{\mathrm{ATPM}}^{\mathrm{LB}}$, $\left(-v_{\mathrm{glc}}^{\mathrm{uptake}}\right)\mu_{\mathrm{EX\_glc(e)}}^{\mathrm{LB}}$ and $\left(-v_{\mathrm{O_2}}^{\mathrm{uptake}}\right)\mu_{\mathrm{EX\_O_2(e)}}^{\mathrm{LB}}$. From this analysis, Constraint 8.12 simplifies as the following linear equality devoid of any binary variables:

$$
\begin{aligned}
v_{\mathrm{biomass}} = v_{\mathrm{ATP}}^{\mathrm{maint.}}\mu_{\mathrm{ATPM}}^{\mathrm{UB}} \\
&-\Big[\left(f v_{\mathrm{biomass}}^{\mathrm{max,WT}}\right)\mu_{\mathrm{biomass}}^{\mathrm{LB}} + \left(v_{\mathrm{ATP}}^{\mathrm{maint.}}\right)\mu_{\mathrm{ATPM}}^{\mathrm{LB}} + \left(-v_{\mathrm{glc}}^{\mathrm{uptake}}\right)\mu_{\mathrm{EX\_glc(e)}}^{\mathrm{LB}} \\
&+\left(-v_{\mathrm{O_2}}^{\mathrm{uptake}}\right)\mu_{\mathrm{EX\_O_2(e)}}^{\mathrm{LB}}\Big].
\end{aligned}
\tag{8.13}
$$

Finally, it is important to not forget to translate the constraint that controls switching on or off reaction fluxes using binary variables in the primal problem (Constraint 8.6) to the dual problem. Otherwise, the dual problem will have no way of knowing which multipliers should be set to zero for cases (i) and (iii) noted earlier. This corresponds to enforcing the CS relations (imposed implicitly for μ_j^{LB} and μ_j^{UB} in cases (i) through (v)) using the following two constraints:

$$
\begin{aligned}
&\mu_j^{\mathrm{LB}} \leq \mu_j^{\mathrm{LB,max}}\left(1-y_j\right), \\
&\qquad \forall j \in J - \{\mathrm{biomass, ATPM, EX\_glc(e), EX\_O_2(e)}\} \cup J^{\mathrm{irrev}}
\end{aligned}
\tag{8.14}
$$

$$
\mu_j^{\mathrm{UB}} \leq \mu_j^{\mathrm{UB,max}}\left(1-y_j\right), \quad \forall j \in J - \{\mathrm{ATPM}\}
\tag{8.15}
$$

where J^{irrev} is the set of irreversible reactions and EX_glc(e) and EX_O_2(e) denote exchange reactions for glucose and oxygen, respectively.

It is important to note that the derivation of the simplified dual objective function assumes that no reactions can reach their arbitrarily large upper or lower bounds (i.e., M and $-M$). Therefore, thermodynamically infeasible cycles (TICs) must be resolved and eliminated before using this *a priori* elimination of terms in the dual objective function. All limiting resources and performance constraints that may become limiting must be included in the objective function to guarantee consistency with the strong duality theorem requirements. Alternatively, all bilinear terms can be exactly linearized as described in Chapter 4 without having to perform any preprocessing or removal of TICs.

The single-level MILP representation of OptKnock upon pre-processing is as follows:

maximize $z = v_{\mathrm{EX\_P(e)}}$

subject to

Primal constraints:

$$\sum_{j \in J} S_{ij} v_j = 0, \quad \forall i \in I \qquad (8.1)$$

$$LB_j y_j \le v_j \le UB_j y_j, \quad \forall j \in J \qquad (8.6)$$

Dual constraints:

$$\sum_{i \in I} S_{ij} \lambda_i + \mu_j^{\mathrm{UB}} - \mu_j^{\mathrm{LB}} = 0, \quad \forall j \in J - \{\mathrm{biomass}\} \qquad (8.8)$$

$$\sum_{i \in I} S_{i,\mathrm{biomass}} \lambda_i + \mu_{\mathrm{biomass}}^{\mathrm{UB}} - \mu_{\mathrm{biomass}}^{\mathrm{LB}} = 1 \qquad (8.9)$$

$$0 \le \mu_j^{\mathrm{LB}} \le \mu_j^{\mathrm{LB,max}}, \quad \forall j \in J \qquad (8.10)$$

$$0 \le \mu_j^{\mathrm{UB}} \le \mu_j^{\mathrm{UB,max}}, \quad \forall j \in J \qquad (8.11)$$

$$\mu_j^{\mathrm{LB}} \le \mu_j^{\mathrm{LB,max}} (1 - y_j), \quad \forall j \in J - \{\mathrm{biomass, ATPM, EX\_glc(e), EX\_O_2(e)}\} \cup J^{\mathrm{irrev}} \qquad (8.14)$$

$$\mu_j^{\mathrm{UB}} \le \mu_j^{\mathrm{UB,max}} (1 - y_j), \quad \forall j \in J - \{\mathrm{ATPM}\} \qquad (8.15)$$

Strong duality:

$$v_{\mathrm{biomass}} = v_{\mathrm{ATP}}^{\mathrm{maint.}} \mu_{\mathrm{ATPM}}^{\mathrm{UB}} - \Big[\left(f v_{\mathrm{biomass}}^{\mathrm{max,WT}} \right) \mu_{\mathrm{biomass}}^{\mathrm{LB}} + \left(v_{\mathrm{ATP}}^{\mathrm{maint.}} \right) \mu_{\mathrm{ATPM}}^{\mathrm{LB}} + \left(-v_{\mathrm{glc}}^{\mathrm{uptake}} \right) \mu_{\mathrm{EX\_glc(e)}}^{\mathrm{LB}} + \left(-v_{\mathrm{O_2}}^{\mathrm{uptake}} \right) \mu_{\mathrm{EX\_O_2(e)}}^{\mathrm{LB}} \Big] \qquad (8.13)$$

Allowable number of knockouts:

$$\sum_{j \in J} (1 - y_j) \le K \qquad (8.7)$$

$$y_j \in \{0,1\}, \quad v_j \in \mathbb{R}, \quad \forall j \in J$$

$$\lambda_i \in \mathbb{R}, \quad \forall i \in I$$

Starting from a relatively low number (i.e., two to three) of knockouts, the OptKnock MILP formulation is usually iteratively solved by successively increasing K to explore the impact of additional interventions on improving the product yield. This allows the prioritization of the interventions whereby the interventions having the largest impact on improving the product yield are identified first. Integer cuts (see Chapter 4) can be used to identify alternate interventions or to explore the suboptimal space of knockouts for a given value of K. The OptKnock formulation and solution procedure were presented here for the special case of minimal medium with glucose as the sole carbon source and under the aerobic condition. More complex production scenarios could be modeled with multiple substrates (e.g., H_2 and CO_2) or with complex performance requirements on ATP maintenance, biomass and byproducts by adjusting the procedure presented here. Note that if a nonlinear but convex objective function is used to represent the cellular fitness, the inner problem can be transformed to a set of regular constraints by enforcing the KKT conditions (see Chapters 9 and 10) in place of the strong duality condition used for linear objective functions.

8.2.2 Improving the Computational Efficiency of OptKnock

The computational efficiency of the algorithm can be improved by fixing the binary variables to one for reactions that must not (or cannot) be eliminated. These reactions include the following:

(i) Biomass, ATPM
(ii) *In silico* and *in vivo* essential reactions
(iii) Blocked reactions
(iv) Exchange reactions
(v) Internal reactions with no gene–protein–reaction (GPR) associations as well as spontaneous and diffusion reactions.
(vi) All reactions in a fully coupled set except for one representative reaction (see Chapter 3 for details of the flux coupling analysis).

In addition to these restrictions, working experience with OptKnock has revealed that the following types of reactions [16] are unlikely to contribute meaningful knockouts:

(i) Transport reactions (i.e., inner membrane transport, outer membrane transport, outer membrane porin transport and tRNA charging)
(ii) Periphery metabolic pathways (e.g., cell envelope biosynthesis, glycerophospholipid metabolism, inorganic ion transport and metabolism, lipopolysaccharide biosynthesis and recycling, membrane lipid metabolism, murein biosynthesis and murein recycling)
(iii) Reactions involving macromolecules and high-molecular-weight metabolites when targeting the production of low-molecular-weight metabolites in central metabolism.

Finally, similarly to essential reactions, one needs to avoid the simultaneous deletion of reactions forming synthetic lethal sets by appending new constraints. For example, if reactions j_1 and j_2 form a synthetic lethal pair, the following constraint can be used to avoid their simultaneous deletion.

$$y_{j_1} + y_{j_2} \geq 1. \tag{8.16}$$

8.2.3 Connecting Reaction Eliminations with Gene Knockouts

OptKnock is a reaction-based formulation and does not directly identify manipulations at the gene level. One thus needs to manually inspect the GPR associations (see Chapter 3) for the eliminated reactions to discern the genes that need to be knocked out. Care must be exercised here as in some cases knocking out a gene may have secondary and negative effects on metabolism. This happens when one gene encodes multiple reactions. For example as shown in Figure 8.2, if OptKnock suggests the elimination of reaction $R1$ but $R2$ is an essential reaction then gene G should not be knocked out.

8.2.4 Impact of Knockouts on the Biomass vs. Product Trade-Off

As discussed in Chapter 3, the trade-off between biomass and product formation can be graphically illustrated by plotting the solutions of the optimization problems that sequentially maximize and minimize the exchange flux of the target product for varying levels of biomass production. Imposing gene knockouts leads to optimization formulations with tighter constraint sets implying more restricted ranges in the trade-off plot. Therefore, the trade-off plot for a knockout mutant would be fully contained within the one for the wild-type strain for the same uptake scenarios. In general, two possible behaviors are manifested upon the imposition of knockouts:

(i) *Maximum biomass and product formation are decoupled*: In this case, the product formation flux can range between zero and a positive value when the biomass flux is fixed at its maximum (see Fig. 8.3a). This is not an ideal scenario for strain design as in the presence of imposed knockouts the mutant strain can escape the requirement for overproduction. Note that in OptKnock the maximization of the product formation flux will yield the high-production solution. Therefore, it is always a good practice to assess OptKnock suggested mutants by generating their trade-off plots.

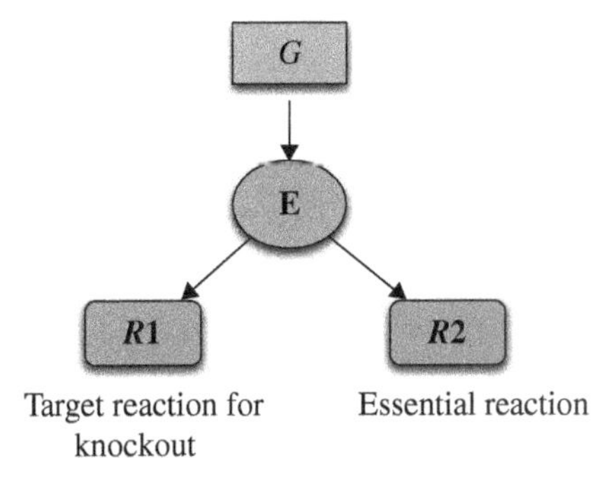

FIGURE 8.2 Example demonstrating that GPR associations need to be carefully scrutinized to ensure that OptKnock suggested reaction eliminations do not eliminate essential reactions. Here, OptKnock suggests eliminating reaction $R1$. However gene G cannot be knocked-out because it is coding for the essential reaction $R2$ in addition to $R1$.

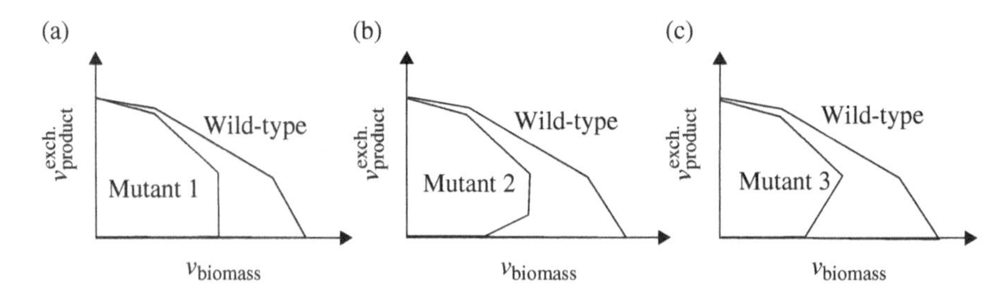

FIGURE 8.3 Trade-off plots for biomass and product formation in wild-type and mutant strains. (a) Product formation is decoupled from biomass formation in Mutant 1. (b) Product formation is coupled with biomass formation in Mutant 2, but it can vary between a positive minimum and maximum value. (c) Product formation is fully coupled with biomass formation in Mutant 3.

(ii) *Maximum biomass and product formation are coupled*: In this case, the minimum product formation flux is nonzero when biomass is fixed at its maximum (see Fig. 8.3b or c). Ideally, there is no variation in the product formation flux (see Fig. 8.3c) implying that the product is a direct by-product of biomass formation. These scenarios are preferred for strain design as a given level of growth fixes the amount of product formation implying that growth selection schemes can be used to boost productivity.

Example 8.1

Use OptKnock to identify a pair of gene knockouts leading to the overproduction of acetate in *E. coli* using the *i*AF1260 metabolic model [18] under the aerobic condition in a minimal medium with glucose as the carbon source. Use $10\frac{\text{mmol}}{\text{gDW}\cdot\text{h}}$ of glucose uptake as the basis and 10% of theoretical maximum as the minimum required level of biomass formation in the network. Check Table 3.1 of Chapter 3 for reaction flux bounds. Generate the trade-off plot for the wild-type and identified mutant strain(s). Is acetate production coupled with biomass formation in the mutant strain(s)?

Solution: OptKnock suggests the elimination of ATP synthase (ATPS) and pyruvate dyhydrogenase (PDH) to reach an acetate production level of $18.04\frac{\text{mmol}}{\text{gDW}\cdot\text{h}}$ for $10\frac{\text{mmol}}{\text{gDW}\cdot\text{h}}$ glucose fed corresponding to 71.02% of the theoretical maximum. The trade-off plot for this mutant is given in Figure 8.4a. As shown in this figure, acetate production can vary between zero and $18.04\frac{\text{mmol}}{\text{gDW}\cdot\text{h}}$ at maximum biomass formation. Therefore, acetate production in this mutant is decoupled from biomass formation. By using integer cuts to identify the next best solution, OptKnock suggests the elimination of ATPS and enolase (ENO) for an acetate production level of $17.83\frac{\text{mmol}}{\text{gDW}\cdot\text{h}}$,

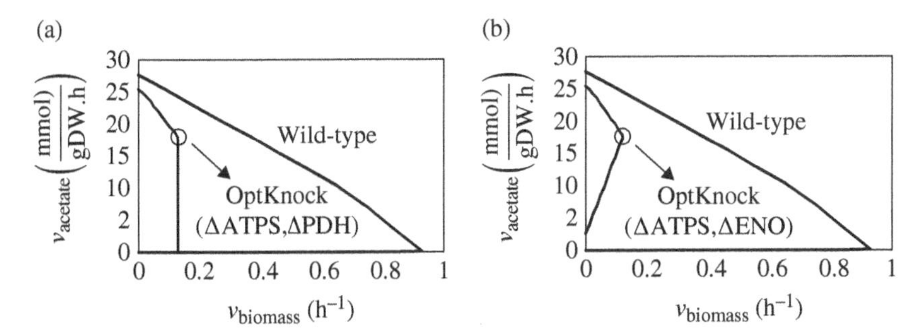

FIGURE 8.4 Trade-off plots for the wild-type and two mutant strains designed by OptKnock for acetate overproduction in *E. coli* (see Example 8.1). (a) An optimal design where biomass and product formation are decoupled. (b) A slightly suboptimal design where biomass and product formation are partially coupled.

which is 70.20% of the theoretical maximum. The trade-off plot for this mutant shows that acetate production varies between $17.40\frac{\text{mmol}}{\text{gDW}\cdot\text{h}}$ and $17.83\frac{\text{mmol}}{\text{gDW}\cdot\text{h}}$ (see Fig. 8.4b), implying that acetate production and biomass formation are partially coupled. Note that having a small, or even better, no variation in the acetate production flux at maximum biomass is the ideal scenario for strain design. A GAMS implementation of this example is available on the book's website. □

8.3 OPTKNOCK MODIFICATIONS

8.3.1 RobustKnock

A potential limitation of OptKnock arises when the suggested mutant involves a range of product formation yields under maximum biomass conditions (see Fig. 8.3a and b). This is because OptKnock always selects the most optimistic (i.e., the highest) flux for the product formation under the maximum biomass condition even though the actual product formation flux can be less than this maximum or even zero (i.e., uncoupled from growth). RobustKnock is a modified version of OptKnock that was developed to address this issue by optimizing the worst-case scenario for the product formation while maximizing the biomass flux [2]. To this end, the objective function of the outer problem in OptKnock is modified as

$$\text{maximize}\left(\text{minimize}\, v_{\text{EX\_P(e)}}\right)$$

which leads to a trilevel optimization problem as follows:

$$\text{maximize}\left(\text{minimize}\right) \quad v_{\text{EX\_P(e)}} \quad \left[\text{RobustKnock}\right]$$

subject to

$$\begin{bmatrix} \text{maximize } v_{\text{biomass}} & \\ \text{subject to} & \\ \sum_{j \in J} S_{ij} v_j = 0, & \forall i \in I \\ LB_j y_j \leq v_j \leq UB_j y_j, & \forall j \in J \end{bmatrix} \quad \begin{matrix} (8.1) \\ (8.6) \end{matrix}$$

$$\sum_{j \in J} \left(1 - y_j\right) \leq K \tag{8.7}$$

$$y_j \in \{0,1\}, \quad v_j \in \mathbb{R}, \quad \forall j \in J$$

The outer problem here is a max-min optimization problem that searches for a set of reaction deletions that maximize the minimum guaranteed production flux of the target metabolite. The inner-level problem is similar to OptKnock and maximizes the cellular objectives subject to reaction deletions identified by the outer problem (note that all constraints representing the uptake conditions are assumed to be embedded in Constraint 8.6 in this representation). This optimization problem can be recast as a standard MILP in two steps: (i) transforming the problem into a standard max-min problem by replacing the inner-level problem with regular constraints and (ii) transforming the resulting max-min problem into a standard MILP by replacing the inner minimization problem with regular constraints. Both steps involve replacing the inner optimization problems with primal and dual constraints and imposing the strong duality condition similarly to OptKnock (see Ref. 2 for more details). This leads to a more complex and generally more computationally demanding optimization problem to solve.

8.3.2 Tilting the Objective Function

A simpler and more computationally tractable alternative to RobustKnock involves using a *tilted objective function* to simulate the worst-case scenario for product formation [16]. In essence, this method modifies the inner objective function in OptKnock to maximize the biomass with a small penalty for target product formation.

$$\text{maximize } v_{\text{biomass}} - \varepsilon v_{\text{EX\_P(e)}}$$

Here ε is a very small weight ensuring that upon maximization of the objective function the smallest possible product formation rate is identified consistent with biomass maximization.

8.4 OTHER STRAIN DESIGN ALGORITHMS

Since the development of OptKnock, a wide array of other optimization-based strain design algorithms have been introduced [3–15]. The detailed description of all these algorithms is out of the scope of this book (see Ref. 19 for a comprehensive review). Briefly, OptGene uses a genetic algorithm to solve the OptKnock optimization formulation [3], whereas SimOptStrain simultaneously identifies reaction deletions and non-native reaction additions [4]. BiMOMA uses the minimization of metabolic adjustment (MOMA) [17] as the cellular objective in the inner problem [4].

All strain design algorithms described so far mainly identify gene knockouts (through reaction eliminations) to enhance the production of a target metabolite. A number of other methods have been developed that enable the simultaneous identification of multiple types of interventions including knockouts, overexpressions and downregulations. OptReg is one such method where flux modulations are modeled as upward and downward deviations from steady-state flux values in the wild-type [5]. Three sets of binary variables are defined for each reaction to determine whether it should be eliminated, upregulated or downregulated. It uses a bilevel optimization structure similar to OptKnock, where the outer problem maximizes the production flux of the target metabolite and the inner problem maximizes the cellular objective (e.g., the biomass formation) while also minimizing the network trafficking (by adding the sum of all fluxes in the objective function multiplied by a small negative scalar). OptORF is a bilevel optimization problem that directly pinpoints optimal metabolic and regulatory gene deletions as well as metabolic gene overexpressions coupling the biomass production and product formation [6].

OptForce is a computational strain design framework that relies on a multistep procedure to identify multiple types of interventions [7]. A key feature of this framework is the ability to incorporate all available flux data for the wild-type or mutant strains. It starts with characterizing the phenotypic space of the wild-type (or a reference) strain by using the available metabolic flux measurements. This is followed by identifying the set of changes in reaction fluxes that must occur in the network consistent with a user-specified yield of the desired product. This is achieved by contrasting the flux range of each reaction in the reference strain (identified in the first step) with those in the overproduction phenotype. Subsequently, one identifies the minimal set of engineering strategies, selected from the changes identified in the second step that must be *directly* imposed to achieve the desired yield. To this end, OptForce uses a max-min bilevel optimization problem to optimize the worst-case scenario for the product formation. The inner objective minimizes the production flux of the target metabolite (i.e., the worst-case scenario) whereas the outer problem identifies direct interventions (including knockouts and up-/down-regulations) that maximize the minimum product formation in the network (see Ref. 20 for a comparison between OptKnock and OptForce). OptForce was also recently extended to include kinetic descriptions for some of the reaction steps (k-OptForce) [8]. Note that the incorporation of kinetic expressions leads to bilevel mixed-integer nonlinear optimization problems (see Chapter 11).

Finally, CosMos is a strain design protocol that identifies continuous modifications in reaction fluxes to enhance the production of a target chemical [9]. Instead of relying on binary variables to flag genetic interventions, it uses continuous variables to capture the required changes in the lower and upper bounds of fluxes. Thus, Constraint 8.6 is replaced with

$$LB_j + \alpha_j \leq v_j \leq UB_j - \beta_j, \quad \forall j \in J \tag{8.16}$$

where $\alpha_j, \beta_j \geq 0$. Observe that a reaction deletion can be modeled by setting $\alpha_j = -LB_j$ (assuming $LB_j \leq 0$) and $\beta_j = UB_j$. Similar to OptForce, the inner problem minimizes the product formation flux to simulate the worst-case scenario whereas the outer problem maximizes the production flux while also minimizing the number of flux modifications. Binary variables are needed here to determine whether a reaction flux has been modified. In addition, imposing the strong duality condition to transform the inner optimization problem into a set of regular constraints leads to a nonlinear and nonconvex optimization problem due to the presence of bilinear terms (the product of α_j or β_j and dual variables). To avoid this issue, KKT conditions are imposed in order to transform the bilevel optimization problem into an MILP, however, at the expense of introducing new sets of binary variables to enforce the complementary slackness condition.

EXERCISES

8.1 Using the dual representation, formulate and solve OptKnock to identify reaction eliminations leading to succinate overproduction using the *i*AF1260 *E. coli* model [18] under the same uptake conditions as that in Example 8.1 but assuming anaerobic condition.

- **(a)** Demonstrate that the dual matches the primal solution for wild-type *E. coli* and some pre-specified knockout mutants.
- **(b)** Identify the best reaction elimination combinations (including alternate reaction removal strategies, if available).
- **(c)** Link the identified reaction deletions with the corresponding gene knockouts using the supplied GPR associations in the model. Comment on their feasibility and compare with literature experimental evidence.
- **(d)** Generate the succinate vs. biomass trade-off plot for all mutants and choose one design that you deem the most promising.

8.2 Consider Exercise 3.2

- **(a)** Biocontainment concerns require that the production strain cannot produce lysine. Derive the formulation whose solution identifies the minimal number of reaction deletions ensuring lysine auxotrophy while allowing for the production of all other 19 amino acids at levels at least 5% of their theoretical maximum yields before the deletions.
- **(b)** The team developed a strain for the production of a proprietary product i^*. Unfortunately, significant flux is diverted toward co-product b. Derive

an optimization formulation for identifying all reaction eliminations that minimize the production of b while keeping unaffected the maximum theoretical yield for i^*.

REFERENCES

1. Burgard AP, Pharkya P, Maranas CD: OptKnock: a bilevel programming framework for identifying gene knockout strategies for microbial strain optimization. *Biotechnol Bioeng* 2003, **84**(6):647–657.
2. Tepper N, Shlomi T: Predicting metabolic engineering knockout strategies for chemical production: accounting for competing pathways. *Bioinformatics* 2010, **26**(4):536–543.
3. Patil KR, Rocha I, Förster J, Nielsen J: Evolutionary programming as a platform for in silico metabolic engineering. *BMC Bioinformat* 2005, **6**:308.
4. Kim J, Reed JL, Maravelias CT: Large-scale bi-level strain design approaches and mixed-integer programming solution techniques. *PLoS One* 2011, **6**(9):e24162.
5. Pharkya P, Maranas CD: An optimization framework for identifying reaction activation/inhibition or elimination candidates for overproduction in microbial systems. *Metab Eng* 2006, **8**(1):1–13.
6. Kim J, Reed JL: OptORF: Optimal metabolic and regulatory perturbations for metabolic engineering of microbial strains. *BMC Syst Biol* 2010, **4**:53.
7. Ranganathan S, Suthers PF, Maranas CD: OptForce: an optimization procedure for identifying all genetic manipulations leading to targeted overproductions. *PLoS Comput Biol* 2010, **6**(4):e1000744.
8. Chowdhury A, Zomorrodi AR, Maranas CD: k-OptForce: integrating kinetics with flux balance analysis for strain design. *PLoS Comput Biol* 2014, **10**(2):e1003487.
9. Cotten C, Reed JL: Constraint-based strain design using continuous modifications (CosMos) of flux bounds finds new strategies for metabolic engineering. *Biotechnol J* 2013, **8**(5):595–604.
10. Pharkya P, Burgard AP, Maranas CD: OptStrain: a computational framework for redesign of microbial production systems. *Genome Res* 2004, **14**(11):2367–2376.
11. Lun DS, Rockwell G, Guido NJ, Baym M, Kelner JA, Berger B, Galagan JE, Church GM: Large-scale identification of genetic design strategies using local search. *Mol Syst Biol* 2009, **5**:296.
12. Choi HS, Lee SY, Kim TY, Woo HM: *In silico* identification of gene amplification targets for improvement of lycopene production. *Appl Environ Microbiol* 2010, **76**(10): 3097–3105.
13. Yang L, Cluett WR, Mahadevan R: EMILiO: a fast algorithm for genome-scale strain design. *Metab Eng* 2011, **13**(3):272–281.
14. Zhuang K, Yang L, Cluett WR, Mahadevan R: Dynamic strain scanning optimization: an efficient strain design strategy for balanced yield, titer, and productivity. DySScO strategy for strain design. *BMC Biotechnol* 2013, **13**:8.
15. Rockwell G, Guido NJ, Church GM: Redirector: designing cell factories by reconstructing the metabolic objective. *PLoS Comput Biol* 2013, **9**(1):e1002882.

16. Feist AM, Zielinski DC, Orth JD, Schellenberger J, Herrgard MJ, Palsson B: Model-driven evaluation of the production potential for growth-coupled products of *Escherichia coli*. *Metab Eng* 2010, **12**(3):173–186.
17. Segrè D, Vitkup D, Church GM: Analysis of optimality in natural and perturbed metabolic networks. *Proc Natl Acad Sci U S A* 2002, **99**(23):15112–15117.
18. Feist AM, Henry CS, Reed JL, Krummenacker M, Joyce AR, Karp PD, Broadbelt LJ, Hatzimanikatis V, Palsson B: A genome-scale metabolic reconstruction for *Escherichia coli* K-12 MG1655 that accounts for 1260 ORFs and thermodynamic information. *Mol Syst Biol* 2007, **3**:121.
19. Zomorrodi AR, Suthers PF, Ranganathan S, Maranas CD: Mathematical optimization applications in metabolic networks. *Metab Eng* 2012, **14**(6):672–686.
20. Chowdhury A, Zomorrodi AR,Maranas CD: Bilevel optimization techniques in computational strain design. *Comput Chem Eng* 2015, **72**:363–372.

9

NLP FUNDAMENTALS

Nonlinear optimization problems often arise in the analysis of metabolic networks due to nonlinear objective functions and/or constraints. Examples include the minimization of metabolic adjustment (MOMA) [1], kinetic and dynamic models of metabolism [2–5] and metabolic flux analysis (MFA) using isotope-labeled atoms [6–8] (see Chapter 10 for a review of these applications). In this chapter, we outline the fundamentals of unconstrained and constrained nonlinear programming (NLP) and relevant solution procedures.

9.1 UNCONSTRAINED NONLINEAR OPTIMIZATION

Unconstrained nonlinear optimization problems involve the maximization or minimization of a nonlinear function without any constraints:

$$\underset{\boldsymbol{x}\in\mathbb{R}^N}{\text{minimize}}\, f(\boldsymbol{x}) \quad \text{[UncNLP]}$$

Even though such problems rarely arise in practical applications, the study of their solution procedures is important as they lay the foundation for those of constrained optimization problems. In general, solution methods for this class of problems rely on improving an initial solution point in an iterative manner until no further improvements are achievable. This is typically done in two separate phases for each iteration k:

(i) Choose a search direction $\boldsymbol{d}_k$

Optimization Methods in Metabolic Networks, First Edition. Costas D. Maranas and Ali R. Zomorrodi.

(ii) Obtain an optimal step size α_k to determine how far to move along the search direction $\boldsymbol{d}_k$. The optimal step size is found by solving a subproblem called *line search,* which is an unconstrained optimization problem with α_k as the only variable $\min_{\alpha_k} f(\boldsymbol{x}_k + \alpha_k \boldsymbol{d}_k)$

The new point is then equal to

$$\boldsymbol{x}_{k+1} = \boldsymbol{x}_k + \alpha_k \boldsymbol{d}_k \tag{9.1}$$

Before moving into the details of solution methods, we start with the optimality conditions for unconstrained optimization problems.

9.1.1 Optimality Conditions for Unconstrained Optimization Problems

Optimality conditions can be *necessary* or *sufficient*. Necessary conditions are those that must be satisfied if a given solution point is a local minimum (i.e., $\boldsymbol{x}^*$ is a local minimum $\Rightarrow$ *necessary conditions* must hold). Sufficient conditions, on the other hand, are conditions that if satisfied for a given point would imply that the point is a local minimum (i.e., *sufficient conditions* hold for a point $\boldsymbol{x}^* \Rightarrow \boldsymbol{x}^*$ is a local minimum). Mathematical proofs for these conditions can be found elsewhere (e.g., see Refs. [9, 10]).

Descent Direction Consider the unconstrained optimization problem [UncNLP] and a current iteration step k. A *descent* direction is a search direction along which the objective function improves, that is $f(\boldsymbol{x}_{k+1}) < f(\boldsymbol{x}_k)$.

Theorem 9.1
Let $f : \mathbb{R}^N \rightarrow \mathbb{R}$ be a continuous function differentiable at a point $\bar{\boldsymbol{x}} \in \mathbb{R}^N$. If there exists a vector $\boldsymbol{d}$ such that $\nabla^{\mathrm{T}} f(\bar{\boldsymbol{x}})\boldsymbol{d} < 0$ (see Chapter 1 for the definition of the gradient vector) then $\boldsymbol{d}$ is a descent direction of f at $\bar{\boldsymbol{x}}$, if for all sufficiently small $\alpha > 0$, $f(\bar{\boldsymbol{x}} + \alpha \boldsymbol{d}) < f(\bar{\boldsymbol{x}})$.

Proof: Given that f is differentiable at $\bar{\boldsymbol{x}}$, we have (see Chapter 1):

$$f(\bar{\boldsymbol{x}} + \alpha \boldsymbol{d}) = f(\bar{\boldsymbol{x}}) + \alpha \nabla^{\mathrm{T}} f(\bar{\boldsymbol{x}})\boldsymbol{d} + \alpha \|\boldsymbol{d}\| u(\bar{\boldsymbol{x}} + \alpha \boldsymbol{d})$$

or

$$\frac{f(\bar{\boldsymbol{x}} + \alpha \boldsymbol{d}) - f(\bar{\boldsymbol{x}})}{\alpha} = \nabla^{\mathrm{T}} f(\bar{\boldsymbol{x}})\boldsymbol{d} + \|\boldsymbol{d}\| u(\bar{\boldsymbol{x}} + \alpha \boldsymbol{d})$$

Since $u(\bar{\boldsymbol{x}} + \alpha \boldsymbol{d}) \rightarrow \boldsymbol{0}$ as $\alpha \rightarrow 0$ and given that $\nabla^{\mathrm{T}} f(\bar{\boldsymbol{x}})\boldsymbol{d} < \boldsymbol{0}$, we have $f(\bar{\boldsymbol{x}} + \alpha \boldsymbol{d}) - f(\bar{\boldsymbol{x}}) < \boldsymbol{0}$ for all sufficiently small $\alpha > 0$. □

First-Order Necessary Optimality Condition Let $f:\mathbb{R}^N \to \mathbb{R}$ be a continuous function differentiable at a point $\bar{\boldsymbol{x}} \in \mathbb{R}^N$. If $\bar{\boldsymbol{x}}$ is a local minimum then $\nabla f(\bar{\boldsymbol{x}}) = \mathbf{0}$.

This condition is called a *first-order* necessary condition because it relies on only information from the first-order derivatives. The necessary optimality condition can also be expressed using the Hessian matrix for twice differentiable functions, which are referred to as the *second-order* optimality conditions.

Second-Order Necessary Optimality Condition Let $f:\mathbb{R}^N \to \mathbb{R}$ be a continuous function twice differentiable at a point $\bar{\boldsymbol{x}} \in \mathbb{R}^N$. If $\bar{\boldsymbol{x}}$ is a local minimum then $\nabla f(\bar{\boldsymbol{x}}) = \mathbf{0}$ and $\boldsymbol{H}(\bar{\boldsymbol{x}})$ is positive semidefinite.

The conditions described so far are necessary meaning that if a point is a local minimum these conditions must hold, however, if a given point satisfies these conditions that point need not be a local minimum.

Second-Order Sufficient Optimality Condition Let $f:\mathbb{R}^N \to \mathbb{R}$ be a continuous function twice differentiable at a point $\bar{\boldsymbol{x}} \in \mathbb{R}^N$. If $\nabla f(\bar{\boldsymbol{x}}) = \mathbf{0}$ and $\boldsymbol{H}(\bar{\boldsymbol{x}})$ is *positive definite* then $\bar{\boldsymbol{x}}$ is a *strict* local minimum.

Theorem 9.2

Let $f:\mathbb{R}^N \to \mathbb{R}$ be a continuous function differentiable at a point $\bar{\boldsymbol{x}} \in \mathbb{R}^N$. If $\nabla f(\bar{\boldsymbol{x}}) = \mathbf{0}$ and $\boldsymbol{H}(\bar{\boldsymbol{x}})$ is indefinite (see Chapter 1) then $\bar{\boldsymbol{x}}$ is a saddle point.

Note that if the Hessian is positive semidefinite at a critical point $\bar{\boldsymbol{x}}$, that point need not be a local minimum. An example is $x = 0$, which is a critical and saddle point for $f(x) = x^3$. However, if the Hessian is positive semidefinite for *all* $\bar{\boldsymbol{x}} \in \mathbb{R}^N$ (i.e., if f is convex) then a critical point $\bar{\boldsymbol{x}} \in \mathbb{R}^N$ will be also a *global* minimum.

Theorem 9.3

Let $f:\mathbb{R}^N \to \mathbb{R}$ be a continuous function differentiable at a point $\bar{\boldsymbol{x}} \in \mathbb{R}^N$. If $\nabla f(\bar{\boldsymbol{x}}) = \mathbf{0}$ and $\boldsymbol{H}$ is positive semidefinite in $\mathbb{R}^N$ then $\bar{\boldsymbol{x}}$ is a global minimum.

It follows from this theorem that if f is strictly convex $\mathbb{R}^N$, then any critical point is a *unique* global minimum. In the following, we describe a more general case for a critical point $\bar{\boldsymbol{x}}$ to be a global minimum, where f need not be convex.

First-Order Sufficient Optimality Condition Let $f:\mathbb{R}^N \to \mathbb{R}$ be pseudoconvex at a point $\bar{\boldsymbol{x}} \in \mathbb{R}^N$, then $\bar{\boldsymbol{x}}$ is a global minimum if $\nabla f(\bar{\boldsymbol{x}}) = \mathbf{0}$. (See Chapter 1 for the definition of the pseudoconvexity.)

By the first-order necessary optimality condition, if $\bar{\boldsymbol{x}}$ is a global minimum then $\nabla f(\bar{\boldsymbol{x}}) = \mathbf{0}$. Therefore, $\nabla f(\bar{\boldsymbol{x}}) = \mathbf{0}$ is both a necessary and sufficient condition if f is pseudoconvex at $\bar{\boldsymbol{x}}$. A summary of all optimality conditions is given in Table 9.1.

TABLE 9.1 Summary of the (A) Necessary and (B) Sufficient Optimality Conditions for Unconstrained Optimization Problems

(A)	Necessary Optimality Conditions
	$\bar{x}$ is a local minimum $\Rightarrow \nabla f(\bar{x}) = \mathbf{0}$
	$\bar{x}$ is a local minimum $\Rightarrow \nabla f(\bar{x}) = \mathbf{0}$ and $H(\bar{x})$ is positive semidefinite
(B)	Sufficient Optimality Conditions
	$\nabla f(\bar{x}) = \mathbf{0}$ and $H(\bar{x})$ is positive definite. $\Rightarrow \bar{x}$ is a strict local minimum
	$\nabla f(\bar{x}) = \mathbf{0}$ and H is positive semidefinite $\forall x \in \mathbb{R}^N \Rightarrow \bar{x}$ is a global minimum
	$\nabla f(\bar{x}) = \mathbf{0}$ and H is positive definite $\forall x \in \mathbb{R}^N \Rightarrow \bar{x}$ is a unique global minimum
	$\nabla f(\bar{x}) = \mathbf{0}$ and f pseudoconvex $\Rightarrow \bar{x}$ is a global minimum
	$\nabla f(\bar{x}) = \mathbf{0}$ and $H(\bar{x})$ is *indefinite* $\Rightarrow \bar{x}$ is a saddle point

9.1.2 An Overview of the Solution Methods for Unconstrained Optimization Problems

Unconstrained optimization solution methods fall into two major categories depending on whether the updating scheme (i.e., finding the search direction and the step size) requires derivative information:

(i) *Methods using only function evaluations*: Updated solution points are determined solely by using objective function evaluations. Iterations are repeated until no further improvement in the objective function is achievable or the maximum number of iterations is reached. These methods may not always be efficient but are easy to implement.

(ii) *Methods using derivative information*: Rely on the information contained in the first- and/or second-order derivatives to determine the search direction. Iterations continue until a critical point is identified. Sufficient optimality conditions need to be assessed to ensure that the identified critical point is indeed a local minimum. These methods can be further categorized as follows:

   (a) *Gradient methods*: Require only first-order derivatives to update the solution point, that is $\mathbf{x}_{k+1} = f(\mathbf{x}_k, f(\mathbf{x}_k), \nabla f(\mathbf{x}_k))$.

   (b) *Second-order methods*: Require both first- and second-order derivatives to update the solution point, that is $\mathbf{x}_{k+1} = f(\mathbf{x}_k, f(\mathbf{x}_k), \nabla f(\mathbf{x}_k), \mathbf{H}_f(\mathbf{x}_k))$.

In this book, we only briefly describe the methods using derivatives. Interested readers are encouraged to refer to optimization textbooks (e.g., Refs. [9, 10]) for details and methods that do not use derivative information.

9.1.3 Steepest Descent (Cauchy or Gradient) Method

The idea of this method is to approximate the objective function by a line (i.e., first-order Taylor's series approximation) and move iteratively along the steepest descent direction (which is orthogonal to the linear approximation of the function) until convergence is achieved. Recall from Chapter 1 that the direction of steepest descent

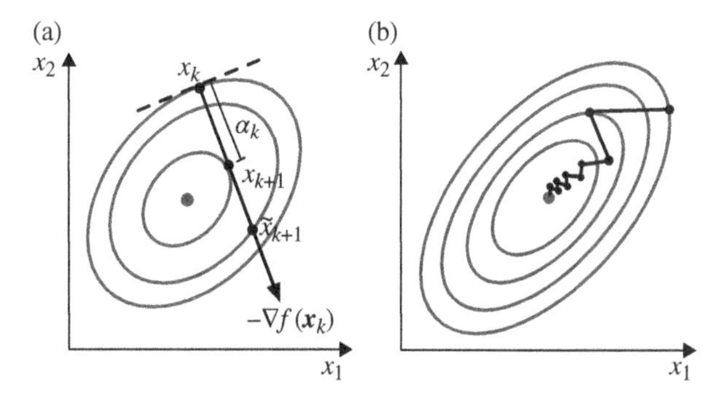

FIGURE 9.1 (a) Calculation of the optimal step size, α_k in the steepest descent method. The objective is to find α_k that minimizes $f(\boldsymbol{x}_{k+1})$. (b) Oscillating iterations (zigzagging) in the steepest descent algorithm as approaching the minimum.

at a point $\boldsymbol{x}_k$ is given by $-\nabla f(\boldsymbol{x}_k)$. The updating scheme for the steepest descent method is:

$$\boldsymbol{x}_{k+1} = \boldsymbol{x}_k + \alpha_k \boldsymbol{d}_k = \boldsymbol{x}_k + \alpha_k \left(-\nabla f\left(\boldsymbol{x}_k\right)\right) = \boldsymbol{x}_k - \alpha_k \nabla f\left(\boldsymbol{x}_k\right) \tag{9.2}$$

where the step size $(\alpha_k \geq 0)$ in the direction of steepest descent is determined by a line search, that is $\alpha_k = \arg\min\ f(\boldsymbol{x}_k + \alpha_k \boldsymbol{d}_k)$ (see Fig. 9.1a). Iterations stop when $\|\nabla f(\boldsymbol{x}_k)\| < \varepsilon$ where ε is a pre-specified small scalar.

The steepest descent method typically performs well in early iterations but leads to oscillations (or zigzagging) closer to the minimum, particularly when the objective function is poorly scaled (see Fig. 9.1b). This is because higher-order derivatives, which are ignored in the steepest descent algorithm become comparable in magnitude to $\|\nabla f(\boldsymbol{x}_k)\|$ as it approaches zero.

9.1.4 Newton's Method

The method of steepest descent uses information from the first-order derivatives to move in the direction orthogonal to the linear approximation of the objective function at each iteration point. Newton's method uses information from both the first- and second-order derivatives and moves along a deflected form of the steepest descent search direction shaped by a quadratic approximation of the objective function. Let q be the quadratic approximation of the objective function f at an iteration point $\boldsymbol{x}_k$. Assuming that f is twice differentiable, from the Taylor's expansion of f at $\boldsymbol{x}_k$ and by ignoring derivatives higher than second-order we have

$$q\left(\boldsymbol{x}\right) = f\left(\boldsymbol{x}_k\right) + \nabla^{\mathrm{T}} f\left(\boldsymbol{x}_k\right)\left(\boldsymbol{x} - \boldsymbol{x}_k\right) + \frac{1}{2}\left(\boldsymbol{x} - \boldsymbol{x}_k\right)^{\mathrm{T}} \boldsymbol{H}\left(\boldsymbol{x}_k\right)\left(\boldsymbol{x} - \boldsymbol{x}_k\right) \tag{9.3}$$

A necessary and sufficient condition for the minimum of the quadratic approximation of f is to set $\nabla q\left(\boldsymbol{x}\right) = \boldsymbol{0}$. From Equation 9.3, we have

$$\nabla q\left(\boldsymbol{x}\right) = \nabla f\left(\boldsymbol{x}_k\right) + \boldsymbol{H}\left(\boldsymbol{x}_k\right)\left(\boldsymbol{x} - \boldsymbol{x}_k\right) = \boldsymbol{0} \tag{9.4}$$

This gives the recursive equation for Newton's method:

$$x_{k+1} = x_k - H^{-1}(x_k)\nabla f(x_k) \tag{9.5}$$

Therefore, Newton's method moves along the direction $-H^{-1}(x_k)\nabla f(x_k)$ with a step size of one. Similarly to the steepest descent method, iterations of Newton's method stop when $\| \nabla f(x_k) \| < \varepsilon$. If the objective function is quadratic, Newton's method arrives at the minimum in only one iteration.

Note that the sequence of the points generated by Newton's method may not converge to a local minimum due to a number of reasons. First, the search direction $-H^{-1}(x_k)\nabla f(x_k)$ may not necessarily be a descent direction if the function is nonconvex. In addition, using a step size of one may not lead to a decrease in f, however if an optimal step size is obtained by a line search, this problem may be resolved:

$$x_{k+1} = x_k - \alpha_k H^{-1}(x_k)\nabla f(x_k) \tag{9.6}$$

where $\alpha_k = \arg\min\, f(x_{k+1})$ and $\alpha_k \geq 0$. A more significant problem with Newton's method is when the inverse of the Hessian matrix does not exist (i.e., the Hessian matrix is singular). However, if the initial point is close enough to a local minimum $\bar{x}$ (i.e., $\nabla f(\bar{x}) = 0$) then $H(x_k)$ will be positive definite at points close to $\bar{x}$ and thus the next point is well defined and Newton's method converges quadratically [9]. If the objective function is convex in $\mathbb{R}^N$ convergence is guaranteed since the Hessian matrix will be nonsingular and positive semidefinite at all points. Furthermore, there are modifications of the Newton's method that resolve the issue of Hessian matrix singularity for a general nonconvex objective function and guarantee convergence regardless of the initial point selection [9].

Interestingly, the Newton–Raphson's method for solving a system of nonlinear equations $w(y) = 0$, where $w(y) = [w_1(y), w_2(y), \ldots, w_L(y)]^T$ uses a linear (i.e., first-order Taylor's series) approximation of $w(y)$ at a given point $y = y_k$:

$$w(y_k) + \nabla w(y_k)(y - y_k) = 0 \tag{9.7}$$

which gives rise to the following recursive equation:

$$y_{k+1} = y_k - \left[\nabla w(y_k)\right]^{-1} w(y_k) \tag{9.8}$$

Comparing Equations 9.5 and 9.8 reveals that Newton's method for solving [UncNLP] can be viewed as the Newton–Raphson's method for solving the system of equations implied by $\nabla f(x) = 0$.

9.1.5 Quasi-Newton Methods

Newton's method requires second-derivative information that can become computationally prohibitive for larger problems. Quasi-Newton methods approximate $H^{-1}(x_k)$ by a positive definite symmetric matrix A_k. This matrix is initialized as a positive definite symmetric matrix (e.g., the identity or a diagonal matrix) and is successively updated at each iteration without a complete recalculation of all its elements. This modifies Newton's step as follows:

$$\boldsymbol{x}_{k+1} = \boldsymbol{x}_k - \boldsymbol{A}_k \nabla f(\boldsymbol{x}_k) \tag{9.9}$$

Positive definiteness of $\boldsymbol{A}_k$ is required to make sure that $-\boldsymbol{A}_k \nabla f(\boldsymbol{x}_k)$ is a descent direction whenever $\nabla f(\boldsymbol{x}_k) \neq \boldsymbol{0}$. In addition, all quasi-Newton methods must satisfy the following property:

$$\lim_{k\to\infty} \boldsymbol{A}_k = \boldsymbol{H}^{-1}(\boldsymbol{x}_k) \tag{9.10}$$

One of the most efficient and widely used methods for updating $\boldsymbol{A}_k$ is the Broyden–Fletcher–Goldfarb–Shanno (BFGS) method:

$$\boldsymbol{A}_{k+1} = \boldsymbol{A}_k + \frac{\Delta \boldsymbol{g}_k (\Delta \boldsymbol{g}_k)^{\mathrm{T}}}{(\Delta \boldsymbol{x}_k)^{\mathrm{T}} \Delta \boldsymbol{g}_k} - \frac{\boldsymbol{A}_k \Delta \boldsymbol{x}_k (\boldsymbol{A}_k \Delta \boldsymbol{x}_k)^{\mathrm{T}}}{(\Delta \boldsymbol{x}_k)^{\mathrm{T}} \boldsymbol{A}_k \Delta \boldsymbol{x}_k} \tag{9.11}$$

where

$$\begin{aligned} \Delta \boldsymbol{x}_k &= \boldsymbol{x}_k - \boldsymbol{x}_{k-1} \\ \Delta \boldsymbol{g}_k &= \nabla f(\boldsymbol{x}_k) - \nabla f(\boldsymbol{x}_{k-1}) \end{aligned} \tag{9.12}$$

Note that $\boldsymbol{A}_k$ is updated by using the information of the previous two steps. If $\boldsymbol{A}_k$ is positive definite and $(\Delta \boldsymbol{x}_k)^{\mathrm{T}} \Delta \boldsymbol{g}_k > \boldsymbol{0}$, it can be shown that $\boldsymbol{A}_{k+1}$ is also positive definite [9]. For a quadratic function of N variables the BFGS method converges in N iterations. However for a general non-quadratic function, it usually requires more iterations than Newton's method but each iteration is less computationally taxing.

9.1.6 Conjugate Gradients (CG) Methods

The convergence characteristics of the steepest descent method can be improved greatly by modifying the new search directions using information from earlier iterations. The search direction for the first iteration is the same as in steepest descent, that is $\boldsymbol{d}_k = -\nabla f(\boldsymbol{x}_k)$ for $k = 1$. For subsequent iterations, CG methods use:

$$\boldsymbol{d}_k = -\nabla f(\boldsymbol{x}_k) + \beta_k \boldsymbol{d}_{k-1} \tag{9.13}$$

where β_k is a deflection parameter specific to the CG variant. For the Hestenes and Stiefel's (SF) method $\beta_k = \dfrac{\|\nabla f(\boldsymbol{x}_k)\|^2}{\|\nabla f(\boldsymbol{x}_{k-1})\|^2}$. Similarly to the quasi-Newton method, N iterations of CG are equivalent to one Newton step. Generally, CG outperforms other methods for large problems. For smaller problems both the original Newton method and BFGS can be competitive. Note that none of the methods described here is guaranteed to find the global minimum for problems with multiple local minima and multistart searches are typically employed.

9.2 CONSTRAINED NONLINEAR OPTIMIZATION

Constrained nonlinear optimization problems consist of a nonlinear objective function and/or at least one nonlinear constraint. A general constrained nonlinear optimization problem can be described in the following form (see Fig. 9.2):

$$\begin{aligned}
&\text{minimize } f(\boldsymbol{x}) \quad \text{[NLP]} \\
&\text{subject to} \\
&g_j(\boldsymbol{x}) \leq 0, \quad j = 1,2,\ldots,M \\
&h_l(\boldsymbol{x}) = 0, \quad l = 1,2,\ldots,L \\
&\boldsymbol{x} \in \mathbb{R}^N
\end{aligned}$$

where $f(\boldsymbol{x})$, $g_j(\boldsymbol{x})$ and $h_l(\boldsymbol{x})$ are functions defined in $\mathbb{R}^N$ denoting the objective function, inequality constraints and equality constraints, respectively.

9.2.1 Equality-Constrained Nonlinear Problems

A nonlinear optimization problem consisting of only equality constraints can be stated as follows:

$$\begin{aligned}
&\text{minimize } f(\boldsymbol{x}) \quad \text{[NLPEQ]} \\
&\text{subject to} \\
&h_l(\boldsymbol{x}) = 0, \quad l = 1,2,\ldots,L \\
&\boldsymbol{x} \in \mathbb{R}^N
\end{aligned}$$

An intuitive way of solving this problem is by converting it to an unconstrained optimization problem. For example, this can be accomplished by a variable elimination through direct substitution.

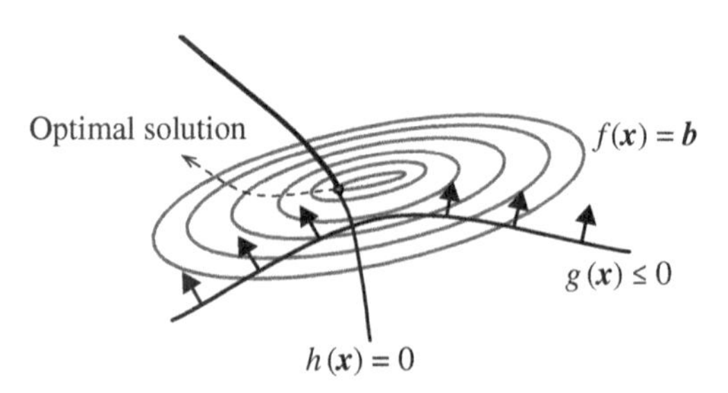

FIGURE 9.2 Schematic illustration of a nonlinear optimization problem with one inequality constraint $g(\boldsymbol{x}) \leq 0$, one equality constraint $h(\boldsymbol{x}) = 0$ and level sets $f(\boldsymbol{x}) = \boldsymbol{b}$.

Example 9.1
Solve the following optimization problem using variable elimination with direct substitution:

$$\begin{aligned} &\text{minimize} \quad z = x_1 x_2 x_3 \\ &\text{subject to} \\ &x_1 + x_2 + x_3 = 1 \\ &x_1, x_2, x_3 \in \mathbb{R} \end{aligned}$$

Solution: By eliminating variable x_3 using the equality constraint, the problem can be seemingly simplified to an unconstrained optimization problem consisting of only two variables:

$$\begin{aligned} &\text{minimize} \ x_1 x_2 (1 - x_1 - x_2) \\ &\text{subject to} \\ &x_1, x_2 \in \mathbb{R} \end{aligned}$$

However, it is very important not to forget that we need to include in the reformulated problem all constraints imposed on the variable that was eliminated. This usually makes the direct substitution impractical for the conversion of the original problem to an unconstrained problem. For example, if x_1, x_2 and x_3 are restricted to be nonnegative in the original problem ($x_1, x_2, x_3 \geq 0$), then the nonnegativity of x_3 has to be imposed as a constraint:

$$\begin{aligned} &\text{minimize} \ x_1 x_2 (1 - x_1 - x_2) \\ &\text{subject to} \\ &1 - x_1 - x_2 \geq 0 \\ &x_1, x_2 \geq 0 \end{aligned}$$

Moreover, variable elimination may also become impractical when multiple equality constraints are present or when complex nonlinear equality constraints such as $x_1^3 x_2 + x_1 x_2^2 + x_2 x_3^2 = 0$ are present. □

Method of Lagrange Multipliers for Equality-Constrained Problems A more effective alternative to formally transform an equality-constrained problem into an unconstrained onc is the method of Lagrange multipliers. This method transforms [NLPEQ] to the following unconstrained optimization problem:

$$\underset{\boldsymbol{x},\boldsymbol{\lambda}}{\text{minimize}} \, L(\boldsymbol{x},\boldsymbol{\lambda}) = f(\boldsymbol{x}) + \sum_{l=1}^{L} \lambda_l h_l(\boldsymbol{x}) \tag{9.14}$$

Here element λ_l of vector $\boldsymbol{\lambda} = [\lambda_1, \lambda_2, \ldots, \lambda_L]^{\mathrm{T}}$ is called the *Lagrangian multiplier* associated with equality constraint $h_l(\boldsymbol{x}) = 0$. $L(\boldsymbol{x},\boldsymbol{\lambda})$ is also referred to as the *Lagrange function*. The critical point of the Lagrange function can be identified by solving the following system of equations:

$$\nabla L = \begin{bmatrix} \nabla_x L \\ \nabla_\lambda L \end{bmatrix} = \mathbf{0} \tag{9.15}$$

or equivalently

$$\frac{\partial L}{\partial x_i} = 0, \quad i = 1,\ldots,N \tag{9.16}$$

$$h_l(\boldsymbol{x}) = 0, \quad l = 1,2,\ldots,L \tag{9.17}$$

These form a system of $N+L$ equations that can be solved for $x_1, x_2, \ldots, x_N$ and $\lambda_1, \lambda_2, \ldots, \lambda_L$. These equations imply that a critical point of the Lagrange function is also a critical point of the [NLPEQ]. The next step after finding the solution for $x_1, x_2, \ldots, x_N$ is to check whether this critical point is indeed a local optimum. To this end, we need to test whether the sufficient optimality conditions hold (e.g., check for the positive definiteness of the Hessian matrix of the Lagrange function).

Example 9.2
Solve the following nonlinear optimization problem using the Lagrange multipliers method:

$$\begin{aligned}
&\text{minimize} \quad (x_1 - 5)^2 + (x_2 - 5)^2 \\
&\text{subject to} \\
&x_1 - x_2 = p \\
&x_1, x_2 \geq 0
\end{aligned}$$

Solution: A graphical representation of this problem is given in Figure 9.3. The Lagrange function for this problem is

$$L(\boldsymbol{x}, \lambda) = (x_1 - 5)^2 + (x_2 - 5)^2 + \lambda(x_1 - x_2 - p)$$

Identifying the critical point of the Lagrange function requires solving the following system of equations:

$$\begin{aligned}
\frac{\partial L}{\partial x_1} &= 2(x_1 - 5) + \lambda = 0 \\
\frac{\partial L}{\partial x_2} &= 2(x_2 - 5) - \lambda = 0 \\
\frac{\partial L}{\partial \lambda} &= x_1 - x_2 - p = 0
\end{aligned}$$

By solving this system of equations, we obtain:

$$x_1^* = \frac{10+p}{2}, \quad x_2^* = \frac{10-p}{2}, \quad \lambda^* = -p$$

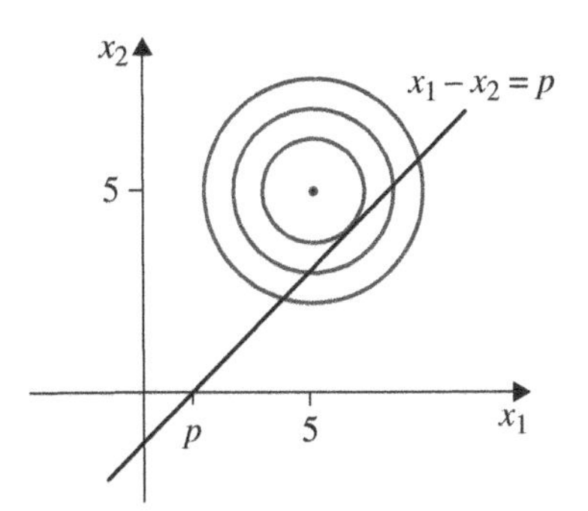

FIGURE 9.3 Graphical representation of the optimization problem in Example 9.2.

Next we need to check the sufficient optimality conditions. The Lagrange function is twice differentiable, so we only have to test for the positive definiteness of the Hessian matrix:

$$H_{L(x,\lambda)}(\boldsymbol{x}) = \begin{bmatrix} \frac{\partial}{\partial x_1}\left(\frac{\partial L(\boldsymbol{x},\lambda)}{\partial x_1}\right) & \frac{\partial}{\partial x_1}\left(\frac{\partial L(\boldsymbol{x},\lambda)}{\partial x_2}\right) \\ \frac{\partial}{\partial x_2}\left(\frac{\partial L(\boldsymbol{x},\lambda)}{\partial x_1}\right) & \frac{\partial}{\partial x_2}\left(\frac{\partial L(\boldsymbol{x},\lambda)}{\partial x_2}\right) \end{bmatrix}$$

$$= \begin{bmatrix} \frac{\partial}{\partial x_1}(2(x_1-5)+\lambda) & \frac{\partial}{\partial x_1}(2(x_2-5)-\lambda) \\ \frac{\partial}{\partial x_2}(2(x_1-5)+\lambda) & \frac{\partial}{\partial x_2}(2(x_2-5)-\lambda) \end{bmatrix} = \begin{bmatrix} 2 & 0 \\ 0 & 2 \end{bmatrix}$$

The Hessian matrix is indeed positive definite and the sufficient optimality conditions are held. Therefore, the identified critical point is a local minimum. □

Economic Interpretation of the Lagrange Multipliers The meaning of the Lagrange multipliers (λ_l) is revealed by perturbing the right-hand side of the equality constraints by a small amount using perturbation vector $\boldsymbol{p}$.

$$\begin{aligned} &\text{minimize } f(\boldsymbol{x}) \\ &\text{subject to} \\ &h_l(\boldsymbol{x}) = p_l, \quad l = 1,2,\ldots,L \\ &\boldsymbol{x} \in \mathbb{R}^N \end{aligned}$$

The Lagrange function associated with this optimization problem is:

$$L(\boldsymbol{x},\lambda) = f(\boldsymbol{x}) + \sum_{l=1}^{L} \lambda_l \left(h_l(\boldsymbol{x}) - p_l\right) \tag{9.18}$$

The critical point of the objective function is found by solving the following equalities (see Eq. 9.16 and 9.17):

$$\frac{\partial L}{\partial x_i} = \frac{\partial f}{\partial x_i} + \sum_{l=1}^{L} \lambda_l \frac{\partial h_l}{\partial x_i} = 0, \quad i = 1, \ldots N \tag{9.19}$$

$$h_l(\boldsymbol{x}) - p_l = 0, \quad l = 1, \ldots, L \tag{9.20}$$

Equation 9.19 for the unperturbed problem can be expanded in a matrix form as follows:

$$\begin{bmatrix} \frac{\partial f}{\partial x_1} \\ \frac{\partial f}{\partial x_2} \\ \vdots \\ \frac{\partial f}{\partial x_N} \end{bmatrix} + \begin{bmatrix} \frac{\partial h_1}{\partial x_1} & \frac{\partial h_2}{\partial x_1} & \cdots & \frac{\partial h_L}{\partial x_1} \\ \frac{\partial h_1}{\partial x_2} & \frac{\partial h_2}{\partial x_2} & \cdots & \frac{\partial h_L}{\partial x_2} \\ \vdots & \vdots & \ddots & \vdots \\ \frac{\partial h_1}{\partial x_N} & \frac{\partial h_2}{\partial x_N} & \cdots & \frac{\partial h_L}{\partial x_N} \end{bmatrix} \begin{bmatrix} \lambda_1 \\ \lambda_2 \\ \vdots \\ \lambda_L \end{bmatrix} = \begin{bmatrix} 0 \\ 0 \\ \vdots \\ 0 \end{bmatrix} \tag{9.21}$$

which can be rewritten in a condensed form as:

$$\nabla f(\boldsymbol{x}) + \nabla^{\mathrm{T}} \boldsymbol{h}(\boldsymbol{x}) \boldsymbol{\lambda} = \boldsymbol{0} \tag{9.22}$$

where $\nabla f(\boldsymbol{x})$ and λ are $N \times 1$ and $L \times 1$ vectors and $\nabla \boldsymbol{h}(\boldsymbol{x})$ is a $L \times N$ matrix as shown in Equation 9.20. We next define

$$\underset{N \times L}{(\nabla_p \boldsymbol{x})} = \begin{bmatrix} \frac{\partial x_1}{\partial p_1} & \frac{\partial x_1}{\partial p_2} & \cdots & \frac{\partial x_1}{\partial p_L} \\ \frac{\partial x_2}{\partial p_1} & \frac{\partial x_2}{\partial p_2} & \cdots & \frac{\partial x_2}{\partial p_L} \\ \vdots & \vdots & \ddots & \vdots \\ \frac{\partial x_N}{\partial p_1} & \frac{\partial x_N}{\partial p_2} & \cdots & \frac{\partial x_N}{\partial p_L} \end{bmatrix} \tag{9.23}$$

By multiplying both sides of Equation 9.22 by $\nabla_{\mathrm{p}}^{T} \boldsymbol{x}$, we get

$$\underset{L \times N}{(\nabla_{\mathrm{p}}^{T} \boldsymbol{x})} \underset{N \times 1}{(\nabla f(\boldsymbol{x}))} + \underset{L \times N}{(\nabla_{\mathrm{p}}^{T} \boldsymbol{x})} \underset{N \times L}{(\nabla^{\mathrm{T}} \boldsymbol{h}(\boldsymbol{x}))} \underset{L \times 1}{\boldsymbol{\lambda}} = \boldsymbol{0} \tag{9.24}$$

According to the chain rule for differentiation we have

$$\frac{\partial f(\boldsymbol{x})}{\partial p_l} = \sum_{i=1}^{N} \frac{\partial f(\boldsymbol{x})}{\partial x_i} \frac{\partial x_i}{\partial p_l}, \quad l = 1, \ldots, L \tag{9.25}$$

$$\frac{\partial h_{l'}(\boldsymbol{x})}{\partial p_l} = \sum_{i=1}^{N} \frac{\partial h_{l'}(\boldsymbol{x})}{\partial x_i} \frac{\partial x_i}{\partial p_l}, \quad l, l' = 1, \ldots, L \tag{9.26}$$

The same relations in matrix form are as follows:

$$\left(\nabla_p^T \boldsymbol{x}\right)\left(\nabla f(\boldsymbol{x})\right) = \nabla_p f(\boldsymbol{x}) \tag{9.27}$$

$$\left(\nabla_p^T \boldsymbol{x}\right)\left(\nabla^T \boldsymbol{h}(\boldsymbol{x})\right) = \nabla_p \boldsymbol{h}(\boldsymbol{x}) \tag{9.28}$$

Therefore, Equation 9.24 can be rewritten as follows:

$$\underset{L\times 1}{\left(\nabla_p f(\boldsymbol{x})\right)} + \underset{L\times L}{\left(\nabla_p \boldsymbol{h}(\boldsymbol{x})\right)} \underset{L\times 1}{\boldsymbol{\lambda}} = \boldsymbol{0} \tag{9.29}$$

It follows from Equation 9.20 that

$$\nabla_p \boldsymbol{h}(\boldsymbol{x}) = I \tag{9.30}$$

where I is the $L \times L$ identity matrix. Equation 9.29 thus becomes

$$\nabla_p f(\boldsymbol{x}) + \boldsymbol{\lambda} = \boldsymbol{0} \tag{9.31}$$

or equivalently,

$$-\frac{\partial f(\boldsymbol{x})}{\partial p_l} = \lambda_l, \quad l = 1, \ldots, L \tag{9.32}$$

This means that the Lagrange multipliers indicate the rate of decrease in the optimal value of the objective function upon a unit increase in the right-hand side of the equality constraints. In other words, they quantify the sensitivity of the objective function optimum value with respect to perturbations in the right-hand side of the equality constraints.

Revisiting Example 9.2, we have

$$-\frac{\partial f(\boldsymbol{x})}{\partial p} = \lambda^* = -p \quad \text{or} \quad \frac{\partial f(\boldsymbol{x})}{\partial p} = p > 0$$

indicating that $f(\boldsymbol{x}^*)$ increases (gets worse) as p increases (see Fig. 9.3). Note that if the Lagrange multiplier is equal to zero, then the solution to the constrained problem would be insensitive to the equality constraint and identical to that of the unconstrained problem.

Existence of the Lagrange Multipliers for NLPs with Equality Constraints Lagrange multipliers exist only when the square set of equations (Eqs. 9.16 and 9.17) has a solution. This system of equations can be expressed in vector form as follows:

$$\nabla f(\boldsymbol{x}) + \nabla^T \boldsymbol{h}(\boldsymbol{x})\boldsymbol{\lambda} = \boldsymbol{0} \tag{9.33}$$

$$\boldsymbol{h}(\boldsymbol{x}) = \boldsymbol{0} \tag{9.34}$$

It can be shown that the Lagrange multipliers may not exist if the rows of $\nabla \boldsymbol{h}(\boldsymbol{x})$ are linearly dependent [9]. Therefore, the existence condition for Lagrange multipliers

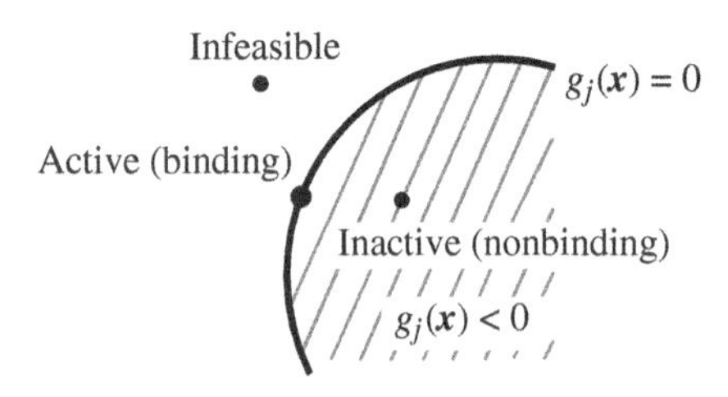

FIGURE 9.4 An example of an infeasible, active and inactive inequality constraint for different points.

requires that matrix $\nabla \boldsymbol{h}(\boldsymbol{x})$ is of full-rank demanding that the set of equations defined by $\nabla \boldsymbol{h}(x)$ must be linearly independent. Any feasible point for which this condition holds is called a *regular* point.

9.2.2 Nonlinear Problems with Equality and Inequality Constraints

Problem [NLP] is the general form of a nonlinear optimization problem with both equality and inequality constraints. Inequality constraints are either active (binding) in which case they can be treated as an equality constraint or inactive (nonbinding) which implies that they can be disregarded (Fig. 9.4):

$$g_j(x) \leq 0 \quad \text{is} \begin{cases} \text{Active (binding) at } \boldsymbol{x} = \overline{\boldsymbol{x}} \text{ if } g_j(\overline{\boldsymbol{x}}) = 0 \\ \text{Inactive (nonbinding) at } \boldsymbol{x} = \overline{\boldsymbol{x}} \text{ if } g_j(\overline{\boldsymbol{x}}) < 0 \end{cases}$$

If the activity or inactivity of the inequality constraints at the optimum solution point was known *a priori*, then the inactive constraints could simply be dropped and the active constraints could be set as equalities. This would transform the problem to a nonlinear optimization problem with only equality constraints.

The general Lagrange function associated problem [NLP] is defined as follows:

$$L(\boldsymbol{x},\boldsymbol{\lambda},\boldsymbol{\mu}) = f(\boldsymbol{x}) + \sum_{j=1}^{M} \mu_j g_j(\boldsymbol{x}) + \sum_{l=1}^{L} \lambda_l h_l(\boldsymbol{x}) \tag{9.35}$$

Here element μ_j of vector $\boldsymbol{\mu} = [\mu_1, \mu_2, \ldots, \mu_M]^{\mathrm{T}}$ is the *Lagrange multiplier* associated with inequality constraint $g_j(\boldsymbol{x}) = 0$. The following two cases must be considered when assessing the existence of the contribution to the Lagrange function by the inequality constraints at a point $\boldsymbol{x} = \overline{\boldsymbol{x}}$:

(i) If $g_j(\overline{\boldsymbol{x}}_j) < 0$ then multiplier μ_j is set to zero and $\mu_j g_j(\overline{\boldsymbol{x}}) = 0$
(ii) If $g_j(\overline{\boldsymbol{x}}) = 0$ then $\mu_j > 0$ and again $\mu_j g_j(\overline{\boldsymbol{x}}) = 0$

This implies that for every feasible point $\boldsymbol{x}$, term $\mu_j g_j(\boldsymbol{x})$ is equal to zero. This is referred to as the *complementary slackness* condition.

Economic Interpretation of the Lagrange Multipliers for Inequality Constraints In analogy to equality constraints, the local sensitivity of the objective function with respect to a perturbation p_j on the right-hand side of the jth inequality constraint is equal to

$$\frac{\partial f(x)}{\partial p_j} = -\mu_j, \quad j = 1,\ldots,M \tag{9.36}$$

Therefore, the Lagrange multipliers for both equality and inequality constraints represent the *shadow prices* for one unit of increase on the right-hand sides of the constraints. This is akin to the interpretation of the dual variables for an LP problem (see Chapter 2). If the right-hand side of the jth inequality constraint $g_j(x) \le 0$ is perturbed by a small positive value $p_j > 0$ (see Fig. 9.5), then the minimum of $f(x)$ cannot increase in value as the feasible region expands in size. Therefore,

$$\frac{\partial f(x)}{\partial p_j} \le 0, \quad j = 1,\ldots,M \tag{9.37}$$

Equation 9.36 together with Inequality 9.37 imply that the Lagrange multiplier associated with the jth inequality constraint μ_j must be nonnegative $(\mu_j \ge 0)$. Nonnegativity of Lagrange multipliers associated with inequality constraints is a requirement arising only for inequalities.

9.2.3 Karush–Kuhn–Tucker Optimality Conditions

Karush–Kuhn–Tucker (KKT) conditions generalize the method of Lagrange multipliers to establish the optimality conditions for problems involving both equality and inequality constraints.

KKT Necessary Conditions Let $\bar{x}$ be a local minimum of problem [NLP] where $f(x)$, $g_j(x)$ and $h_l(x)$ are continuously differentiable at $\bar{x}$. Also, assume that $\bar{x}$ is a regular point (i.e., $\nabla g_j(\bar{x})$, $j = 1,2,\ldots,M$ and $\nabla h_l(\bar{x})$, $l = 1,2,\ldots,L$ are

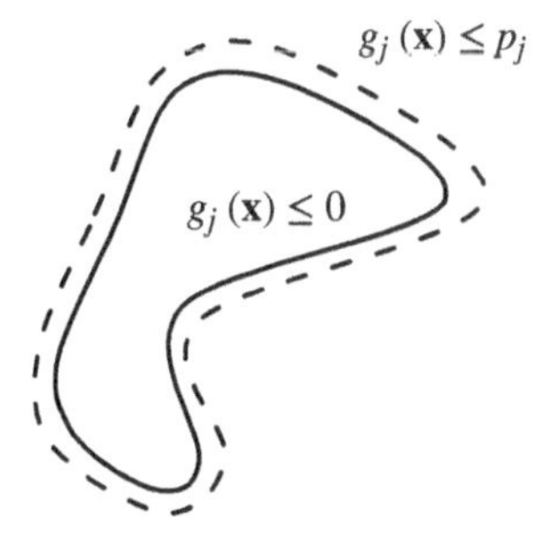

FIGURE 9.5 Perturbation of the right-hand side of an inequality constraint with a positive value p_j expands the size of the feasible region.

linearly independent). Then there exist unique multipliers μ_j, $j = 1,2,\ldots,M$ and λ_l, $l = 1,2,\ldots,L$ such that

$$\nabla f(\bar{x}) + \sum_{j=1}^{M} \mu_j \nabla g_j(\bar{x}) + \sum_{l=1}^{L} \lambda_l \nabla h_l(\bar{x}) = 0 \tag{9.38}$$

$$g_j(\bar{x}) \le 0, \quad j = 1,\ldots,M \tag{9.39}$$

$$h_l(\bar{x}) = 0, \quad l = 1,\ldots,L \tag{9.40}$$

$$\mu_j g_j(\bar{x}) = 0, \quad j = 1,\ldots,M \tag{9.41}$$

$$\mu_j \ge 0, \quad j = 1,\ldots,M \tag{9.42}$$

Constraints 9.39 and 9.40 require $\bar{x}$ to be feasible for problem [NLP] and are called *primal feasibility* (PF) conditions. Constraints 9.38 and 9.42 are referred to as *dual feasibility* (DF) conditions. Constraint 9.41 is the *complementary slackness* (CS) condition. Conditions PF, DF and CS together form the KKT necessary conditions. Any point $\bar{x}$ that satisfies PF, DF and CS is called a *KKT point*. Similarly to a critical point in unconstrained NLP, not every KKT point is necessarily a local minimum of [NLP]. KKT sufficient conditions describe the required conditions for the optimality of a KKT point.

KKT Sufficient Conditions Let $\bar{x}$ be a feasible solution to problem [NLP] where $f(x)$, $g_j(x)$ and $h_l(x)$ are continuously differentiable at $\bar{x}$. Also, assume that $f(x)$ and $g_j(x)$, $j = 1,2,\ldots,M$ are convex for all x and that $h_l(x)$, $l = 1,2,\ldots,L$ are linear. If $\bar{x}$ satisfies the KKT conditions, then it is a global minimum of problem [NLP]. Furthermore, if the convexity assumptions of $f(x)$ and $g_j(x)$ are restricted to an ε. Neighborhood of $\bar{x}$ (i.e., $N_\varepsilon(\bar{x})$ for some $\varepsilon > 0$), then $\bar{x}$ is a local minimum of [NLP].

Note that Equation 9.38 can be rearranged as follows:

$$-\nabla f(x) = \sum_{j=1}^{M} \mu_j \nabla g_j(x) + \sum_{l=1}^{L} \lambda_l \nabla h_l(x) \tag{9.43}$$

This implies that at an optimal point, the direction of steepest descent $-\nabla f(x)$ can be expressed as a linear combination of the gradients of the binding inequality constraints and the gradients of the equality constraints. It is interesting to note that the DF conditions for an LP problem correspond to the constraints of its dual formulation. In addition, imposing the PF and CS conditions for an LP problem requires that the solution of the resulting system of equations be optimal for both primal and dual problems.

There is a variety of methods for solving constrained NLPs. In this book, we focus on two popular classes of deterministic methods. The first one solves the KKT conditions using an active set strategy, while the second relies on the use of reduced gradients.

9.2.4 Sequential (Successive) Quadratic Programming

Sequential (successive) quadratic programming (SQP) is an iterative algorithm, which employs Newton–Raphson's method to solve the system of equations representing the KKT point of the NLP problem. The iterative solution of the KKT conditions proceeds through the minimization of the quadratic approximation of the Lagrange function subject to the linear approximation of the constraints of the original problem. At each iteration, a subset of inequality constraints is deemed to be active and treated as equality constraints while the remaining are judged inactive and therefore are disregarded. Here, we first present the SQP algorithm for an equality-only constrained optimization problem. An extension for problems containing both equality and inequality constraints is described next.

Problems Containing Only Equality Constraints Consider problem [NLPEQ] composed of functions that are continuous and twice differentiable. The KKT conditions for this problem are as follows in vector form:

$$\nabla_x L(\boldsymbol{x},\boldsymbol{\lambda}) = \nabla f(\boldsymbol{x}) + \nabla^{\mathrm{T}} \boldsymbol{h}(\boldsymbol{x})\boldsymbol{\lambda} = \boldsymbol{0} \tag{9.44}$$

$$\nabla_\lambda L(\boldsymbol{x},\boldsymbol{\lambda}) = \boldsymbol{h}(\boldsymbol{x}) = \boldsymbol{0} \tag{9.45}$$

This system of equations is solved using Newton–Raphson's method involving the following update formula for iteration point $\boldsymbol{y}_k = \begin{bmatrix} \boldsymbol{x}_k \\ \boldsymbol{\lambda}_k \end{bmatrix}$ (see Eq. 9.7):

$$\nabla L(\boldsymbol{y}_k) + \nabla^2 L(\boldsymbol{y})(\boldsymbol{y} - \boldsymbol{y}_k) = \boldsymbol{0} \tag{9.46}$$

where

$$\nabla L(\boldsymbol{y}) = \begin{bmatrix} \nabla_x L \\ \nabla_\lambda L \end{bmatrix} = \begin{bmatrix} \nabla f(\boldsymbol{x}) + \nabla^{\mathrm{T}} \boldsymbol{h}(\boldsymbol{x})\boldsymbol{\lambda}_k \\ h(\boldsymbol{x}) \end{bmatrix} \tag{9.47}$$

$$\nabla^2 L(\boldsymbol{y}) = \begin{bmatrix} \nabla_x(\nabla_x L(\boldsymbol{x},\boldsymbol{\lambda})) & \nabla_x(\nabla_\lambda L(\boldsymbol{x},\boldsymbol{\lambda})) \\ \nabla_\lambda(\nabla_x L(\boldsymbol{x},\boldsymbol{\lambda})) & \nabla_\lambda(\nabla_\lambda L(\boldsymbol{x},\boldsymbol{\lambda})) \end{bmatrix} = \begin{bmatrix} \nabla^2 L(\boldsymbol{x},\boldsymbol{\lambda}) & \nabla^{\mathrm{T}} \boldsymbol{h}(\boldsymbol{x}) \\ \nabla \boldsymbol{h}(\boldsymbol{x}) & 0 \end{bmatrix} \tag{9.48}$$

Here, $\nabla^2 L(\boldsymbol{x},\boldsymbol{\lambda})$ denotes the Hesian matrix of the Lagrange function. Therefore, Equation 9.46 can be expanded as follows:

$$\begin{bmatrix} \nabla f(\boldsymbol{x}_k) + \nabla^{\mathrm{T}} \boldsymbol{h}(\boldsymbol{x}_k)\boldsymbol{\lambda}_k \\ \boldsymbol{h}(\boldsymbol{x}_k) \end{bmatrix} + \begin{bmatrix} \nabla^2 L(\boldsymbol{x}_k,\boldsymbol{\lambda}_k) & \nabla^{\mathrm{T}} \boldsymbol{h}(\boldsymbol{x}) \\ \nabla \boldsymbol{h}(\boldsymbol{x}_k) & \boldsymbol{0} \end{bmatrix} \begin{bmatrix} \boldsymbol{x} - \boldsymbol{x}_k \\ \boldsymbol{\lambda} - \boldsymbol{\lambda}_k \end{bmatrix} = \begin{bmatrix} \boldsymbol{0} \\ \boldsymbol{0} \end{bmatrix} \tag{9.49}$$

Note that the dimensions of $\boldsymbol{x}$, $\boldsymbol{\lambda}$, $\boldsymbol{h}(\boldsymbol{x}_k)$, $\nabla f(\boldsymbol{x}_k)$, $\nabla \boldsymbol{h}(\boldsymbol{x}_k)$ and $\nabla^2 L(\boldsymbol{x}_k,\boldsymbol{\lambda}_k)$ are $N \times 1$, $L \times 1$, $L \times 1$, $N \times 1$, $L \times N$ and $N \times N$, respectively. Equation 9.49 can be expressed as the following system of equations:

$$\nabla^2 L(\boldsymbol{x}_k,\boldsymbol{\lambda}_k)(\boldsymbol{x} - \boldsymbol{x}_k) + \nabla^{\mathrm{T}} \boldsymbol{h}(\boldsymbol{x})(\boldsymbol{\lambda} - \boldsymbol{\lambda}_k) = -\nabla f(\boldsymbol{x}_k) - \nabla^{\mathrm{T}} \boldsymbol{h}(\boldsymbol{x})\boldsymbol{\lambda}_k \tag{9.50}$$

$$\nabla \boldsymbol{h}(\boldsymbol{x}_k)(\boldsymbol{x} - \boldsymbol{x}_k) = -\boldsymbol{h}(\boldsymbol{x}_k) \tag{9.51}$$

By defining direction $\boldsymbol{d} = \boldsymbol{x} - \boldsymbol{x}_k$, the system of equations is simplified as follows:

$$\nabla^2 L(\boldsymbol{x}_k, \boldsymbol{\lambda}_k)\boldsymbol{d} + \nabla^{\mathrm{T}}\boldsymbol{h}(\boldsymbol{x})\boldsymbol{\lambda} = -\nabla f(\boldsymbol{x}_k) \tag{9.52}$$

$$\nabla \boldsymbol{h}(\boldsymbol{x}_k)\boldsymbol{d} = -\boldsymbol{h}(\boldsymbol{x}_k) \tag{9.53}$$

Therefore, the recursive equation for solving the KKT conditions in matrix form are as follows:

$$\begin{bmatrix} \nabla^2 L(\boldsymbol{x}_k, \boldsymbol{\lambda}_k) & \nabla^{\mathrm{T}}\boldsymbol{h}(\boldsymbol{x}) \\ \nabla \boldsymbol{h}(\boldsymbol{x}_k) & \boldsymbol{0} \end{bmatrix} \begin{bmatrix} \boldsymbol{d}_k \\ \boldsymbol{\lambda}_{k+1} \end{bmatrix} = \begin{bmatrix} -\nabla f(\boldsymbol{x}_k) \\ -\boldsymbol{h}(\boldsymbol{x}_k) \end{bmatrix} \tag{9.54}$$

This system of linear equations is solved for $(\boldsymbol{d}_k, \boldsymbol{\lambda}_{k+1})$ and the new point is found by setting $\boldsymbol{x}_{k+1} = \boldsymbol{x}_k + \boldsymbol{d}_k$ and increasing k until $\|\boldsymbol{d}_k\| < \varepsilon$.
Now consider the following quadratic programming problem:

$$\underset{\boldsymbol{d}}{\text{minimize}} \quad L(\boldsymbol{x}_k) + \nabla_{\mathrm{x}}^{\mathrm{T}} L(\boldsymbol{x}_k)\boldsymbol{d} + \frac{1}{2}\boldsymbol{d}^{\mathrm{T}}\nabla^2 L(\boldsymbol{x}_k)\boldsymbol{d} \quad [\text{QPEQ}]$$

subject to

$$\boldsymbol{h}(\boldsymbol{x}_k) + \nabla \boldsymbol{h}(\boldsymbol{x}_k)\boldsymbol{d} = \boldsymbol{0}$$

where $\boldsymbol{d} = \boldsymbol{x} - \boldsymbol{x}_k$ as before. Observe that the objective function of this problem is the quadratic approximation of the Lagrange function for [NLPEQ] and the constraint is the linear approximation of the equality constraints of [NLPEQ] at $\boldsymbol{x} = \boldsymbol{x}_k$. By substituting the expressions for $L(\boldsymbol{x}_k)$ and $\nabla_{\mathrm{x}} L(\boldsymbol{x}_k)$ into the objective function of [QPEQ], we obtain:

$$\begin{aligned} & L(\boldsymbol{x}_k) + \nabla_{\mathrm{x}}^{\mathrm{T}} L(\boldsymbol{x}_k)\boldsymbol{d} + \frac{1}{2}\boldsymbol{d}^{\mathrm{T}}\nabla^2 L(\boldsymbol{x}_k)\boldsymbol{d} \\ & = f(\boldsymbol{x}_k) + \boldsymbol{\lambda}^{\mathrm{T}}\boldsymbol{h}(\boldsymbol{x}_k) + \left[\nabla f(\boldsymbol{x}_k) + \nabla^{\mathrm{T}}\boldsymbol{h}(\boldsymbol{x})\boldsymbol{\lambda}\right]^{\mathrm{T}} \boldsymbol{d} + \frac{1}{2}\boldsymbol{d}^{\mathrm{T}}\nabla^2 L(\boldsymbol{x}_k)\boldsymbol{d} \\ & = f(\boldsymbol{x}_k) + \nabla^{\mathrm{T}} f(\boldsymbol{x}_k)\boldsymbol{d} + \boldsymbol{\lambda}^{\mathrm{T}}\left[\boldsymbol{h}(\boldsymbol{x}_k) + \nabla \boldsymbol{h}(\boldsymbol{x}_k)\boldsymbol{d}\right] \\ & + \frac{1}{2}\boldsymbol{d}^{\mathrm{T}}\nabla^2 L(\boldsymbol{x}_k)\boldsymbol{d} \end{aligned} \tag{9.55}$$

Since $\boldsymbol{h}(\boldsymbol{x}_k) + \nabla \boldsymbol{h}(\boldsymbol{x}_k)\boldsymbol{d} = \boldsymbol{0}$, problem [QPEQ] can be simplified as:

$$\underset{\boldsymbol{d}}{\text{minimize}} \quad f(\boldsymbol{x}_k) + \nabla^{\mathrm{T}} f(\boldsymbol{x}_k)\boldsymbol{d} + \frac{1}{2}\boldsymbol{d}^{\mathrm{T}}\nabla^2 L(\boldsymbol{x}_k)\boldsymbol{d} \quad [\text{QPEQ}']$$

subject to

$$\boldsymbol{h}(\boldsymbol{x}_k) + \nabla \boldsymbol{h}(\boldsymbol{x}_k)\boldsymbol{d} = \boldsymbol{0}$$

The Lagrange function for this problem is as follows:

$$L(\boldsymbol{d}) = f(\boldsymbol{x}_k) + \nabla^{\mathrm{T}} f(\boldsymbol{x}_k)\boldsymbol{d} + \frac{1}{2}\boldsymbol{d}^{\mathrm{T}}\nabla^2 L(\boldsymbol{x}_k)\boldsymbol{d} + \left[\boldsymbol{h}(\boldsymbol{x}_k) + \nabla \boldsymbol{h}(\boldsymbol{x}_k)\boldsymbol{d}\right]^{\mathrm{T}} \boldsymbol{\lambda} \tag{9.56}$$

The KKT optimality conditions for this problem are thus

$$\nabla_{\mathrm{d}} L(\boldsymbol{d}) = \nabla f(\boldsymbol{x}_k) + \nabla^2 L(\boldsymbol{x}_k)\boldsymbol{d} + \nabla^{\mathrm{T}} \boldsymbol{h}(\boldsymbol{x}_k)\boldsymbol{\lambda} = \boldsymbol{0} \tag{9.57}$$

$$\nabla_{\lambda} L(\boldsymbol{d}) = \boldsymbol{h}(\boldsymbol{x}_k) + \nabla \boldsymbol{h}(\boldsymbol{x}_k)\boldsymbol{d} = \boldsymbol{0} \tag{9.58}$$

which are equivalent to conditions 9.52 and 9.53. Hence, each iteration of the Newton–Raphson's method for solving the KKT condition for [NLPEQ] is equivalent to solving the quadratic program [QPEQ′]. This means that the original problem [NLPEQ] can be solved by iteratively solving its quadratic approximation [QPEQ′].

Problems Containing Equality and Inequality Constraints Quadratic problem [QPEQ′] for an NLP problem involving both equality and inequality constraints can be stated as follows:

$$\underset{\boldsymbol{d}}{\text{minimize}} \quad f(\boldsymbol{x}_k) + \nabla^{\mathrm{T}} f(\boldsymbol{x}_k)\boldsymbol{d} + \frac{1}{2}\boldsymbol{d}^{\mathrm{T}}\nabla^2 L(\boldsymbol{x}_k)\boldsymbol{d} \quad [\text{QPNLP}]$$

subject to

$$\boldsymbol{g}(\boldsymbol{x}_k) + \nabla \boldsymbol{g}(\boldsymbol{x}_k)\boldsymbol{d} \le \boldsymbol{0}$$
$$\boldsymbol{h}(\boldsymbol{x}_k) + \nabla \boldsymbol{h}(\boldsymbol{x}_k)\boldsymbol{d} = \boldsymbol{0}$$

The KKT conditions for this problem are as follows:

$$\nabla f(\boldsymbol{x}_k) + \nabla^2 L(\boldsymbol{x}_k)\boldsymbol{d} + \nabla^{\mathrm{T}} \boldsymbol{g}_A(\boldsymbol{x}_k)\boldsymbol{\mu} + \nabla^{\mathrm{T}} \boldsymbol{h}(\boldsymbol{x}_k)\boldsymbol{\lambda} = \boldsymbol{0} \tag{9.59}$$

$$\boldsymbol{g}(\boldsymbol{x}_k) + \nabla \boldsymbol{g}(\boldsymbol{x}_k)\boldsymbol{d} = \boldsymbol{0} \tag{9.60}$$

$$\boldsymbol{h}(\boldsymbol{x}_k) + \nabla \boldsymbol{h}(\boldsymbol{x}_k)\boldsymbol{d} = \boldsymbol{0} \tag{9.61}$$

Here, $\boldsymbol{g}_A(\boldsymbol{x}_k)$ denotes the set of active inequality constraints at $\boldsymbol{x} = \boldsymbol{x}_k$:

$$A = \{ j \mid g_j(\boldsymbol{x}_k) = 0 \}$$

The CS condition is redundant here as active inequality constraints are predetermined and enforced as equalities. It is important to keep in mind that the set of constraints that are active must be re-adjusted at every iteration. As before, the termination criterion is met if $\|\boldsymbol{d}_k\|$ becomes less than a user-specified threshold. The solution found by SQP will be an optimal solution to [NLP] if the KKT sufficient optimality conditions hold. If the inequality constraints are nonconvex, potential solutions may be lost during the SQP process as shown in Figure 9.6. The SQP algorithm generally requires fewer iterations than the reduced gradient class of algorithms (see next section).

Note that problem [QPNLP] might become unbounded or infeasible even if [NLP] is not. The former can be addressed by imposing bounds on $\boldsymbol{d}$. The latter occurs when the Hessian of the objective function of problem [QPNLP] is not positive

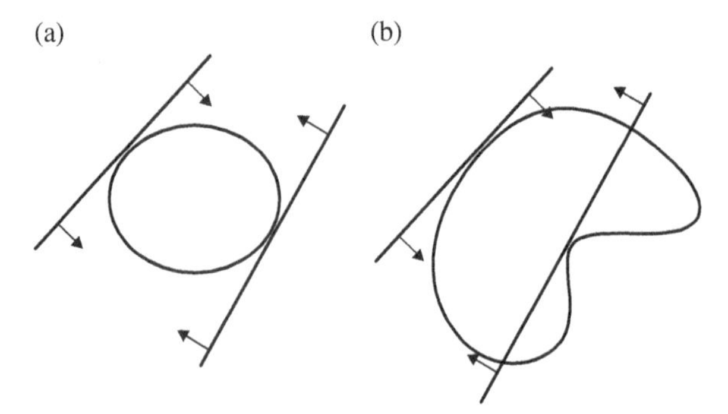

FIGURE 9.6 (a) The feasible region is conserved when using the SQP algorithm for problems with convex inequality constraints. (b) Using the SQP algorithm to solve problems involving nonconvex inequality constraints can lead to a partial loss of the feasible region.

definite (in general $\nabla^2 L(\boldsymbol{x}_k)$ need not be pd). This can be addressed by replacing $\nabla^2 L(\boldsymbol{x}_k)$ in the objective function with a quasi-Newton positive definite approximation (see Ref. [11] for details). In addition, the SQP procedure may not necessarily converge with a step size of one ($\boldsymbol{x}_{k+1} = \boldsymbol{x}_k + \boldsymbol{d}$), even if combined with the quasi-Newton method. This situation can be remedied by performing a line search after finding the improving direction $\boldsymbol{d}$. An efficient implementation of the SQP algorithm is accessible through the fmincon function (under the active-set algorithms) in the nonlinear optimization toolbox of MATLAB and through the NLP solvers SNOPT and CONOPT in GAMS.

9.2.5 Generalized Reduced Gradient

The generalized reduced gradient (GRG) method was originally proposed for NLP problems with equality constraints and was later generalized for problems with both equality and inequality constraints. The basic idea of this method is to convert the original problem into an unconstrained optimization subproblem in a reduced dimensional space. To this end, a linear approximation of the nonlinear constraints at each iteration is used to express a subset of variables as a function of another subset of independent variables. The solution of this subproblem in the reduced dimensional space can be shown to be a KKT point of the original problem. This subproblem is solved iteratively until a KKT point is identified.

Consider problem [NLPEQ] in the matrix form as follows:

$$
\begin{aligned}
&\text{minimize}\, f(\boldsymbol{x}) \quad [\text{NLPEQ}] \\
&\text{subject to} \\
&\boldsymbol{h}(\boldsymbol{x}) = 0 \\
&\boldsymbol{x} \in \mathbb{R}^N
\end{aligned}
$$

where $\boldsymbol{h}(\boldsymbol{x}) = [h_1(\boldsymbol{x}), h_2(\boldsymbol{x}), \ldots, h_L(\boldsymbol{x})]^{\mathrm{T}}$. All inequality constraints are assumed to have been transformed to equality constraints using slack variables (see Chapter 2). Now consider the linear approximation of the equality constraints at a feasible point $\boldsymbol{x}_k$:

$$\begin{aligned}&\text{minimize } f(\boldsymbol{x}) \quad [\text{NLPEQ}'] \\ &\text{subject to} \\ &\boldsymbol{h}(\boldsymbol{x}) = \boldsymbol{h}(\boldsymbol{x}_k) + \nabla \boldsymbol{h}(\boldsymbol{x}_k)(\boldsymbol{x} - \boldsymbol{x}_k) = \boldsymbol{0} \\ &\boldsymbol{x} \in \mathbb{R}^N\end{aligned} \tag{9.62}$$

Note that $\boldsymbol{h}(\boldsymbol{x}_k) = \boldsymbol{0}$ since $\boldsymbol{x}_k$ is a feasible solution. This simplifies Equation 9.62 to

$$\nabla \boldsymbol{h}(\boldsymbol{x}_k)\boldsymbol{x} = \nabla \boldsymbol{h}(\boldsymbol{x}_k)\boldsymbol{x}_k \tag{9.63}$$

which is of the form $\boldsymbol{A}\boldsymbol{x} = \boldsymbol{b}$. Consider the partitioning of $\boldsymbol{x}_k$ as $\boldsymbol{x} = \begin{bmatrix} \boldsymbol{x}_B \\ \boldsymbol{x}_N \end{bmatrix}$ and of the gradient $\nabla \boldsymbol{h}(\boldsymbol{x})$ as $[\boldsymbol{B} \quad \boldsymbol{N}]$. Here, $\boldsymbol{x}$, $\boldsymbol{x}_B$ and $\boldsymbol{x}_N$ are $N \times 1$, $L \times 1$ and $(N-L) \times 1$ vectors and $\nabla \boldsymbol{h}(\boldsymbol{x})$, $\boldsymbol{B}$ and $\boldsymbol{N}$ are $L \times N$, $L \times L$ and $L \times (N-L)$ matrices, respectively. In analogy to the terminology used in LP problem (see Chapter 2), $\boldsymbol{x}_B$ and $\boldsymbol{x}_N$ are referred to as basic (dependent) and nonbasic (independent) variables, respectively.

Problem [NLPEQ′] can be converted into an unconstrained optimization problem by expressing the basic variables as a function of the nonbasic variables. To this end, we can rewrite Equation 9.63 as follows:

$$[\boldsymbol{B} \quad \boldsymbol{N}]\begin{bmatrix} \boldsymbol{x}_B \\ \boldsymbol{x}_N \end{bmatrix} = [\boldsymbol{B} \quad \boldsymbol{N}]\begin{bmatrix} \boldsymbol{x}_{B,k} \\ \boldsymbol{x}_{N,k} \end{bmatrix} \tag{9.64}$$

This results in

$$\boldsymbol{x}_B = \boldsymbol{x}_{B,k} - \boldsymbol{B}^{-1}\boldsymbol{N}(\boldsymbol{x}_N - \boldsymbol{x}_{N,k}) \tag{9.65}$$

By substituting $\boldsymbol{x}_B$ from this equation in the objective function of [NLPEQ′], it is converted to the following unconstrained NLP problem in the reduced $(N-L)$ dimensional space

$$\underset{\boldsymbol{x}_N \in \mathbb{R}^{N-L}}{\text{minimize}} f\left(\left[\boldsymbol{x}_B = \boldsymbol{x}_{B,k} - \boldsymbol{B}^{-1}\boldsymbol{N}(\boldsymbol{x}_N - \boldsymbol{x}_{N,k}) \quad \boldsymbol{x}_N\right]\right) \quad [\text{NLPEQ}'']$$

The critical point of the objective function for this unconstrained problem with $\boldsymbol{x}_N$ as independent variables is found by setting the gradient of f with respect to $\boldsymbol{x}_N$ to zero:

$$\nabla_N f(\boldsymbol{x}_N) = \boldsymbol{0} \tag{9.66}$$

From the chain rule for differentiation:

$$\nabla_B^{\mathrm{T}} f(\boldsymbol{x})(\nabla_N \boldsymbol{x}_B) + \nabla_N^{\mathrm{T}} f(\boldsymbol{x})(\nabla_N \boldsymbol{x}_N) = \boldsymbol{0} \tag{9.67}$$

where $\boldsymbol{x} = [\boldsymbol{x}_B \; \boldsymbol{x}_N]$ and accordingly $\nabla = [\nabla_{\mathrm{B}} \; \nabla_{\mathrm{N}}]$. From Equation 6.65, we have $\nabla_{\mathrm{N}} \boldsymbol{x}_B = -\boldsymbol{B}^{-1}\boldsymbol{N}$. Therefore, Equation 9.67 is reduced to

$$\nabla_N^{\mathrm{T}} f(\boldsymbol{x}) - \nabla_B^{\mathrm{T}} f(\boldsymbol{x}) \boldsymbol{B}^{-1}\boldsymbol{N} = \boldsymbol{0} \tag{9.68}$$

The left-hand side expression of Equation 9.68 is called the *reduced gradient* $(\boldsymbol{r}_g)$, which is in essence the projection of the gradient of the original N-dimensional problem onto the $(N-L)$ dimensional space:

$$\boldsymbol{r}_g = \nabla_N^{\mathrm{T}} f(\boldsymbol{x}) - \nabla_B^{\mathrm{T}} f(\boldsymbol{x}) \boldsymbol{B}^{-1}\boldsymbol{N} \tag{9.69}$$

Next, we show that a critical point of the unconstrained problem [NLPEQ″] in the reduced dimensional space (i.e., a point satisfying $\boldsymbol{r}_g = \boldsymbol{0}$) is equivalent to a KKT point of [NLPEQ]. The KKT conditions for [NLPEQ] can be written as follows:

$$\nabla f(\boldsymbol{x}) + \nabla^{\mathrm{T}} \boldsymbol{h}(\boldsymbol{x}) \boldsymbol{\lambda} = \boldsymbol{0} \tag{9.70}$$

Upon partitioning, $\nabla f(\boldsymbol{x}) = \begin{bmatrix} \nabla_B f(\boldsymbol{x}) \\ \nabla_N f(\boldsymbol{x}) \end{bmatrix}$ where $\nabla_B f(\boldsymbol{x})$ and $\nabla_N f(\boldsymbol{x})$ are $L \times 1$ and $(N-L) \times 1$ vectors denoting the gradient of f with respect to $\boldsymbol{x}_B$ and $\boldsymbol{x}_N$, respectively. Equation 9.70 becomes

$$\begin{bmatrix} \nabla_B f(\boldsymbol{x}) \\ \nabla_N f(\boldsymbol{x}) \end{bmatrix} + \begin{bmatrix} \boldsymbol{B}^{\mathrm{T}} \\ \boldsymbol{N}^{\mathrm{T}} \end{bmatrix} \boldsymbol{\lambda} = \begin{bmatrix} \boldsymbol{0} \\ \boldsymbol{0} \end{bmatrix} \tag{9.71}$$

From the equation in the first row it follows that

$$\boldsymbol{\lambda} = -\left(\boldsymbol{B}^{\mathrm{T}}\right)^{-1} \nabla_B f(\boldsymbol{x}) \tag{9.72}$$

By substituting $\boldsymbol{\lambda}$ into the equation in the second row of Equation 9.71, we derive

$$\nabla_N f(\boldsymbol{x}) + \boldsymbol{N}^{\mathrm{T}} \left(-\left(\boldsymbol{B}^{\mathrm{T}}\right)^{-1} \nabla_B f(\boldsymbol{x}) \right) = \boldsymbol{0} \tag{9.73}$$

By transposing both terms in Equation 9.73, we have

$$\nabla_{\mathrm{N}}^{\mathrm{T}} f(\boldsymbol{x}) - \nabla_{\mathrm{B}}^{\mathrm{T}} f(\boldsymbol{x}) \boldsymbol{B}^{-1}\boldsymbol{N} = \boldsymbol{0} \tag{9.74}$$

whose left-hand side is identical to $\boldsymbol{r}_g$ (see Eqs. 9.68 and 9.69). Hence, we demonstrated that imposing $\boldsymbol{r}_g = \boldsymbol{0}$ for the unconstrained problem [NLPEQ″] in the reduced dimensional space is equivalent to the KKT condition for problem [NLPEQ″]. This implies that [NLPEQ] can be solved by solving [NLPEQ″]. Usually, a steepest descent method is used to solve the unconstraint optimization problem [NLPEQ″]. Iterations end when $\boldsymbol{d} = \boldsymbol{r}_g = \boldsymbol{0}$ or $\|\boldsymbol{d}\| = \|\boldsymbol{r}_g\|$ is less than a user-specified threshold. Note that a line search may be employed at each iteration to obtain the optimal step size. The GRG algorithm is particularly effective for problems with many linear equality constraints and only a few nonlinear constraints. It is accessible through the NLP solvers CONOPT, LINDO and LGO in GAMS.

9.3 LAGRANGIAN DUALITY THEORY

In this section, we introduce the Lagrangian duality theory, which plays an important role in the theoretical development of many NLP solution algorithms as well as those of mixed-integer nonlinear programming (MINLP) problems (see Chapter 11). Consider problem [NLP] again:

$$\begin{aligned}
&\text{minimize } f(\boldsymbol{x}) \quad [\text{NLP}]\\
&\text{subject to}\\
&g_j(\boldsymbol{x}) \le 0, \quad j = 1,2,\ldots,M\\
&h_l(\boldsymbol{x}) = 0, \quad l = 1,2,\ldots,L\\
&\boldsymbol{x} \in \mathbb{R}^N
\end{aligned}$$

This problem is referred to as the *primal problem*. The *Lagrange relaxation* of this problem minimizes the Lagrange function with respect to $\boldsymbol{x}$:

$$\theta(\boldsymbol{\lambda},\boldsymbol{\mu}) = \underset{\boldsymbol{x}}{\text{minimize}}\ L(\boldsymbol{x},\boldsymbol{\lambda},\boldsymbol{\mu}) = \underset{\boldsymbol{x}}{\text{minimize}}\ f(\boldsymbol{x}) + \sum_{j=1}^{M} \mu_j g_j(\boldsymbol{x}) + \sum_{l=1}^{L} \lambda_l h_l(\boldsymbol{x}) \tag{9.75}$$

Here, $\boldsymbol{\mu} = [\mu_1, \mu_2, \ldots, \mu_M]^{\mathrm{T}}$ and $\boldsymbol{\lambda} = [\lambda_1, \lambda_2, \ldots, \lambda_L]^{\mathrm{T}}$ are the Lagrange multiplier vectors. $\theta(\boldsymbol{\lambda},\boldsymbol{\mu})$ is referred to as the dual function. The Lagrangian dual of the primal problem [NLP] maximizes $\theta(\boldsymbol{\lambda},\boldsymbol{\mu})$ over $\boldsymbol{\lambda}$ and $\boldsymbol{\mu}$ and is formulated as follows:

$$\begin{aligned}
&\underset{\boldsymbol{\lambda},\boldsymbol{\mu}}{\text{maximize}}\ \theta(\boldsymbol{\lambda},\boldsymbol{\mu}) \quad [\text{NLPD}]\\
&\text{subject to}\\
&\boldsymbol{\mu} \ge \boldsymbol{0}
\end{aligned}$$

By replacing $\theta(\boldsymbol{\lambda},\boldsymbol{\mu})$ from Equation 9.75:

$$\begin{aligned}
&\underset{\boldsymbol{\lambda},\boldsymbol{\mu}}{\text{maximize}}\quad \underset{\boldsymbol{x}}{\text{minimize}}\quad f(\boldsymbol{x}) + \sum_{j=1}^{M} \mu_j g_j(\boldsymbol{x}) + \sum_{l=1}^{L} \lambda_l h_l(\boldsymbol{x}) \quad [\text{NLPD}]\\
&\text{subject to}\\
&\boldsymbol{\mu} \ge \boldsymbol{0}
\end{aligned}$$

Note that the inner minimization problem is parametric in both $\boldsymbol{\lambda}$ and $\boldsymbol{\mu}$. It can become unbounded or a minimum may not always exist in which case the operator "minimize" should be replaced with the *infimum* (the greatest lower bound) operator. Similarly, the outer maximization problem can be unbounded or a maximum may not always exist in which case "maximize" should be replaced with *supremum* (the smallest upper bound). The Lagrangian dual problem can be solved using the cutting plane (or outer linearization) method (see Ref. [9]).

9.3.1 Relationships between the Primal and Dual Problems

The primal and dual problems are closely related as seen before for LP problems (see Chapter 2). Their relationships can be expressed in terms of the weak and strong duality theorems.

Theorem 9.4: Weak duality theorem

Let $\boldsymbol{x}$ be a feasible solution to the primal problem [NLP] and a feasible solution to the dual problem [NLPD]. Then the objective function of the primal problem evaluated at $\boldsymbol{x}$ is always greater than or equal to the objective function of the dual problem evaluated at $(\boldsymbol{\lambda},\boldsymbol{\mu})$, that is, $f(\boldsymbol{x}) \geq \theta(\boldsymbol{\lambda},\boldsymbol{\mu})$. If $f(\boldsymbol{x}) > \theta(\boldsymbol{\lambda},\boldsymbol{\mu})$ then the difference $f(\boldsymbol{x}) - \theta(\boldsymbol{\lambda},\boldsymbol{\mu})$ is referred to as the *duality gap*.

Theorem 9.5: Strong duality theorem

Let $\boldsymbol{x} \in X$ and X be a nonempty convex set in $\mathbb{R}^N$, $f(\boldsymbol{x})$ and $\boldsymbol{g}(\boldsymbol{x})$ be convex and $\boldsymbol{h}(\boldsymbol{x})$ be linear functions. Then the optimal objective function values of the primal and dual problems are equal.

Note that if the conditions of the strong duality theorem hold, then it is possible to solve the primal problem indirectly by solving the dual problem. The following are direct implications of the weak and strong duality theorems [9, 12]:

- If the objective function value of the primal problem $f(\boldsymbol{x})$ at a feasible point $\boldsymbol{x}$ is equal to the objective function value of the dual $\theta(\boldsymbol{\lambda},\boldsymbol{\mu})$ at a feasible point $(\boldsymbol{\lambda},\boldsymbol{\mu})$, then $\boldsymbol{x}$ solves the primal problem and $(\boldsymbol{\lambda},\boldsymbol{\mu})$ solves the dual problem.
- If $(\boldsymbol{\lambda},\boldsymbol{\mu})$ are optimal for the dual problem, then $\boldsymbol{x}$ is optimal to the primal problem, if and only if $(\boldsymbol{x}, \boldsymbol{\lambda}, \boldsymbol{\mu})$ satisfies the KKT optimality conditions of the primal problem.
- If the dual problem is unbounded then the primal problem is infeasible.
- If the primal problem is unbounded (the optimal objective function value diverges to $-\infty$) then $\theta(\boldsymbol{\lambda},\boldsymbol{\mu})$ also diverges to $-\infty$ for every $\boldsymbol{\mu} > \mathbf{0}$.

EXERCISES

9.1 Consider the problem

$$\text{minimize} \quad z = \sum_{i=1}^{N} x_i$$

subject to

$$\sum_{i=1}^{N} \frac{x_i^2}{K_i} \leq D$$

$$x_i \geq 0, \quad i = 1,2,\ldots,N$$

where K_i and D are positive constants.

(a) Write the KKT conditions for this problem.

(b) Show that a solution to the KKT conditions gives an optimal solution to the NLP problem.

(c) Find the optimal solution. (Is it unique?)

9.2 Consider a hyperplane with equation $\boldsymbol{Ax} = \boldsymbol{b}$, where $\boldsymbol{x}$ is an N-dimensional vector, $\boldsymbol{A}$ is a $M \times N$ matrix with $M < N$ and $\boldsymbol{b}$ is a $M \times 1$ vector. Consider a *given* point z outside the hyperplane.

(a) Formulate the task of identifying the projection of point z on the hyperplane as a constrained optimization problem.

(b) By applying the Lagrange multipliers method, identify the coordinates of the projection point $\boldsymbol{x}^*$ of point z on the hyperplane with equation $\boldsymbol{Ax} = \boldsymbol{b}$.

9.3 BioGen Inc. is planning to install a number of fermentors in an existing square floorplan whose side is L meters. Each prefabricated fermentor has a circular footprint with a diameter of $0.29L$. Is it possible to fit 10 reactors in the plant? How about 11? Plot the arrangements and give your recommendations to your floor manager.

REFERENCES

1. Segrè D, Vitkup D, Church GM: Analysis of optimality in natural and perturbed metabolic networks. *Proc Natl Acad Sci U S A* 2002, **99**(23):15112–15117.
2. Pozo C, Miro A, Guillen-Gosalbez G, Sorribas A, Alves R, Jimenez L: Gobal optimization of hybrid kinetic/FBA models via outer-approximation. *Computers & Chemical Engineering* 2015, **72**:325–333.
3. Mahadevan R, Edwards JS, Doyle FJ: Dynamic flux balance analysis of diauxic growth in *Escherichia coli*. *Biophys J* 2002, **83**(3):1331–1340.
4. Sorribas A, Pozo C, Vilaprinyo E, Guillén-Gosálbez G, Jiménez L, Alves R: Optimization and evolution in metabolic pathways: global optimization techniques in Generalized Mass Action models. *J Biotechnol* 2010, **149**(3):141–153.
5. Pozo C, Marín-Sanguino A, Alves R, Guillén-Gosálbez G, Jiménez L, Sorribas A: Steady-state global optimization of metabolic non-linear dynamic models through recasting into power-law canonical models. *BMC Syst Biol* 2011, **5**:137.
6. Suthers PF, Burgard AP, Dasika MS, Nowroozi F, Van Dien S, Keasling JD, Maranas CD: Metabolic flux elucidation for large-scale models using ^{13}C labeled isotopes. *Metab Eng* 2007, **9**(5–6):387–405.
7. Chang Y, Suthers PF, Maranas CD: Identification of optimal measurement sets for complete flux elucidation in metabolic flux analysis experiments. *Biotechnol Bioeng* 2008, **100**(6): 1039–1049.
8. Riascos CAM, Gombert AK, Pinto JM: A global optimization approach for metabolic flux analysis based on labeling balances. *Comput Chem Eng* 2005, **29**(3):447–458.
9. Bazaraa MS, Sherali HD, Shetty CM: *Nonlinear programming: theory and algorithms*, 3rd edn. Hoboken, N.J.: Wiley-Interscience; 2006.

10. Rardin RL: *Optimization in operations research.* Upper Saddle River, N.J.: Prentice Hall; 1998.
11. Edgar TF, Himmelblau DM, Lasdon LS: *Optimization of chemical processes*, 2nd edn. New York: McGraw-Hill; 2001.
12. Floudas CA: *Nonlinear and mixed-integer optimization: fundamentals and applications.* New York: Oxford University Press; 1995.

10

NLP APPLICATIONS IN METABOLIC NETWORKS

In this chapter, we discuss three examples of nonlinear optimization in the analysis of metabolic networks. The first example replaces the biomass flux maximization in the flux balance analysis (FBA) of metabolic networks (see Chapter 3) with a nonlinear objective function enforcing the minimization of metabolic adjustment (MOMA). The presence of the nonlinear objective function requires the use of the Karush–Kuhn–Tucker (KKT) optimality conditions for establishing a set of constraints describing the optimal solution. The second example introduces kinetic expressions to quantify the flux of a subset of the reactions in a stoichiometric metabolic model. The last example considers a least squares minimization problem arising in metabolic flux analysis (MFA) involving a nonlinear objective function and constraints.

10.1 MINIMIZATION OF THE METABOLIC ADJUSTMENT

The minimization of metabolic adjustment (MOMA) method relies on the hypothesis that the metabolic flux distribution in a knockout mutant strain undergoes minimal redistribution upon perturbation with respect to original fluxes in the wild-type strain [1] (see Fig. 10.1). Predicted flux distributions by MOMA have a generally higher agreement with experimental flux and growth rate data in knockout mutant strains of *Escherichia coli* compared to those predicted by FBA [1]. Following MOMA, a number of other hypothesized objective functions have been introduced to better explain flux redistribution in mutants using optimality principles [2–5].

MOMA can be formulated as a quadratic programming (QP) problem, where the sum of the squared differences between the fluxes after a gene knockout and

Optimization Methods in Metabolic Networks, First Edition. Costas D. Maranas and Ali R. Zomorrodi.

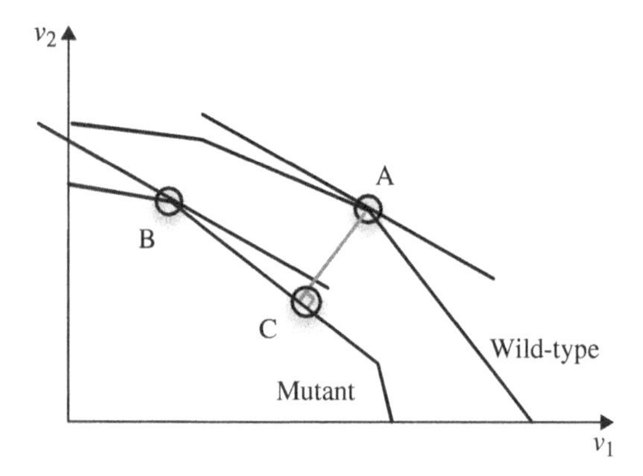

FIGURE 10.1 Comparison of FBA and MOMA for a simple network with only two reactions. The area below the piecewise linear lines represents the feasible regions of the wild-type and a knockout mutant strain. Points A and B denote the maximum biomass solution from FBA for the wild-type and mutant strain, respectively. Point C, derived by MOMA, is the projection of point A onto the feasible region of the mutant strain. Source: Adapted from Ref. 1 with permission of National Academy of Sciences, USA. @ 2002.

the steady-state fluxes in the wild-type strain is minimized. The constraints are the same as those in FBA yielding the following MOMA formulation:

$$\text{minimize} \sum_{j \in J} \left(v_j - v_j^{\text{WT}} \right)^2 \qquad [\text{MOMA}]$$

subject to

$$\sum_{j \in J} S_{ij} v_j = 0, \qquad \forall i \in I \tag{10.1}$$

$$LB_j \leq v_j \leq UB_j, \qquad \forall j \in J \tag{10.2}$$

Here, I and J denote the set of metabolites and reactions. In addition, v_j and v_j^{WT} are the fluxes of reaction j in the mutant and wild-type strains, respectively. The latter can be estimated from FBA or a combination of FBA and experimental flux measurements. Therefore, MOMA can be viewed as an extension of FBA as most flux values remain at or very close to the values calculated for the wild-type strain using FBA. Constraint 10.1 represents the steady-state mass balance for each metabolite i and Constraint 10.2 imposes lower and upper bounds on reaction fluxes. Consider the case of aerobic growth in a minimal medium with glucose as the carbon source. The following constraints are needed to impose this uptake condition (see also Chapter 3):

$$v_{\text{EX\_glc(e)}} \geq -v_{\text{glc}}^{\text{uptake}} \tag{10.3}$$

$$v_{\text{EX\_O}_2\text{(e)}} \geq -v_{\text{O}_2}^{\text{uptake}} \tag{10.4}$$

$$v_j \geq -M, \quad \forall j \in J^{\text{excess}} \tag{10.5}$$

$$v_{\text{ATPM}} = v_{\text{ATP}}^{\text{maint.}} \tag{10.6}$$

Here, $J^{\text{excess}} \subset J$ denotes the set of exchange reactions for metabolites present in excess in the growth medium (e.g., water) and M is a large scalar (e.g., 1000). Constraints 10.3 and 10.4 specify the maximum allowable uptake rates for glucose and oxygen, respectively, by setting the lower bound on the corresponding exchange reactions. Constraint 10.5 allows unlimited uptake of chemical species in excess J^{excess} and Constraint 10.6 sets the non-growth-associated ATP maintenance requirements. Note that all these constraints can be condensed into Constraint 10.2, where appropriate lower and upper bounds are assigned. The lower and upper bounds on the rest of the reactions in the network are the same as in FBA (see Table 3.1).

The Lagrange function associated with problem $[\text{MOMA}]$ can be written as follows:

$$\begin{aligned}
L(v,\lambda,\mu) = {} & \sum_{j\in J}\left(v_j - v_j^{\text{WT}}\right)^2 + \sum_{i\in I}\left(\sum_{j\in J} S_{ij}v_j\right)\lambda_i \\
& + \left(-v_{\text{EX_glc(e)}} - v_{\text{glc}}^{\text{uptake}}\right)\mu_{\text{EX_glc(e)}}^{\text{LB}} \\
& + \left(-v_{\text{EX_O}_2\text{(e)}} - v_{\text{O}_2}^{\text{uptake}}\right)\mu_{\text{EX_O}_2\text{(e)}}^{\text{LB}} \\
& + \left(-v_{\text{ATPM}} + v_{\text{ATP}}^{\text{maint.}}\right)\mu_{\text{ATPM}}^{\text{LB}} + \left(v_{\text{ATPM}} - v_{\text{ATP}}^{\text{maint.}}\right)\mu_{\text{ATPM}}^{\text{UB}} \\
& + \sum_{j\in J-\{\text{EX_glc(e)},\text{EX_O}_2\text{(e)},\text{ATPM}\}} \left(-v_j + LB_j\right)\mu_j^{\text{LB}} \\
& + \sum_{j\in J-\{\text{ATPM}\}} \left(v_j - UB_j\right)\mu_j^{\text{UB}}
\end{aligned} \tag{10.7}$$

The KKT optimality conditions (see Chapter 9) are as follows:

***Dual feasibility*:**

$$2\left(v_j - v_j^{\text{WT}}\right) + \sum_{i\in I}\left(\sum_{j\in J} S_{ij}\right)\lambda_i + \mu_j^{\text{UB}} - \mu_j^{\text{LB}} = 0, \quad \forall j \in J \tag{10.8}$$

***Primal feasibility*:**

Constraints 10.1 to 10.6.

***Complementary slackness*:**

$$\left(-v_j + LB_j\right)\mu_j^{\text{LB}} = 0, \quad \forall j \in J - \{\text{EX\_glc(e)},\text{EX\_O}_2\text{(e)},\text{ATPM}\} \tag{10.9}$$

$$\left(v_j - UB_j\right)\mu_j^{\text{UB}} = 0, \quad \forall j \in J - \{\text{ATPM}\} \tag{10.10}$$

$$\left(-v_{\text{EX}_{\text{glc(e)}}} + \left(-v_{\text{glc}}^{\text{uptake}}\right)\right)\mu_{\text{EX\_glc(e)}}^{\text{LB}} = 0 \tag{10.11}$$

$$\left(-v_{\mathrm{EX}_{\mathrm{O_2(e)}}}+\left(-v_{\mathrm{O_2}}^{\mathrm{uptake}}\right)\right)\mu_{\mathrm{EX\_O_2(e)}}^{\mathrm{LB}}=0 \tag{10.12}$$

If [MOMA] is used in the inner problem of a bilevel optimization problem, we can use the KKT optimality conditions to convert the bilevel problem into a regular optimization problem with explicit constraints. This situation may arise, for example, in OptKnock if one chooses MOMA to represent the cellular objective instead of the maximization of the biomass formation [6] (see Exercise 10.1). If a binary variable y_j is used to enforce reaction eliminations, as in OptKnock (see Chapter 8), the CS conditions can be rewritten as follows:

$$\left(-v_j+LB_j y_j\right)\mu_j^{\mathrm{LB}}=0,\quad \forall j\in J-\left\{\mathrm{EX\_glc(e)},\mathrm{EX\_O_2(e)},\mathrm{ATPM}\right\} \tag{10.13}$$

$$\left(v_j-UB_j y_j\right)\mu_j^{\mathrm{UB}}=0,\quad \forall j\in J-\left\{\mathrm{ATPM}\right\} \tag{10.14}$$

$$\left(-v_{\mathrm{EX}_{\mathrm{glc(e)}}}+\left(-v_{\mathrm{glc}}^{\mathrm{uptake}}\right)\right)\mu_{\mathrm{EX\_glc(e)}}^{\mathrm{LB}}=0 \tag{10.15}$$

$$\left(-v_{\mathrm{EX}_{\mathrm{O_2(e)}}}+\left(-v_{\mathrm{O_2}}^{\mathrm{uptake}}\right)\right)\mu_{\mathrm{EX\_O_2(e)}}^{\mathrm{LB}}=0 \tag{10.16}$$

Note that ATPM, EX_glc(e) and EX_O_2(e) are not candidates for knockout implying that $y_{\mathrm{ATPM}}=y_{\mathrm{EX\_glc(e)}}=y_{\mathrm{EX\_O_2(e)}}=1$. Constraints 10.13 to 10.16 contain bilinear terms, which are the product of reaction fluxes and Lagrange multipliers. Some of the equations containing these bilinear terms can be eliminated after a careful inspection of the CS conditions. Two cases can be considered here:

(i) If a reaction j is eliminated ($y_j=0$), both terms $(-v_j+LB_j y_j)$ and $(v_j-UB_j y_j)$ appearing in Constraints 10.13 and 10.14 are zero. Therefore, μ_j^{LB} and μ_j^{UB} are free to assume any nonnegative value.

(ii) If a reaction j is not eliminated (y_j = 1) and the reaction flux does not hit its wide lower or upper bound $-M$ and M, the Lagrange multipliers will be equal to zero:

$$\mu_j^{\mathrm{LB}}=0,\quad \forall j\in J-\left\{\mathrm{EX\_glc(e)},\mathrm{EX\_O_2(e)},\mathrm{ATPM}\right\}\cup J^{\mathrm{irrev}}\cup J^{\mathrm{exch,not\ media}}$$
$$\mu_j^{\mathrm{UB}}=0,\quad \forall j\in J-\left\{\mathrm{ATPM}\right\}$$

whereJ^{irrev} and $J^{\mathrm{exch,not\ media}}$ denote the set of irreversible reactions. and the set of exchange reactions for any metabolite that is not present in the growth medium, respectively.

Conditions (i) and (ii) can be enforced by using the following constraints:

$$\mu_j^{\mathrm{LB}}\le\left(1-y_j\right)\mu_j^{\mathrm{LB,max}},\ \forall j\in J-\{\mathrm{EX\_glc(e)},\ \mathrm{EX\_O_2(e)},\ \mathrm{ATPM}\}\cup J^{\mathrm{irrev}}\cup J^{\mathrm{exch,not\,media}} \tag{10.17}$$

$$\mu_j^{\text{UB}} \le (1-y_j)\mu_j^{\text{UB,max}}, \quad \forall j \in J - \{\text{ATPM}\} \tag{10.18}$$

where $\mu_j^{\text{LB,max}}$ and $\mu_j^{\text{UB,max}}$ denote the upper bounds on μ_j^{LB} and μ_j^{UB}, respectively. By imposing these conditions, Equation 10.13 is reduced to

$$v_j\mu_j^{\text{LB}} = 0, \quad \forall j \in J^{\text{irrev}} \cup J^{\text{exch,not media}} \tag{10.19}$$

Therefore, all that remains is to linearize Equations 10.15, 10.16 and 10.19 as they contain bilinear terms. They can be linearized by introducing binary variables w_j:

$$\mu_j^{\text{LB}} \le \mu_j^{\text{LB,max}} w_j, \quad \forall j \in J^{\text{irrev}} \cup J^{\text{exch,not media}} \tag{10.20}$$

$$v_j \le UB_j(1-w_j), \quad \forall j \in J^{\text{irrev}} \tag{10.21}$$

$$\mu_{\text{EX\_glc(e)}}^{\text{LB}} \le \mu_{\text{EX\_glc(e)}}^{\text{LB,max}} w_{\text{EX\_glc(e)}} \tag{10.22}$$

$$v_{\text{EX\_glc(e)}} - \left(-v_{\text{glc}}^{\text{uptake}}\right) \le M\left(1 - w_{\text{EX}_{\text{glc(e)}}}\right) \tag{10.23}$$

$$\mu_{\text{EX\_O}_2\text{(e)}}^{\text{LB}} \le \mu_{\text{EX\_O}_2\text{(e)}}^{\text{LB,max}} w_{\text{EX\_O}_2\text{(e)}} \tag{10.24}$$

$$v_{\text{EX\_O}_2\text{(e)}} - \left(-v_{\text{O}_2}^{\text{uptake}}\right) \le M\left(1 - w_{\text{EX\_O}_2\text{(e)}}\right) \tag{10.25}$$

where $\mu_{\text{EX\_glc(e)}}^{\text{LB,max}}$ and $\mu_{\text{EX\_O}_2\text{(e)}}^{\text{LB,max}}$ denote upper bounds on $\mu_{\text{EX\_glc(e)}}^{\text{LB}}$ and $\mu_{\text{EX\_O}_2\text{(e)}}^{\text{LB}}$, respectively and M is a large positive scalar. When a binary variable w_j assumes a value of one, the constraint on reaction j is active (i.e., $v_j = 0, \forall j \in J^{\text{irrev}}$ and $v_{\text{EX\_glc(e)}} = -v_{\text{glc}}^{\text{uptake}}$ or $v_{\text{EX\_O}_2\text{(e)}} = -v_{\text{O}_2}^{\text{uptake}}$) and the corresponding Lagrange multiplier can assume any nonnegative value. On the other hand, if it assumes a value of zero, the constraint on reaction j becomes inactive and the corresponding Lagrange multiplier is forced to assume a value of zero. This example shows how KKT conditions can be invoked to convert a convex nonlinear minimization problem into a set of constraints.

10.2 INCORPORATION OF KINETIC EXPRESSIONS IN STOICHIOMETRIC MODELS

Stoichiometric metabolic models alone cannot quantitatively capture the effect of concentration levels and enzyme saturation on reaction throughput and regulation. This may lead to flux distribution predictions that are inconsistent with allowable ranges for intracellular metabolite concentrations, enzymatic activities and substrate-level regulation. Kinetic models of metabolism directly describe the time evolution of metabolite concentrations, enzyme activities and reaction fluxes using a system of ordinary differential equations (ODEs) (see Chapter 3). However, in contrast to stoichiometric models, reconstruction of large-scale kinetic models has been impeded by the paucity of experimental data on fluxes and concentrations to enable robust model parameterization [7, 8]. Kinetic models are currently available for a short list of organisms spanning only a fraction of known metabolism [7–12]. Several efforts have been made in recent years for improving the prediction of metabolic phenotypes using genome-scale stoichiometric-based models supplemented with kinetic information [13–17]. In these hybrid models, the flux of reactions with available kinetic information is identified by solving the system of ODEs representing the component balances and FBA is used to infer the flux for the remaining of the reactions in the network.

In Chapter 3, we introduced an LP formulation for identifying the maximum product yield (MPY) capacity of an organism. Here, we revisit formulation [MPY] to assess how the predicted product yield is affected when incorporating some kinetic information. The set of reactions in the network J is divided into two subsets: reactions with kinetic information J^{kin} and those having only stoichiometric information J^{stoic}. For reactions in J^{kin} the maximum enzymatic reaction rate $v_j^{\max}$ is allowed to vary between zero (denoting removal of the reaction activity) to a maximum allowable up-regulation f from the wild-type (reference) strain enzyme activity $v_j^{\max,\text{ref}}$. The following NLP identifies the maximum production yield of a product P:

$$\underset{C_i, v_j}{\text{maximize}} \quad z = v_{\text{EX\_P(e)}} \quad [\text{k-MPY}]$$

subject to

$$v_j = v_j\left(v_j^{\max}, C_i, P_j\right), \qquad \forall j \in J^{\text{kin}} \tag{10.26}$$

$$0 \le v_j^{\max} \le f\, v_j^{\max,\text{ref}}, \quad \forall j \in J^{\text{kin}} \tag{10.27}$$

$$C_i^{\text{LB}} \le C_i \le C_i^{\text{UB}}, \qquad \forall i \in I^{\text{kin}} \tag{10.28}$$

$$\sum_{j \in J} S_{ij} v_j = 0, \qquad \forall i \in I \tag{10.29}$$

$$LB_j \le v_j \le UB_j, \qquad \forall j \in J^{\text{stoic}} \tag{10.30}$$

$$C_i \ge 0, \qquad \forall i \in I^{\text{kin}}$$

$$v_j \in \mathbb{R}, \qquad \forall j \in J$$

Here, $v_{\text{EX\_P(e)}}$ is the exchange flux of the target product P, C_i is the concentration of a metabolite i in the kinetic part of the model (I^{kin}), C_i^{LB} and C_i^{UB} denote lower and upper bounds on the concentration of a metabolite i and P_j is the set of kinetic parameters for reaction $j \in J^{\text{kin}}$.

The objective function of this optimization problem maximizes the exchange flux of a target metabolite consistent with reaction kinetics and stoichiometry. Constraint 10.26 determines the reaction rates using the kinetic rate laws for reactions in J^{kin}. Constraint 10.27 imposes the allowable range of enzymatic manipulations in the kinetic part of the network, while Constraint 10.28 imposes lower and upper bounds on the concentration of metabolites $i \in I^{\text{kin}}$. Finally, Constraints 10.29 and 10.30 enforce the steady-state mass balance and a lower and upper bound on reaction fluxes as in FBA.

Observe that while the flux of reactions in J^{stoic} is constrained only by the steady-state mass balance (Constraint 10.29) and bounds (Constraint 10.30) those in J^{kin} are determined by the kinetic rate expressions (Constraints 10.26–10.28). Kinetic expressions (Constraint 10.26), are typically nonlinear functions of the metabolite concentrations.

Example 10.1

Use a hybrid stoichiometric and kinetic modeling description to generate the trade-off plot for the production of acetate in *E. coli* under aerobic conditions in minimal medium with glucose as the carbon source. Use the *i*AF1260 stoichiometric model [18] and a kinetic model of *E. coli* central metabolism presented in Ref. 8 for the calculations. Allow for up to twofold upregulation of $v_j^{\text{max,ref}}$ for all reactions in J^{kin} (i.e., $f = 2$). Furthermore, allow departure from the wild-type concentrations by at most 1.5-fold ($0.67C_i^{\text{WT}} \leq C_i \leq 1.5C_i^{\text{WT}}, \quad \forall i \in I^{\text{kin}}$). Compare the trade-off plots for the purely stoichiometric and hybrid models.

Solution: The maximum theoretical biomass for the hybrid model can be obtained by replacing the objective function of problem [k-MPY] with v_{biomass} and solving the resulting problem. To generate the trade-off plot, problem [k-MPY] must be solved multiple times while fixing the biomass flux at different fractions (from zero to one) of its theoretical maximum as performed in Chapter 3. The problem can be solved with a local NLP solver (e.g., CONOPT) using multiple starting points or a global solver (e.g., BARON). A GAMS implementation of this example using the former approach (i.e., CONOPT with multiple starting points) is available through the book's website.

Figure 10.2 compares the trade-off plots for acetate vs. biomass production with and without the inclusion of kinetic constraints. Overall, the imposition of kinetic expressions reduces the size of the feasible acetate vs. biomass production flux space compared to that for a purely stoichiometric model. In particular, it results in a 14.6% reduction in maximum theoretical biomass flux (from 0.963 to 0.822 h^{-1}). The reason for this can be inferred by inspecting the identified flux distributions and metabolite concentrations in the kinetic part of the model. The imposed upper limits on the concentrations of pentose phosphate (PP) pathway metabolites 6-phosphogluconate (6pgc), ribulose-5-phosphate (ru5p) and sedo-heptulose-7-phosphate (s7p) restrict the maximum flux through the PP pathway

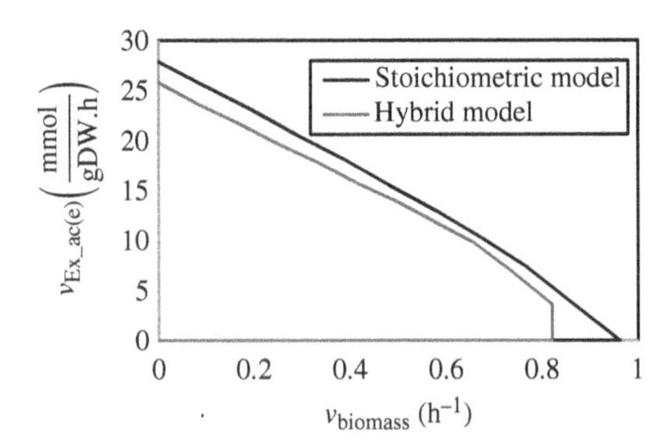

FIGURE 10.2 Comparison of the trade-off plots for acetate production using a pure stoichiometric and a hybrid stoichiometric and kinetic model (see Example 10.1).

(from 4.8 to 1.64 $\frac{\text{mmol}}{\text{gDW}\cdot\text{h}}$ per 10 $\frac{\text{mmol}}{\text{gDW}\cdot\text{h}}$ of glucose uptake). This results in a reduction in the synthesis of key biomass precursors ribose-5-phosphate (r5p) and erythrose-4-phosphate (e4p) thus lowering the maximum biomass flux. The additional flux that is not routed through PP pathway and biomass production can now be directed toward acetate or lower glycolysis/TCA cycle products (such as serine, succinate, fatty acids) thereby allowing for the acetate production flux to vary between zero and 3.64 $\frac{\text{mmol}}{\text{gDW}\cdot\text{h}}$ (per 10 $\frac{\text{mmol}}{\text{gDW}\cdot\text{h}}$ of glucose uptake) at maximum biomass (see Fig. 10.2). This observation is consistent with flux measurements reporting a non-zero acetate production flux (up to 7 $\frac{\text{mmol}}{\text{gDW}\cdot\text{h}}$ per 10 $\frac{\text{mmol}}{\text{gDW}\cdot\text{h}}$ of glucose uptake) in the exponential phase of growth for wild-type *E. coli* [19]. In addition, a reduction in the maximum theoretical acetate yield (from 2.783 to 2.577 $\frac{\text{mmol acetate}}{\text{mmol glucose}}$) is observed as the metabolic flux toward acetate from nucleotide metabolism is restricted due to the reduced availability of precursor r5p. □

10.3 METABOLIC FLUX ANALYSIS (MFA)

Inferring the true metabolic state of a biological system requires accurate measurements of intracellular metabolic fluxes. The current state-of-the-art technique for the quantification of intracellular fluxes is isotope labeling–based metabolic flux analysis (MFA). While here we focus only on ^{13}C isotope labeling, MFA using other tracers such as ^{18}O, ^{2}H and ^{15}N follows the same basic principles. The central idea of MFA is to trace labeling patterns of metabolites (often the amino acid pool) after feeding cells

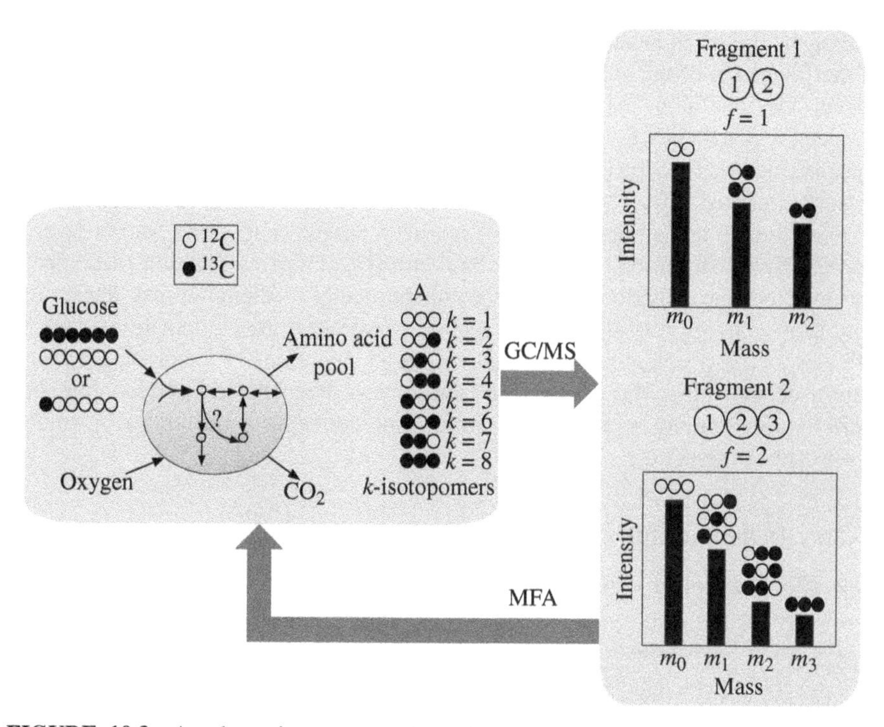

FIGURE 10.3 A schematic representation of metabolic flux analysis (MFA). First, a carbon labeled substrate such as glucose (e.g. labeled at the first carbon, [1-^{13}C]), or a mixture of uniformly labeled [U-^{13}C] and unlabeled glucose is used as the carbon source. The patterns of isotope incorporation into intermediate metabolites (usually amino acids) are then analyzed using GC/MS measurements. Molecules are derivatized during GC/MS and may be fragmented into different-sized species. For example, metabolite A here yields a two-carbon ($f = 1$, including carbons 1 and 2) and a three-carbon ($f = 2$) fragment. GC/MS measures the fraction abundance of fragments with different numbers of labeled atoms (bar graphs), but it cannot identify the position of the labeled atoms. These measurements are then used to computationally infer the labeling patters of the original molecule and reaction fluxes. Source: Adapted from Ref. 22 with permission of Elsevier.

with stable isotopes (e.g., ^{13}C) substrates (Fig. 10.3). For example, glucose molecules labeled only at one carbon position or a mixture of uniformly labeled (i.e., labeled in all carbon positions) and unlabeled glucose molecules is used as the carbon source. The isotope-labeled atoms of glucose are incorporated with distinct labeling distributions into the structure of various metabolites generated through the conversion of glucose in metabolic pathways. These labeling distributions are a function of reaction mechanism/flux and the selected input labeled tracer. Different isotope forms (isomers) of a metabolite are referred to as *isotopomers*. Highly sensitive measurements using nuclear magnetic resonance (NMR) spectroscopy and/or gas chromatography/mass spectrometry (GC/MS) are used to quantify the relative abundance of ^{13}C carbons in the fragments of all analyzed metabolites. The inverse problem of finding the flux of all metabolic reactions consistent with the measured ^{13}C distribution in the fragments of the analyzed metabolites is solved using the metabolic network connectivity and atom mapping information for every reaction in the model. This problem, which defines the core computational task in MFA, involves a nonlinear least squares objective function (as in MOMA) as well as a system of nonlinear isotopomer balance equations [20].

Even though it is straightforward to predict a unique isotopomer pattern given a feasible flux distribution, the inverse problem (i.e., inferring reaction fluxes from the detected mass distributions) is a computationally challenging task due to the nonlinear coupling relations. In this section, we present the computational task of MFA as a NLP problem minimizing the squared deviations of the predicted distribution of the molecular species of isotopomers from the measured ones. We start by introducing a number of important concepts required to build the optimization model.

10.3.1 Definition of the Relevant Parameters and Variables

The following sets are defined:

J = Set of the reactions in the network
I = Set of the metabolites in the network
$I^{\text{meas}} \subseteq I$ = Set of all measured metabolites (e.g., all amino acids)
K_i = Set of the isotopomers for metabolite $i \in I^{\text{meas}}$
F_i = Set of all possible fragments generated upon ionization of a metabolite $i \in I^{\text{meas}}$
M_f^i = Set of the mass fractions observed for a fragment f of a metabolite $i \in I^{\text{meas}}$

Isotopomer Distribution Vector (IDV): Schmidt et al. [21] introduced the idea of representing the labeling patterns of a molecule as binary numbers where labeled and unlabeled carbon atoms are denoted with 1's and 0's, respectively. For example, an unlabeled glucose molecule is represented as 000000_{bin} (subscript 'bin' denotes binary), whereas a glucose molecule labeled with a single ^{13}C-isotope in the first position is shown by 100000_{bin}. The *isotopomer distribution vector* for a metabolite $i \in I^{\text{meas}}$ (denoted by IDV_i) is a vector whose k^{th} element is the molar fraction of metabolite i that exists in isotopomer form $k \in K_i$. It follows from the aforementioned definition that each metabolite consisting of N carbon atoms will have 2^N isotopomers (i.e., IDV_i is a 2^N element vector assuming that impurities in tracers are ignored). The index of each specific labeling pattern in this vector is obtained by converting its binary representation to a decimal number and adding one to it. For example, the fraction of 100000_{bin} glucose can be found at index location $32+1=33$ as 32 is the decimal representation of the 100000_{bin}. The sum of all isotopomer fractions for any metabolite i must be equal to one:

$$\sum_{k \in K_i} \text{IDV}_{ik} = 1, \quad \forall i \in I \tag{10.31}$$

***Mass Distribution Vector (MDV)*:** NMR or GC/MS measured metabolites are derivatized and broken into fragments with a different number of carbons. GC/MS and NMR provide only raw data on the total mass of all isotopomers of a fragment with the same number of labeled carbons (i.e., they do not directly measure IDVs). The *mass distribution vector* of a fragment $f \in F_i$ for a metabolite $i \in I^{\text{meas}}$ (denoted by $\text{MDV}_{\text{f}}^{i}$) is a vector whose m^{th} element represents the fraction of all isotopomers with $m-1$ labeled carbons. For example, all three isotope forms 001_{bin}, 010_{bin} and 100_{bin} of a three-carbon molecule contribute to the second element in the MDV of fragment $f = 2$ as seen in Figure 10.3. Note that, if fragment f has N carbons, $\text{MDV}_{\text{f}}^{i}$ has $N+1$ elements.

***Isotopomer Grouping Matrix (IGM)*:** An isotopomer grouping matrix, denoted by $\text{IGM}_{\text{f}}^{i}$ is a matrix which links isotopomers of molecule i (as rows) and mass distributions in a fragment f (as columns). A nonzero entry at index (k,m) of this matrix indicates that the isotopomer form k of metabolite i appears in the fragment f with $m-1$ labeled carbons. Non-zero elements are generally set to a value of one, however, they can assume a fractional value for molecules with a plane of symmetry where isotopomer k appears in more than one mass isotopomer of a fragment f.

Example 10.2

Construct the isotopomer grouping matrix for the eight isotopomer forms of metabolite A in Figure 10.3.

Solution: Fragment $f = 1$ consists of only carbons 1 and 2 while fragment $f = 2$ consists of all three carbon atoms. The IGMs for fragments $f = 1$ and $f = 2$ can be expressed as follows:

$$\text{IGM}_{f=1}^{\text{A}} = \begin{bmatrix} 1 & 0 & 0 \\ 1 & 0 & 0 \\ 0 & 1 & 0 \\ 0 & 1 & 0 \\ 0 & 1 & 0 \\ 0 & 1 & 0 \\ 0 & 0 & 1 \\ 0 & 0 & 1 \end{bmatrix}, \quad \text{IGM}_{f=2}^{\text{A}} = \begin{bmatrix} 1 & 0 & 0 & 0 \\ 0 & 1 & 0 & 0 \\ 0 & 1 & 0 & 0 \\ 0 & 0 & 1 & 0 \\ 0 & 1 & 0 & 0 \\ 0 & 0 & 1 & 0 \\ 0 & 0 & 1 & 0 \\ 0 & 0 & 0 & 1 \end{bmatrix}$$

For example, consider row $k = 2$ in $\text{IGM}_{f=1}^{\text{A}}$. Given that carbons 1 and 2 are unlabeled for isotopomer $k = 2$, they appear only in bin m_0 of fragment $f = 1$. □

Metabolites with either a plane or point of symmetry require a special treatment. For example, glycerol contains a point of symmetry located at carbon 2, whereas succinate and fumarate have a plane of symmetry passing through the C–C bond between carbons 2 and 3. In such cases, the IGM is computed as the mean contribution from all equivalent forms of metabolite i appearing in fragment f. This is explained with the following example.

Example 10.3

Construct the IGM of metabolite A in Figure 10.4 for fragment $f = 1$ assuming that A is a symmetric molecule with a point of symmetry located at carbon 2.

Solution: Since molecule A has a point of symmetry located at carbon 2, $A_{C_1-C_2-C_3}$ is chemically and biochemically equivalent to $A_{C_3-C_2-C_1}$ (the subscripts of the carbon atoms represent their position). Due to this symmetric structure, fragment $f = 1$ is a mixture of $A_{C_1-C_2}$ and $A_{C_3-C_2}$ while fragment $f = 2$ is a mixture of $A_{C_1-C_2-C_3}$ and $A_{C_3-C_2-C_1}$. The contributions from $A_{C_1-C_2}$ and $A_{C_3-C_2}$ to the IGM are as follows:

$$\text{IGM}_{f=1}^{A_{C_1-C_2}} = \begin{bmatrix} 1 & 0 & 0 \\ 1 & 0 & 0 \\ 0 & 1 & 0 \\ 0 & 1 & 0 \\ 0 & 1 & 0 \\ 0 & 1 & 0 \\ 0 & 0 & 1 \\ 0 & 0 & 1 \end{bmatrix}, \quad \text{IGM}_{f=1}^{A_{C_3-C_2}} = \begin{bmatrix} 1 & 0 & 0 \\ 0 & 1 & 0 \\ 0 & 1 & 0 \\ 0 & 0 & 1 \\ 1 & 0 & 0 \\ 0 & 1 & 0 \\ 0 & 1 & 0 \\ 0 & 0 & 1 \end{bmatrix}$$

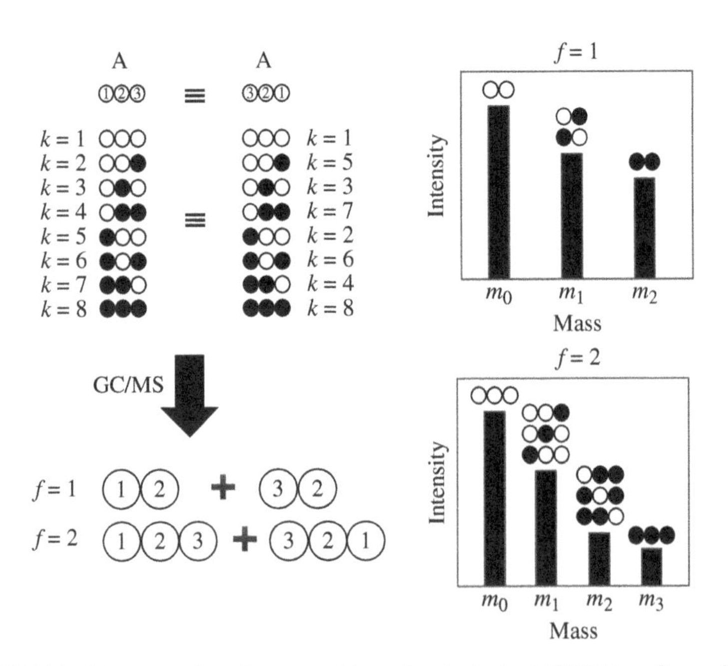

FIGURE 10.4 Fragmentation of a symmetric molecule during GC/MS (see Example 10.3).

The IGM for fragment $f = 1$ is thus defined as the average contribution from $A_{C_1-C_2}$ and $A_{C_3-C_2}$:

$$\text{IGM}_{f=1}^{\text{A}} = \frac{1}{2}\left(\text{IGM}_{f=1}^{\text{A}_{\text{C}_1\text{-C}_2}} + \text{IGM}_{f=1}^{\text{A}_{\text{C}_3\text{-C}_2}}\right) = \begin{bmatrix} 1 & 0 & 0 \\ 0.5 & 0.5 & 0 \\ 0 & 1 & 0 \\ 0 & 0.5 & 0.5 \\ 0.5 & 0.5 & 0 \\ 0 & 1 & 0 \\ 0 & 0.5 & 0.5 \\ 0 & 0 & 1 \end{bmatrix}.$$

For example, consider the second row of this matrix corresponding to $k = 2$. This isotopomer form contributes to mass m_0 of $f = 1$ for $A_{C_1-C_2}$ and to its mass m_1 for $A_{C_3-C_2}$. Therefore, the average contribution from each form to m_0 and m_1 is 0.5. □

The mass distribution vector is linked with the isotopomer distribution vector through the isotopomer grouping matrix as follows:

$$\text{MDV}_{\text{fm}}^{i} = \sum_{k\in K_i} \text{IGM}_{f,k\rightarrow m}^{i} \text{IDV}_{ik}, \quad \forall i \in I^{\text{meas}}, \quad f \in F_i, \quad m \in M_j^{i} \tag{10.32}$$

which can be recast in matrix form as follows:

$$\text{MDV}_{\text{f}}^{i} = \left(\text{IGM}_{\text{f}}^{i}\right)^{\text{T}} \text{IDV}_i, \quad \forall i \in I^{\text{meas}}, \quad f \in F_i \tag{10.33}$$

Isotopomer Mapping Matrix (IMM): Isotopomer mapping matrices (IMMs) identify what product isotopomer a particular reactant isotopomer will assume after a reaction step. An IMM is defined for each pair of reactant and product molecules that involve carbon exchange in a metabolic reaction. Rows and columns in IMMs correspond to the elements in the IDVs of the product and reactant, respectively. The elements of the IMM are usually either zero or one though a fractional value is used in the case of symmetric molecules. Non-zero elements of this matrix are found by identifying all labeling patterns of the product molecule resulting from each isotopomer of the reactant molecule. Therefore, element (k',k) of matrix $\text{IMM}_{j:i'\rightarrow i}$ encodes the transfer of carbons from isotopomer form k' of reactant i' to a product i in isotopomer form k according to reaction j. A zero value implies that no such transfer is possible given the reaction mechanism, whereas a value of one indicates that such a carbon transfer is occurring between reactant and product in the corresponding isotopomer forms. The position of nonzero elements of IMMs can be found by enumerating all labeling patterns of the product molecule resulting from all possible isotopomers of the reactant using the information stored in atom mapping matrices [21]. Note that all reversible reactions in the network have to be decomposed into their forward and backward directions to track the exchange of labeled metabolites in a reversible reaction. This is because both the forward and backward reactions may carry a nonzero flux and as a result involve different labeled carbon exchange even though the net rate is zero. It is easy to verify that the IMMs of the backward reaction can be obtained by transposing those of the forward reaction. The concept of IMMs is graphically explained using Example 10.4.

Example 10.4

Construct the IMMs for the carbon exchange reactions j_1, j_2 and j_3 shown in Figure 10.5.

Solution: One IMM needs to be constructed for each reactant-product pair in each reaction.

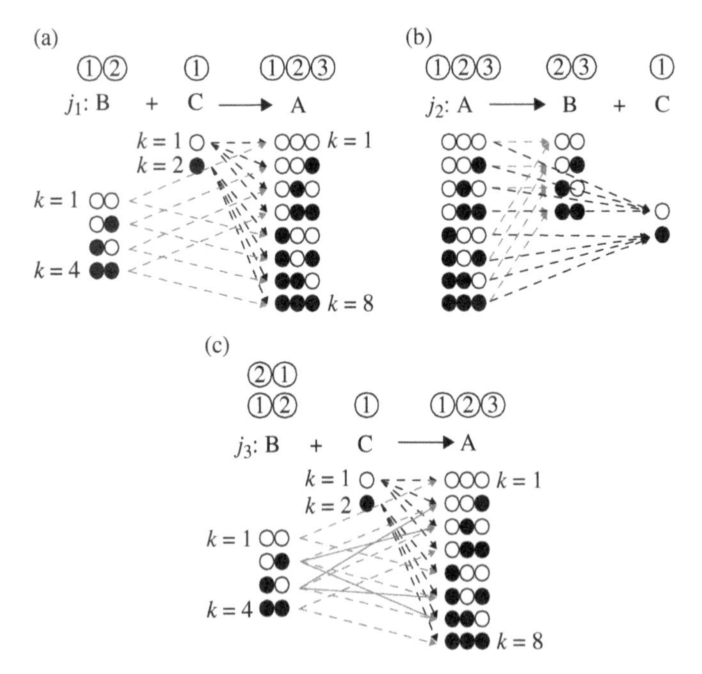

FIGURE 10.5 (a) A carbon exchange reaction involving the conversion of two reactants B (asymmetric) and C to a product A. (b) The dissociation reaction of metabolite A into two products B (asymmetric) and C. (c) The carbon exchange reaction involving the conversion of two reactants B (symmetric) and C to a product A (see Example 10.4). Dashed arrows indicate the mapping of isotopomer forms of asymmetric reactants to those of products and solid arrows in (C) represent additional mappings arising due to the symmetric structure of B.

(i) Reaction j_i: The set of isotopomers for B, C and A are $K_B = \{1,2,3,4\}$, $K_C = \{1,2\}$ and $K_A = \{1,2,\ldots,8\}$, respectively. Therefore, the IMMs are defined as follows:

$$\text{IMM}_{j_1:\text{B}\to\text{A}} = \begin{bmatrix} 1 & 0 & 0 & 0 \\ 0 & 1 & 0 & 0 \\ 0 & 0 & 1 & 0 \\ 0 & 0 & 0 & 1 \\ 1 & 0 & 0 & 0 \\ 0 & 1 & 0 & 0 \\ 0 & 0 & 1 & 0 \\ 0 & 0 & 0 & 1 \end{bmatrix}, \quad \text{IMM}_{j_1:\text{C}\to\text{A}} = \begin{bmatrix} 1 & 0 \\ 1 & 0 \\ 1 & 0 \\ 1 & 0 \\ 0 & 1 \\ 0 & 1 \\ 0 & 1 \\ 0 & 1 \end{bmatrix}$$

(ii) Reaction j_2 j: K_A, K_B and K_C are the same as those in j_1. IMMs are simply transposed for the reverse reaction direction:

$$\text{IMM}_{j_2:\text{A}\to\text{B}} = \begin{bmatrix} 1 & 0 & 0 & 0 & 1 & 0 & 0 & 0 \\ 0 & 1 & 0 & 0 & 0 & 1 & 0 & 0 \\ 0 & 0 & 1 & 0 & 0 & 0 & 1 & 0 \\ 0 & 0 & 0 & 1 & 0 & 0 & 0 & 1 \end{bmatrix}$$

$$\text{IMM}_{j_2:\text{A}\to\text{C}} = \begin{bmatrix} 1 & 1 & 1 & 1 & 0 & 0 & 0 & 0 \\ 0 & 0 & 0 & 0 & 1 & 1 & 1 & 1 \end{bmatrix}$$

(iii) Reaction j_3: K_A, K_B and K_C are the same as in reaction j_1, however, in this case B is symmetric with a plane of symmetry between carbons 1 and 2. The IMMs assume fractional entries for the indistinguishable carbon pairs:

$$\text{IMM}_{j_3:\text{B}\to\text{A}} = \begin{bmatrix} 1 & 0 & 0 & 0 \\ 0 & 0.5 & 0.5 & 0 \\ 0 & 0.5 & 0.5 & 0 \\ 0 & 0 & 0 & 1 \\ 1 & 0 & 0 & 0 \\ 0 & 0.5 & 0.5 & 0 \\ 0 & 0.5 & 0.5 & 0 \\ 0 & 0 & 0 & 1 \end{bmatrix}$$

For example, consider the second row corresponding to $k_A = 2$. The isotopomers of B that map to this isotopomer of A include both $k_B = 2$ and $k_B = 3$ due to the symmetric nature of B thus each contributing 0.5 to the row corresponding to $k_A = 2$. Note that for an asymmetric reactant B (reaction j_1) only $k_B = 2$ is mapped to $k_A = 2$. $\text{IMM}_{j_3:\text{C}\to\text{A}}$ is the same as $\text{IMM}_{j_1:\text{C}\to\text{A}}$. □

Linking the IDV of Reactants and Products Using IMMs The molar fraction of the isotopomers of the product molecules for a given carbon exchange reaction can be determined from the reactant molecules and the reaction mechanism (specified by IMMs) [21]. In particular, Fr_{ik}^{j} the fraction of a product molecule i in isotopomer form k is related to the IDV of reactants in a given carbon exchange reaction j as follows:

$$Fr_k^j = \prod_{i' \in \{i' | S_{i'j} < 0\}} \left(\sum_{k' \in K_{i'}} \mathrm{IMM}_{j:i' \to i, k' \to k} \mathrm{IDV}_{i'k'} \right), \quad \forall i \in I, k \in K_i \tag{10.34}$$

where $\mathrm{IMM}_{j:i' \to i, k' \to k}$ is the (k', k) element of isotopomer mapping matrix $\mathrm{IMM}_{j:i' \to i}$. Equivalently in matrix form this yields

$$Fr_i^j = \prod_{i' \in \{i' | S_{i'j} < 0\}} \mathrm{IMM}_{j:i' \to i} \mathrm{IDV}_{i'}, \quad \forall i \in I \tag{10.35}$$

The summation term in Equation 10.34 accounts for multiple isotopomer forms of a reactant i' that may contribute to isotopomer form k of product i through a reaction j. The multiplication operator (denoted by Π) accounts for multiple reactants contributing to isotopomer form k of product i in a bimolecular reaction mechanism. In this case, the product of the fraction of all reactants in their "appropriate" isotopomer forms that yield isotopomer form k for product i is calculated. It follows from these definitions that the summation term is eliminated, if only one isotopomer form of a metabolite i' contributes to isotopomer k of product i. Similarly, the product term is eliminated if the reaction has only one reactant (e.g., degradation reactions). For example, the fraction of istopomer k of metabolite A in reaction j_1 in Example 10.4 (Fig. 10.5a) is given by

$$Fr_{\mathrm{A},k}^{j_1} = \left(\sum_{k' \in K_\mathrm{B}} \mathrm{IMM}_{j_1:\mathrm{B} \to \mathrm{A}, k' \to k} \mathrm{IDV}_{\mathrm{B}k'} \right) \left(\sum_{k' \in K_\mathrm{C}} \mathrm{IMM}_{j_1:\mathrm{C} \to \mathrm{A}, k' \to k} \mathrm{IDV}_{\mathrm{C}k'} \right), \quad \forall k \in K_\mathrm{A}$$

In particular, $Fr_{\mathrm{A},1}^{j_1} = \mathrm{IDV}_{\mathrm{B},1}\mathrm{IDV}_{\mathrm{C},1}$ and $Fr_{\mathrm{A},3}^{j_1} = \mathrm{IDV}_{\mathrm{B},3}\mathrm{IDV}_{\mathrm{C},1}$. If B is symmetric (e.g., reaction j_3 in Fig. 10.5c), then $Fr_{\mathrm{A},3}^{j_3} = (0.5\,\mathrm{IDV}_{\mathrm{B},2} + 0.5\,\mathrm{IDV}_{\mathrm{B},3})\mathrm{IDV}_{\mathrm{C},1}$.

It is important to not confuse $Fr_{i,k}^{j}$ with IDV_{ik}. The former represents the molar fraction of isotopomer k of product i resulting from a given reaction j whereas the latter represents the molar fraction of the total amount of isotopomer (i,k) in the entire metabolic network which may be derived from multiple producing reactions. These two metrics are related through the mass balance equations for isotopomer (i,k).

10.3.2 Isotopomer Mass Balance

In analogy to the mass balance for each metabolite in the network, we can establish a mass balance for each isotopomer form k of metabolite i. The mass balance equation for isotopomer (i,k) is written as follows:

$$\frac{dC_{ik}}{dt} = \sum_{j \in \{j | S_{ij} > 0\}} S_{ij} v_j Fr_{ik}^j + \sum_{j \in \{j | S_{ij} < 0\}} S_{ij} v_j \mathrm{IDV}_{ik} = 0, \quad \forall i \in I, k \in K_i \tag{10.36}$$

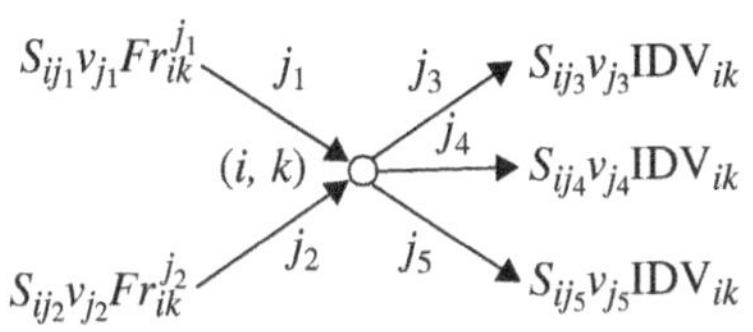

$S_{ij_1}v_{j_1}Fr_{ik}^{j_1}$ j_1 j_3 $S_{ij_3}v_{j_3}\mathrm{IDV}_{ik}$
(i,k) j_4 $S_{ij_4}v_{j_4}\mathrm{IDV}_{ik}$
$S_{ij_2}v_{j_2}Fr_{ik}^{j_2}$ j_2 j_5 $S_{ij_5}v_{j_5}\mathrm{IDV}_{ik}$

FIGURE 10.6 An example of an isotopomer mass balance. Metabolite i in isotopomer form k is generated through reactions j_1 and j_2 and is consumed by reactions j_3, j_4 and j_5. Source: Adapted from Ref. 22.

Here, C_{ik} denotes the concentration of isotopomer (i,k), S_{ij} is the stoichiometric coefficient of metabolite i in reaction j and v_j is the flux through reaction j. The first term on the right-hand side of this equation is the production/generation term for isotopomer (i,k) and the second term is the consumption term. By replacing Fr_{ik}^{j} in the generation term from Equation 10.34 the mass balance equation can be recast as follows under both metabolic and isotopic steady-state conditions:

$$\sum_{j\in\{j|S_{ij}>0\}} S_{ij}v_j\left[\prod_{i'\in\{i'|S_{i'j}<0\}}\left(\sum_{k'\in K_{i'}}\mathrm{IMM}_{j:i'\to i,k'\to k}\mathrm{IDV}_{i'k'}\right)\right]+ \sum_{j\in\{j|S_{ij}<0\}} S_{ij}v_j\mathrm{IDV}_{ik}=0,\quad \forall i\in I,k\in K_i \tag{10.37}$$

Note that the consumption rate of isotopomer (i,k) is directly proportional to its IDV regardless of the product. However, its production rate is dependent upon the IDV of the substrates contributing to the generation of isotopomer (i,k) through possibly multiple reactions j (see Fig. 10.6).

10.3.3 Optimization Formulation for MFA

The problem of elucidating intracellular metabolic fluxes and isotope labeling patterns from a given set of GC-MS data as well as a set of experimentally measured extracellular fluxes (e.g., growth rate, product secretion rates, uptake rates) can be formulated as the following NLP problem [22]:

$$\text{minimize } z=\sum_{i\in I^{meas}}\sum_{f\in F_i}\sum_{m\in M_f^i}\left(\frac{\mathrm{MDV}_{f,m}^{i}-\mathrm{MDV}_{f,m}^{i,\exp}}{\sigma_{f,m}^{i}}\right)^2+\sum_{j\in J^{meas}}\left(\frac{v_j-v_j^{\exp}}{\sigma_j}\right)^2$$

subject to

$$\sum_{j\in J} S_{ij}v_j=0,\quad \forall i\in I \tag{10.38}$$

$$\sum_{j\in\{j|S_{ij}>0\}} S_{ij}v_j\left[\prod_{i'\in\{i'|S_{i'j}<0\}}\left(\sum_{k'\in K_{i'}}\mathrm{IMM}_{j:i'\to i,k'\to k}\mathrm{IDV}_{i'k'}\right)\right]+ \sum_{j\in\{j|S_{ij}<0\}} S_{ij}v_j\mathrm{IDV}_{ik}=0,\quad \forall i\in I,k\in K_i \tag{10.37}$$

$$\mathrm{MDV}^i_{f,m} = \sum_{k\in K_i} \mathrm{IGM}^i_{f,k\to m}\mathrm{IDV}_{ik}, \quad \forall i \in I^{\mathrm{meas}}, \quad f \in F_i, \quad m \in M_f \tag{10.32}$$

$$\sum_{k\in K_i} \mathrm{IDV}_{ik} = 1, \quad \forall i \in I^{\mathrm{meas}} \tag{10.31}$$

$$0 \le v_j \le UB_j, \quad \forall j \in J \tag{10.39}$$

$$0 \le \mathrm{IDV}_{ik} \le 1, \quad \forall i \in I^{\mathrm{meas}}, k \in K_i \tag{10.40}$$

Here, J^{meas} represents the set of reactions with measured fluxes and $\sigma^i_{f,m}$ and σ_j denote the standard deviation of the mass distribution and flux measurements, respectively. The first term in the objective function minimizes the sum of the squared deviations of the calculated mean mass distributions $\mathrm{MDV}^i_{f,m}$ from those measured experimentally $\mathrm{MDV}^{i,\exp}_{f,m}$. The second term in the objective function minimizes the sum of the squared discrepancies between predicted fluxes v_j and experimentally measured (external) fluxes $v^{\exp}_j$. Equation 10.38 represents the steady-state mass balance for all metabolites in the network. Equation 10.39 constrains all fluxes to be nonnegative as all reversible reactions are split into forward and backward reactions. The variables of this optimization problem include the reaction fluxes v_j, mass distributions $\mathrm{MDV}^i_{f,m}$ and isotopomer distributions $\mathrm{IDV}_{i,k}$. The production term in Equation 10.37 is the source of nonlinearities in the constraints. Unimolecular reactions (e.g. B $\to$ A) yield bilinear terms as a result of the multiplication of metabolic fluxes v_j and isotopomer fractions, whereas bimolecular reactions (e.g. B+C $\to$A) contribute trilinear terms.

Due to the nonconvex nature of this problem, identifying a global optimum cannot be guaranteed using standard NLP solvers. Global optimization solvers such as BARON [23] and ANTIGONE [24] can be used instead. However, the large size of the problem and variations in the measured MDVs can make the use of global optimization techniques computationally challenging. One can instead pursue the identification of multiple local optima starting from different initial conditions and reporting the best solution found. The following stepwise procedures can be put forth to implement this strategy:

Step 0: Populate the model parameters S_{ij}, $\mathrm{IMM}^j_{i\to i'}$, $\mathrm{IGM}^i_{f,k\to m}$, UB_j, $\mathrm{MDV}^{i,\exp}_{f,m}$ and $v^{\exp}_j$. Initialize the iteration counter ($t = 0$)

Step 1: Find a random initial feasible flux distribution v^0_j and initialize IDV_{ik}

Step 2: Solve the optimization problem using a local optimization solver (e.g., CONOPT) to find a local solution for v_j, $\mathrm{MDV}^i_{f,m}$ and $\mathrm{IDV}_{i,k}$. Store the current solution and the corresponding objective function value z^t.

Step 3: Update the iteration counter $t \leftarrow t+1$ and go back to Step 1 if the maximum allowable number of iterations has not been reached.

Step 4: Report the solution with the best objective function value.

Schmidt et al. [25] proposed an alternative way of solving this optimization problem by taking advantage of the fact that the IDVs and subsequently MDVs are fully specified given a flux distribution and tracer scheme. This implies that reaction fluxes can be considered as the only variables of the optimization problem with metabolite mass balances (Eq. 10.38) and bounds on reaction fluxes (Eq. 10.39) as the only constraints. Wiechert et al. [26] suggested using the concept of the null-space matrix to express reaction fluxes as a linear combination of a subset of independent (free) fluxes. This eliminates the equality constraints in Equation 10.38. The same procedure can be applied to eliminate any active flux bounds (i.e., inequalities in Equation 10.39 satisfied as equalities). These transformations render the constrained optimization problem with fluxes as the only variables to an unconstrained optimization problem in the space of independent fluxes. This is similar to the projection that the GRG algorithm (see Chapter 9) performs on the degrees of freedom for a problem with equality constraints. The resulting unconstrained optimization problem is then solved using variations of Newton's method. The gradients and Hessian matrix can be obtained using numerical methods or analytical derivations [26, 27]. It is important to note that the optimization problem in the projected space of independent fluxes remains nonconvex due to the nonlinear coupling relations between fluxes and IDVs.

In addition to the projection to the space of free variables, several ways of encoding labeling information beyond the use of isotopomers have been introduced. Cumulative isotopomers or cumomers allow the problem to be solved as a cascade of easier-to-solve subproblems [20]. Elementary metabolite units (EMUs) [33] further simplify the solution procedure by describing isotopomer balances using only MDVs instead of IDVs for all the metabolites within the network. This approach traces target fragments back to all the inputs of the metabolic model, thereby greatly decreasing the number of variables. The EMU algorithm provides a framework to decompose the metabolic network into sub-networks of metabolite fragments of various sizes. Mass balances on the EMU sub-networks form a cascaded square system of linear equations describing MDVs of target EMUs for a given flux distribution and input MDVs. As a result, the least squares MFA problem can be solved as a nonlinear parameter optimization problem with fluxes as the only variables [27]. Implicit in this optimization problem is the calculation of the MDVs for the current values of metabolic fluxes rendering the overall optimization problem nonconvex.

Note that the variability of mass spectrometry measurements may lead to multiple values for $\mathrm{MDV}_{f,m}^{i,\exp}$, which in turn give rise to different flux distribution estimations. It is important to remember that about 1% of all carbon in nature is in the form of ^{13}C. This contribution to the labeling patterns needs to be subtracted from the calculations. Uncertainty in the mass spectrometry measurements implies that not a single value but rather a normally distributed range (characterized by a mean and variance) is reported. Generally, statistical analyses in the form of chi-squared goodness-of-fit tests are performed [26, 27] to assess whether the calculated flux distributions represent the true metabolic state of the system. Several software packages have been developed to solve the MFA optimization problem using different algorithms and approaches to modeling labeling information examples including 13CFLUX2 [28], OpenFLUX [29], Metran [30] and INCA [31].

EXERCISES

10.1 Consider Example 8.1 from Chapter 8 (OptKnock). Repeat the study after replacing the maximization of biomass formation in the inner objection with the minimization of metabolic adjustment (MOMA) criterion (see Ref. 1). Use the fluxes obtained upon maximizing biomass formation (using FBA) as an estimation of the wild-type steady-state metabolic fluxes. Apply the KKT conditions to express the optimality relations for the inner problem. What are the differences in the results?

10.2 Formulate an optimization problem for parameterizing a kinetic model of *E. coli* central metabolism by minimizing the sum of the squared differences between model predictions and the set of measured steady-state reaction fluxes [32] as shown in the following figure and the kinetic expressions detailed in the table.

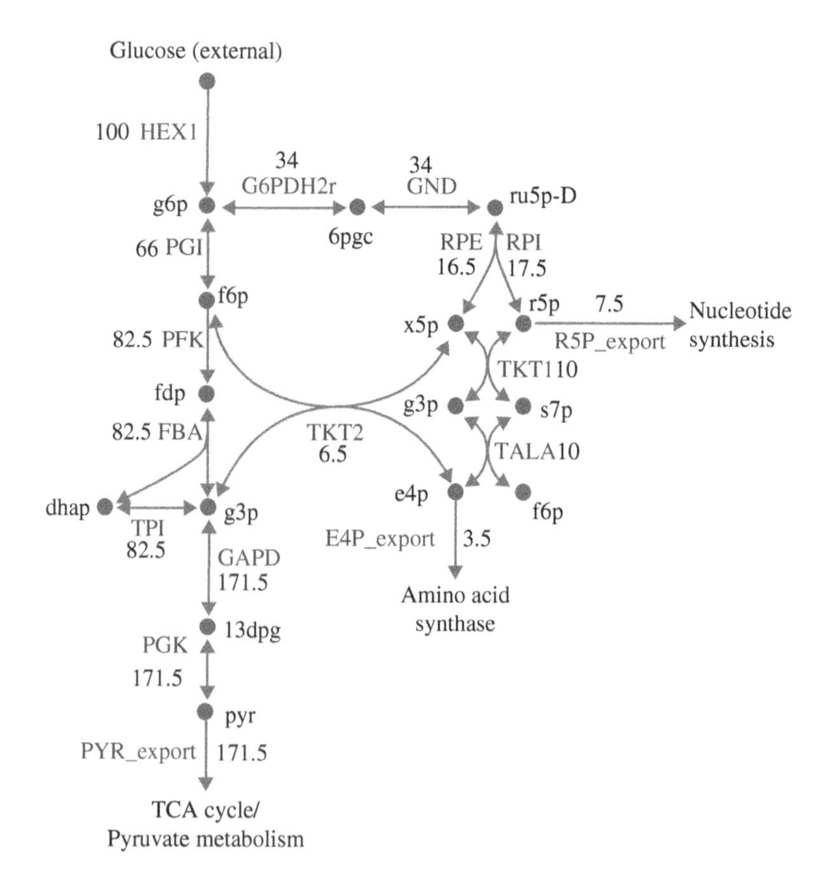

Reaction and metabolite names in this figure follow those of the *i*AF1260 model of *E. coli* [18]. Cofactors are eliminated and some reactions are lumped together (GND and PGL are lumped into GND, and PGK, PGM, ENO and PK are lumped into PGK) in this model for simplification [32]. Export reactions are placeholders for pathways consuming the corresponding metabolites. Numbers on reactions in this figure represent experimentally measured flux values. Choose an appropriate kinetic expression for each reaction in the model from the following table:

Reaction Type	Form of the Kinetic Expression
Irreversible reactions with one substrate: $A \rightarrow P$	$v = \dfrac{v_{max} C_A}{1 + (C_A / K_m)}$
Reversible reactions with one substrate and one product: $A \leftrightarrow P$	$v = \dfrac{v^f_{max} C_A - v^b_{max} C_P}{1 + \dfrac{C_A}{K_{mA}} + \dfrac{C_P}{K_{mP}}}$
Reversible reactions with two substrates and one product: $A + B \leftrightarrow P$	$v = \dfrac{v^f_{max} C_A C_B - v^b_{max} C_P}{\left(1 + \dfrac{C_A}{K_{mA}} + \dfrac{C_B}{K_{mB}}\right)\left(1 + \dfrac{C_P}{K_{mP}}\right)}$
Reversible reactions with one substrate and two products: $A \leftrightarrow P + W$	$v = \dfrac{v^f_{max} C_A - v^b_{max} C_P C_W}{\left(1 + \dfrac{C_A}{K_{mA}}\right)\left(1 + \dfrac{C_P}{K_{mP}} + \dfrac{C_W}{K_{mW}}\right)}$
Reversible reactions with two substrates and two products: $A + B \leftrightarrow P + W$	$v = \dfrac{v^f_{max} C_A C_B - v^b_{max} C_P C_W}{\left(1 + \dfrac{C_A}{K_{mA}} + \dfrac{C_B}{K_{mb}}\right)\left(1 + \dfrac{C_P}{K_{mp}} + \dfrac{C_W}{K_{mw}}\right)}$

Here, C_A, C_B, C_P and C_W represent the concentration of metabolites A, B, P and W, respectively. K_m and v_{max} denote Michaelis–Menten constants and superscript f and b denote the forward and backward reactions of a reversible reaction, respectively. Allow intracellular metabolite concentrations to range between 0.01 and 20 mM [18] and set the external glucose concentration to 0.05 mM.

10.3 For the metabolic network given in the figure below (Source: Adapted from Ref. 33), estimate the flux distribution if the mass isotopomer distribution of D was experimentally measured to be [0.0001 ± 0.003, 0.8008 ± 0.0025, 0.1983 ± 0.002, 0.0009 ± 0.0015], the uptake rate of A was measured to be 100 moles and metabolite A is labeled at position 2. Note that an atom transition scheme such as B(abc) → E(a) + C(bc) defines the atom mapping for a reaction involving the breakup of a three-carbon metabolite A (where each carbon atom is identified using "a," "b" and "c") such that atom "a" (carbon 1) of metabolite A maps to atom "a" (i.e., carbon 1) of metabolite

B and atoms "b" and "c" (carbons 2 and 3 of metabolite A) map to atoms "b" and "c" (carbons 1 and 2) of metabolite C.

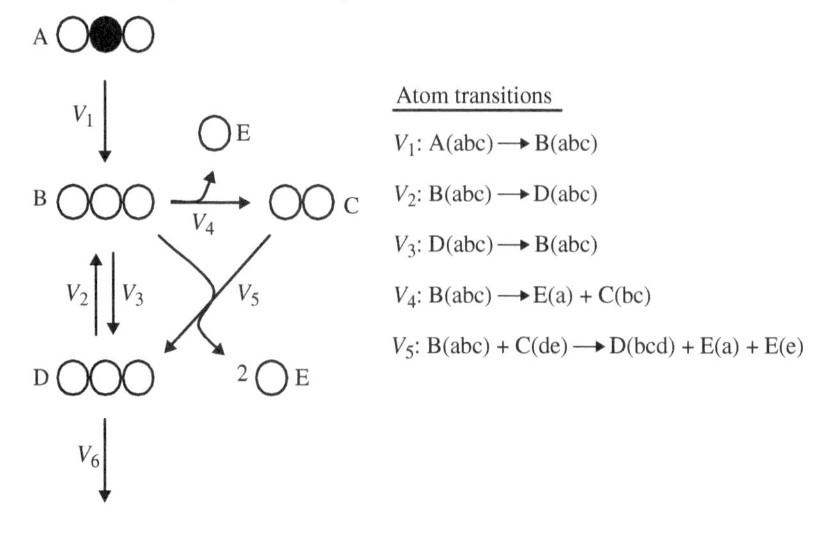

REFERENCES

1. Segrè D, Vitkup D, Church GM: Analysis of optimality in natural and perturbed metabolic networks. *Proc Natl Acad Sci U S A* 2002, **99**(23):15112–15117.
2. Shlomi T, Berkman O, Ruppin E: Regulatory on/off minimization of metabolic flux changes after genetic perturbations. *Proc Natl Acad Sci U S A* 2005, **102**(21):7695–7700.
3. Snitkin ES, Segrè D: Optimality criteria for the prediction of metabolic fluxes in yeast mutants. *Genome Inform* 2008, **20**:123–134.
4. Kim J, Reed JL: RELATCH: relative optimality in metabolic networks explains robust metabolic and regulatory responses to perturbations. *Genome Biol* 2012, **13**(9):R78.
5. Kleessen S, Nikoloski Z: Dynamic regulatory on/off minimization for biological systems under internal temporal perturbations. *BMC Syst Biol* 2012, **6**:16.
6. Kim J, Reed JL, Maravelias CT: Large-scale bi-level strain design approaches and mixed-integer programming solution techniques. *PLoS One* 2011, **6**(9):e24162.
7. Teusink B, Passarge J, Reijenga CA, Esgalhado E, van der Weijden CC, Schepper M, Walsh MC, Bakker BM, van Dam K, Westerhoff HV *et al.*: Can yeast glycolysis be understood in terms of in vitro kinetics of the constituent enzymes? Testing biochemistry. *Eur J Biochem* 2000, **267**(17):5313–5329.
8. Khodayari A, Zomorrodi AR, Liao JC, Maranas CD: A kinetic model of *Escherichia coli* core metabolism satisfying multiple sets of mutant flux data. *Metab Eng* 2014, **25C**:50–62.
9. Rizzi M, Baltes M, Theobald U, Reuss M: In vivo analysis of metabolic dynamics in *Saccharomyces cerevisiae*: II Mathematical model. *Biotechnol Bioeng* 1997, **55**(4):592–608.
10. Won W, Park C, Park C, Lee SY, Lee KS, Lee J: Parameter Estimation and Dynamic Control Analysis of Central Carbon Metabolism in Escherichia coli. *Biotechnol Bioprocess Eng* 2011, **16**(2):216–222.

11. Peskov K, Mogilevskaya E, Demin O: Kinetic modelling of central carbon metabolism in *Escherichia coli*. *FEBS J* 2012, **279**(18):3374–3385.
12. Kadir TA, Mannan AA, Kierzek AM, McFadden J, Shimizu K: Modeling and simulation of the main metabolism in *Escherichia coli* and its several single-gene knockout mutants with experimental verification. *Microb Cell Fact* 2010, **9**:88.
13. Cotten C, Reed JL: Mechanistic analysis of multi-omics datasets to generate kinetic parameters for constraint-based metabolic models. *BMC Bioinformat* 2013, **14**:32.
14. Chowdhury A, Zomorrodi AR, Maranas CD: k-OptForce: integrating kinetics with flux balance analysis for strain design. *PLoS Comput Biol* 2014, **10**(2):e1003487.
15. Pozo C, Miro A, Guillen-Gosalbez G, Sorribas A, Alves R, Jimenez L: Gobal optimization of hybrid kinetic/FBA models via outer-approximation. *Comput Chem Eng* 2015, **72**:325–333.
16. Machado D, Costa RS, Ferreira EC, Rocha I, Tidor B: Exploring the gap between dynamic and constraint-based models of metabolism. *Metab Eng* 2012, **14**(2):112–119.
17. Fleming RM, Thiele I, Provan G, Nasheuer HP: Integrated stoichiometric, thermodynamic and kinetic modelling of steady state metabolism. *J Theor Biol* 2010, **264**(3): 683–692.
18. Feist AM, Henry CS, Reed JL, Krummenacker M, Joyce AR, Karp PD, Broadbelt LJ, Hatzimanikatis V, Palsson B: A genome-scale metabolic reconstruction for *Escherichia coli* K-12 MG1655 that accounts for 1260 ORFs and thermodynamic information. *Mol Syst Biol* 2007, **3**:121.
19. Leighty RW, Antoniewicz MR: COMPLETE-MFA: complementary parallel labeling experiments technique for metabolic flux analysis. *Metab Eng* 2013, **20**:49–55.
20. Schmidt K, Carlsen M, Nielsen J, Villadsen J: Modeling isotopomer distributions in biochemical networks using isotopomer mapping matrices. *Biotechnol Bioeng* 1997, **55**(6):831–840.
21. Schmidt K, Carlsen M, Nielsen J, Villadsen J: Modeling isotopomer distributions in biochemical networks using isotopomer mapping matrices. *Biotechnol Bioeng* 1997, **55**(6):831–840.
22. Suthers PF, Burgard AP, Dasika MS, Nowroozi F, Van Dien S, Keasling JD, Maranas CD: Metabolic flux elucidation for large-scale models using 13C labeled isotopes. *Metab Eng* 2007, **9**(5–6):387–405.
23. Sahinidis N: BARON: A general purpose global optimization software package. *J Global Optim* 1996, **8**(2):201–205.
24. Misener R, Floudas CA: ANTIGONE: Algorithms for coNTinuous/Integer Global Optimization of Nonlinear Equations. *J Global Optim* 2014, **59**(2–3):503–526.
25. Schmidt K, Nielsen J, Villadsen J: Quantitative analysis of metabolic fluxes in *Escherichia coli*, using two-dimensional NMR spectroscopy and complete isotopomer models. *J Biotechnol* 1999, **71**(1–3):175–189.
26. Wiechert W, Siefke C, de Graaf AA, Marx A: Bidirectional reaction steps in metabolic networks: II. Flux estimation and statistical analysis. *Biotechnol Bioeng* 1997, **55**(1):118–135.
27. Antoniewicz MR, Kelleher JK, Stephanopoulos G: Determination of confidence intervals of metabolic fluxes estimated from stable isotope measurements. *Metab Eng* 2006, **8**(4):324–337.

28. Weitzel M, Noh K, Dalman T, Niedenfuhr S, Stute B, Wiechert W: 13CFLUX2--high-performance software suite for ^{13}C-metabolic flux analysis. *Bioinformatics* 2013, **29**(1):143–145.
29. Quek LE, Wittmann C, Nielsen LK, Kromer JO: OpenFLUX: efficient modelling software for ^{13}C-based metabolic flux analysis. *Microbial Cell Factories* 2009, **8**:25.
30. Yoo H, Antoniewicz MR, Stephanopoulos G, Kelleher JK: Quantifying reductive carboxylation flux of glutamine to lipid in a brown adipocyte cell line. *J Biol Chem* 2008, **283**(30):20621–20627.
31. Young JD: INCA: a computational platform for isotopically non-stationary metabolic flux analysis. *Bioinformatics* 2014, **30**(9):1333–1335.
32. Kadir TA, Mannan AA, Kierzek AM, McFadden J, Shimizu K: Modeling and simulation of the main metabolism in *Escherichia coli* and its several single-gene knockout mutants with experimental verification. *Microbial Cell Factories* 2010, **9**:88.
33. Antoniewicz MR, Kelleher JK, Stephanopoulos G: Elementary metabolite units (EMU): a novel framework for modeling isotopic distributions. *Metab Eng* 2007, **9**(1):68–86.

11

MINLP FUNDAMENTALS AND APPLICATIONS

Mixed-integer nonlinear programming (MINLP) problems involve both continuous ($\boldsymbol{x}$) and integer ($\boldsymbol{y}$) variables with nonlinearities in the objective function and/or constraints. The general form of an MINLP problem is as follows:

$$\begin{aligned}
&\underset{\boldsymbol{x},\boldsymbol{y}}{\text{minimize}}\ f(\boldsymbol{x},\boldsymbol{y}) \quad [\text{MINLP}]\\
&\text{subject to}\\
&\boldsymbol{g}(\boldsymbol{x},\boldsymbol{y}) \le 0\\
&\boldsymbol{h}(\boldsymbol{x},\boldsymbol{y}) = 0\\
&\boldsymbol{x} \in X \subset \mathbb{R}^N\\
&\boldsymbol{y} \in Y \subset \mathbb{Z}^P
\end{aligned}$$

where $f(\boldsymbol{x},\boldsymbol{y})$, $\boldsymbol{g}(\boldsymbol{x},\boldsymbol{y}) = [g_1(\boldsymbol{x},\boldsymbol{y}), g_2(\boldsymbol{x},\boldsymbol{y}), \ldots, g_M(\boldsymbol{x},\boldsymbol{y})]^{\mathrm{T}}$ and $\boldsymbol{h}(\boldsymbol{x},\boldsymbol{y}) = [h_1(\boldsymbol{x},\boldsymbol{y}), h_2(\boldsymbol{x},\boldsymbol{y}), \ldots, h_L(\boldsymbol{x},\boldsymbol{y})]^{\mathrm{T}}$ are nonlinear functions of $\boldsymbol{x}$ and $\boldsymbol{y}$. This class of problems arise in computational strain design applications with nonlinear objective functions (e.g., the minimization of metabolic adjustment (MOMA) [1]) or when kinetic expressions are used to describe (some of) the metabolic fluxes [2–6]. In this chapter, we focus on 0–1 MINLP problems (i.e., $\boldsymbol{y} \in \{0,1\}^P$) and outline two popular solution procedures. This is followed by an MINLP-based extension of the OptKnock procedure (see Chapter 8) that allows for kinetic descriptions for some of the reactions in the model.

Optimization Methods in Metabolic Networks, First Edition. Costas D. Maranas and Ali R. Zomorrodi.

11.1 AN OVERVIEW OF THE MINLP SOLUTION PROCEDURES

For any MINLP problem one can simply relax the binary variables by treating them as continuous variables varying between zero and one and solve the resulting nonlinear programming NLP problem. If the optimal values of the relaxed variables are all zero or one, the original MINLP has been solved. In general, however, this relaxation does not yield integer solutions, therefore customized solution strategies need to be invoked. MINLP problems are solved using stochastic or deterministic methods. Stochastic methods are typically reserved for very large and highly nonlinear problems that cannot be solved to optimality within the allotted time using deterministic methods and/or when a provably optimal solution is not necessarily needed. Popular stochastic methods include simulated annealing (SA) [7], generic algorithms (GA) [8], ant colony optimization [9] and particle swarm optimization [10]. In contrast, deterministic methods have well-defined termination criteria and convergence properties. Examples of these methods include, among others, generalized Benders decomposition (GBD) [11], outer approximation (OA) [12] and outer approximation with equality relaxation (OA/ER) [13]. All these methods generate a converging sequence of upper and lower bounds bracketing the optimal solution. The upper bound in all these methods is found by fixing the binary variables and solving the resulting NLP problem in the continuous variables $\boldsymbol{x}$ (also called the *primal problem*). The lower bound is obtained by solving a relaxation of the original problem called the *master problem*. Existing methods differ primarily in the way they formulate the master problem. In the following, we present a brief description of the GBD, OA and OA/ER algorithms. Interested readers should consult with the original publications [11–13] and/or MINLP textbooks [14] for a more thorough treatment.

11.2 GENERALIZED BENDERS DECOMPOSITION

The GBD algorithm, which was proposed by Geoffrion [11] can be used to solve the following class of MINLP problems:

$$\begin{aligned}
&\underset{\boldsymbol{x},\boldsymbol{y}}{\text{minimize}}\ f(\boldsymbol{x},\boldsymbol{y}) \quad [\text{MINLP}_{\text{GBD}}]\\
&\text{subject to}\\
&\boldsymbol{g}(\boldsymbol{x},\boldsymbol{y}) \le \boldsymbol{0}\\
&\boldsymbol{h}(\boldsymbol{x},\boldsymbol{y}) = \boldsymbol{0}\\
&\boldsymbol{x} \in X\\
&\boldsymbol{y} \in Y = \{0,1\}^P
\end{aligned}$$

where X is a nonempty compact and convex subset of $\mathbb{R}^N$ (e.g., $X = \{\boldsymbol{x} \in \mathbb{R}^N | \boldsymbol{x}^{\text{LB}} \le \boldsymbol{x} \le \boldsymbol{x}^{\text{UB}}\}$), $f(\boldsymbol{x})$ and $\boldsymbol{g}(\boldsymbol{x}, \boldsymbol{y}) = [g_1(\boldsymbol{x}, \boldsymbol{y}), g_2(\boldsymbol{x}, \boldsymbol{y}),\ldots, g_M(\boldsymbol{x}, \boldsymbol{y})]^{\text{T}}$ are convex and $\boldsymbol{h}(\boldsymbol{x}, \boldsymbol{y}) = [h_1(\boldsymbol{x}, \boldsymbol{y}), h_2(\boldsymbol{x}, \boldsymbol{y}),\ldots, h_L(\boldsymbol{x}, \boldsymbol{y})]^{\text{T}}$ are linear functions, respectively for fixed values of $\boldsymbol{y}$.

11.2.1 The Primal Problem

The primal problem (i.e., an NLP) provides an upper bound to the optimal solution by fixing the binary variables at a particular 0–1 combination $\boldsymbol{y}_k$ in some iteration k.

$$\begin{aligned}
&\underset{\boldsymbol{x}}{\text{minimize}}\ f(\boldsymbol{x},\boldsymbol{y}_k) \quad \left[\mathrm{P}^k_{\mathrm{GBD}}\right]\\
&\text{subject to}\\
&\boldsymbol{g}(\boldsymbol{x},\boldsymbol{y}_k)\le \boldsymbol{0}\\
&\boldsymbol{h}(\boldsymbol{x},\boldsymbol{y}_k)= \boldsymbol{0}\\
&\boldsymbol{x}\in X
\end{aligned}$$

Problem $[\mathrm{P}^k_{\mathrm{GBD}}]$ may be feasible or infeasible for $\boldsymbol{y}=\boldsymbol{y}_k$.

- *Feasible primal problem*: In this case, the upper bound is updated as the lowest encountered so far, $\mathrm{UB}=\min(\mathrm{UB}, f(\boldsymbol{x}_k, \boldsymbol{y}_k))$ where $\boldsymbol{x}_k$ is the optimal solution to $[\mathrm{P}^k_{\mathrm{GBD}}]$. Solving the primal problem also provides the optimal values of the Lagrange multipliers associated with the equality $\boldsymbol{\lambda}_k$ and inequality $\boldsymbol{\mu}_k$ constraints. These values can be used to define the Lagrange function at iteration k:

$$L_k(\boldsymbol{x}, \boldsymbol{y}, \boldsymbol{\lambda}_k, \boldsymbol{\mu}_k)= f(\boldsymbol{x},\boldsymbol{y})+\sum_{j=1}^{M}\mu_{jk}g_j(\boldsymbol{x},\boldsymbol{y})+\sum_{l=1}^{L}\lambda_{lk}h_l(\boldsymbol{x},\boldsymbol{y}) \tag{11.1}$$

- *Infeasible primal problem*: In this case, the following feasibility problem is formulated by introducing slack variables and relaxing all nonlinear constraints:

$$\begin{aligned}
&\text{minimize}\sum_{j=1}^{M}a_j+\sum_{l=1}^{L}\left(b_l^+ + b_l^-\right) \quad \left[\mathrm{PF}^k_{\mathrm{GBD}}\right]\\
&\text{subject to}\\
&g_j(\boldsymbol{x},\boldsymbol{y}_k)\le a_j, \qquad j=1,2,\ldots,M\\
&h_l(\boldsymbol{x},\boldsymbol{y}_k)= b_l^+ - b_l^-, \quad l=1,2,\ldots,L\\
&a_j, b_l^+, b_l^-\ge 0, \qquad j=1,2,\ldots,M,\ l=1,2,\ldots,L\\
&\boldsymbol{x}\in \boldsymbol{X}
\end{aligned}$$

This problem is solved to determine a new $\boldsymbol{x}_k$ that is the closest possible to feasibility as shown pictorially in Figure 11.1 (note that the objective function value of $[\mathrm{PF}^k_{\mathrm{GBD}}]$ quantifies the degree of infeasibility). Solving this problem also provides the optimal values of the Lagrange multipliers associated with equality ($\boldsymbol{\lambda}'_k$) and inequality ($\boldsymbol{\mu}'_k$) constraints of the infeasible primal problem. The corresponding Lagrange function involves only terms arising from the constraints [14]:

$$L_k^{\text{infeas}}(\boldsymbol{x}, \boldsymbol{y}, \boldsymbol{\lambda}'_k, \boldsymbol{\mu}'_k)= \sum_{j=1}^{M}\mu'_{jk}g_j(\boldsymbol{x},\boldsymbol{y})+\sum_{l=1}^{L}\lambda'_{lk}h_l(\boldsymbol{x},\boldsymbol{y}) \tag{11.2}$$

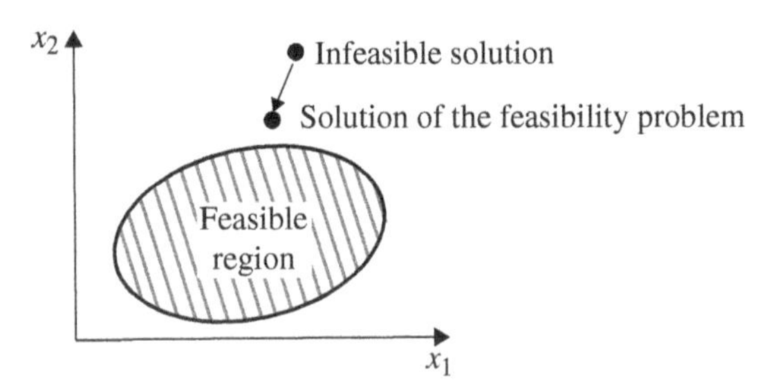

FIGURE 11.1 A feasibility problem is solved when the primal problem is infeasible at iteration k to determine a new $\boldsymbol{x}_k$ which is the closest possible to the feasible region.

Note that problem [$\mathrm{PF}_{\mathrm{GBD}}^k$] does not contain any information on the objective function, therefore the upper bound is not updated after an infeasible primal.

11.2.2 The Master Problem

Problem [$\mathrm{MINLP}_{\mathrm{GBD}}$] can be rewritten as follows by separating the minimization operator for $\boldsymbol{x}$ and $\boldsymbol{y}$:

$$\begin{aligned} &\underset{y}{\text{minimize}}\ \underset{x}{\text{minimize}}\ f(\boldsymbol{x}, \boldsymbol{y}) \quad [\mathrm{MINLP}_{\mathrm{GBD}}] \\ &\text{subject to} \\ &\boldsymbol{g}(\boldsymbol{x}, \boldsymbol{y}) \le 0 \\ &\boldsymbol{h}(\boldsymbol{x}, \boldsymbol{y}) = 0 \\ &\boldsymbol{x} \in X \\ &\boldsymbol{y} \in Y = \{0,1\}^P \end{aligned}$$

Let $v(\boldsymbol{y})$ denote the solution to the inner minimization problem

$$\begin{aligned} &v(\mathbf{y}) = \underset{x}{\text{minimize}}\ f(\boldsymbol{x}, \boldsymbol{y}) \\ &\text{subject to} \\ &\boldsymbol{g}(\boldsymbol{x}, \boldsymbol{y}) \le 0 \\ &\boldsymbol{h}(\boldsymbol{x}, \boldsymbol{y}) = 0 \\ &\boldsymbol{x} \in X \end{aligned}$$

Note that this inner problem is identical to the primal problem for $\boldsymbol{y} = \boldsymbol{y}_k$. Let V be the set of variables $\boldsymbol{y}$ for which a feasible solution to the inner minimization problem exists

$$V = \left\{ \boldsymbol{y} \middle| \boldsymbol{y} \in Y, \boldsymbol{h}(\bar{\boldsymbol{x}}, \boldsymbol{y}) = \mathbf{0}, \boldsymbol{g}(\bar{\boldsymbol{x}}, \boldsymbol{y}) \le \mathbf{0} \ \text{for some}\ \bar{\boldsymbol{x}} \in X \right\}$$

Problem $[\text{MINLP}_{\text{GBD}}]$ can be recast as follows:

$$\begin{aligned}&\underset{y}{\text{minimize}}\ v(\mathbf{y}) \quad [\text{MINLP}_{\text{GBD}}]\\&\text{subject to}\\&\mathbf{y} \in V\end{aligned}$$

The difficulty in solving this problem is that both set V and function $v(\mathbf{y})$ are only known implicitly. A possible treatment is to approximate them as the intersection of a set of explicit constraints. This is facilitated by the following theorem.

Theorem 11.1

Under the conditions stated for f, $\mathbf{g}$ and $\mathbf{h}$ a combination of binary variables $\mathbf{y} \in Y$ belongs to V if and only if it satisfies the following set of constraints [11, 14]:

$$\underset{\mathbf{x}\in X}{\text{minimize}}\ L^{\text{infeas}}(\mathbf{x}, \mathbf{y}, \boldsymbol{\lambda}', \boldsymbol{\mu}') \le 0, \quad \forall \boldsymbol{\lambda}', \boldsymbol{\mu}' \in A \tag{11.3}$$

where

$$A = \left\{ \boldsymbol{\lambda}' \in \mathbb{R}^L, \boldsymbol{\mu}' \in \mathbb{R}^M \,\middle|\, \boldsymbol{\mu}' \ge 0, \sum_{j=1}^{M} \mu_j' = 1 \right\}$$

Constraint 11.3 enables the use of Lagrange multiplier information derived from the infeasible primal to deduce the potential feasibility of a binary variable combination.

Next, we need to recast $v(\mathbf{y})$ using explicit constraints. To this end, we can replace $v(\mathbf{y})$ by its Lagrangian dual (see Chapter 9). Given the conditions stated for f, $\mathbf{g}$ and $\mathbf{h}$, the strong duality theorem holds and therefore there is no duality gap between the primal and dual problems:

$$\begin{aligned}&v(\mathbf{y}) = \underset{\boldsymbol{\lambda},\boldsymbol{\mu}\ge 0}{\text{maximize}}\ \theta(\boldsymbol{\lambda}, \boldsymbol{\mu}, \mathbf{y}) =\\&\underset{\boldsymbol{\lambda},\boldsymbol{\mu}\ge 0}{\text{maximize}}\ \underset{\mathbf{x}\in X}{\text{minimize}}\ L(\mathbf{x}, \mathbf{y}, \boldsymbol{\lambda}, \boldsymbol{\mu}), \quad \forall \mathbf{y} \in V\end{aligned} \tag{11.4}$$

where

$$L(\mathbf{x}, \mathbf{y}, \boldsymbol{\lambda}, \boldsymbol{\mu}) = f(\mathbf{x},\mathbf{y}) + \sum_{j=1}^{M} \mu_j g_j(\mathbf{x},\mathbf{y}) + \sum_{l=1}^{L} \lambda_l h_l(\mathbf{x},\mathbf{y}) \tag{11.5}$$

Having expressed set V and function $v(\mathbf{y})$ using explicit constraints, problem $[\text{MINLP}_{\text{GBD}}]$ can now be rewritten as follows:

$$\begin{aligned}&\underset{y}{\text{minimize}}\ \ \underset{\boldsymbol{\lambda},\boldsymbol{\mu}\ge 0}{\text{maximize}}\ \ \underset{\mathbf{x}\in X}{\text{minimize}}\ L(\mathbf{x}, \mathbf{y}, \boldsymbol{\lambda}, \boldsymbol{\mu})\\&\text{subject to}\\&\underset{\mathbf{x}\in X}{\text{minimize}}\ L^{\text{infeas}}(\mathbf{x}, \mathbf{y}, \boldsymbol{\lambda}', \boldsymbol{\mu}') \le 0, \quad \forall \boldsymbol{\lambda}', \boldsymbol{\mu}' \in A\end{aligned}$$

By introducing variable u to denote the maximum of the Lagrange function over multipliers $\boldsymbol{\lambda}, \boldsymbol{\mu} \geq 0$, this problem can be re-formulated as follows:

$$\underset{y \in Y, u}{\text{minimize}} \; u \qquad [\text{M}_{\text{GBD}}]$$

subject to

$$u \geq \underset{x \in X}{\text{minimize}} \, L(\boldsymbol{x}, \boldsymbol{y}, \boldsymbol{\lambda}, \boldsymbol{\mu}), \qquad \forall \boldsymbol{\lambda} \in \mathbb{R}^L, \boldsymbol{\mu} \geq 0 \tag{11.6}$$

$$0 \geq \underset{x \in X}{\text{minimize}} \, L^{\text{infeas}}(\boldsymbol{x}, \boldsymbol{y}, \boldsymbol{\lambda}', \boldsymbol{\mu}'), \quad \forall \boldsymbol{\lambda}', \boldsymbol{\mu}' \in A \tag{11.7}$$

This is defined as the *master problem* and is denoted by $[\text{M}_{\text{GBD}}]$. Note that both inner minimization problems are parametric in $\boldsymbol{y}$. Solving the master problem directly is not possible for two reasons:

(i) Constraints 11.6 and 11.7 are stated only implicitly as two inner minimization problems over variables $\boldsymbol{x}$. This can be addressed by relaxing these two constraints at each iteration k and evaluating $L(\boldsymbol{x}, \boldsymbol{y}, \boldsymbol{\lambda}, \boldsymbol{\mu})$ and $L^{\text{infeas}}(\boldsymbol{x}, \boldsymbol{y}, \boldsymbol{\lambda}', \boldsymbol{\mu}')$ at fixed values of $\boldsymbol{x}$ identified as the solutions of the primal problems (both feasible and infeasible) in the previous $k-1$ iterations. This approximates the inner minimization problem with the intersection of a set of constraints.

(ii) There is an infinite number of constraints that have to be satisfied as Constraint 11.6 must be imposed for *all* $\boldsymbol{\lambda} \in \mathbb{R}^L, \boldsymbol{\mu} \geq 0$ and Constraint 11.7 for *all* $\boldsymbol{\lambda}', \boldsymbol{\mu}' \in A$. To address this, Constraints 11.6 and 11.7 are relaxed at each iteration k by imposing them only for the values of multipliers $\boldsymbol{\lambda}, \boldsymbol{\mu}, \boldsymbol{\lambda}'$ and $\boldsymbol{\mu}'$ obtained in all previous $k-1$ iterations by the primal (both feasible and infeasible) problems.

Upon employing the two relaxations stated in (i) and (ii), the master problem is converted to an MILP with respect to $\boldsymbol{y}$ and u:

$$\underset{y \in Y, u}{\text{minimize}} \; u \qquad [\text{RM}^k_{\text{GBD}}]$$

subject to

$$u \geq L(\boldsymbol{x}_p, \boldsymbol{y}, \boldsymbol{\lambda}_p, \boldsymbol{\mu}_p), \qquad \forall p \in F \tag{11.8}$$

$$0 \geq L^{\text{infeas}}(\boldsymbol{x}_q, \boldsymbol{y}, \boldsymbol{\lambda}'_q, \boldsymbol{\mu}'_q), \quad \forall q \in IF \tag{11.9}$$

Here, F and IF denote the set of iterations where the primal problem was feasible or infeasible, respectively. The solution of the relaxed master problem $[\text{RM}^k_{\text{GBD}}]$ solves $[\text{M}_{\text{GBD}}]$ if it satisfies all ignored constraints. If not, then the relaxed master problem is resolved by accumulating additional constraints arising from subsequent primal problems. Iterations continue until all ignored constraints are satisfied or until an acceptable optimal solution based on a user-defined termination criterion is found.

Because only relaxed versions of Constraints 11.6 and 11.7 are used, the solution of the relaxed master problem will provide a lower bound on the solution of problem [$\mathrm{M_{GBD}}$] and therefore of [$\mathrm{MINLP_{GBD}}$]. The last remaining check is to ensure the feasibility of the solution of the relaxed master problem with respect to the ignored constraints. Assume that $(\boldsymbol{y}_k, u)$ is the solution to the relaxed master problem. Based on Theorem 11.1 and the definition of set V, $\boldsymbol{y}_k$ satisfies Constraint 11.7 for all $\boldsymbol{\lambda}', \boldsymbol{\mu}' \in A$ if and only if [$\mathrm{P}^k_{\mathrm{GBD}}$] has a feasible solution for $\boldsymbol{y} = \boldsymbol{y}_k$. If [$\mathrm{P}^k_{\mathrm{GBD}}$] is feasible, then it can be shown using the duality theorems [11] that

$$v(\boldsymbol{y}_k) = \theta(\boldsymbol{\lambda}, \boldsymbol{\mu}, \boldsymbol{y}_k) = \underset{\boldsymbol{x} \in X}{\text{minimize}}\, L(\boldsymbol{x}, \boldsymbol{y}_k, \boldsymbol{\lambda}, \boldsymbol{\mu})$$

where $v(\boldsymbol{y}_k)$ is the optimal objective function value of [$\mathrm{P}^k_{\mathrm{GBD}}$] at $\boldsymbol{y} = \boldsymbol{y}_k$ and $\theta(\boldsymbol{\lambda}, \boldsymbol{\mu}, \boldsymbol{y}_k)$ is that for its dual. This implies that $(\boldsymbol{y}_k, u)$ satisfies Constraint 11.6, if and only if $u \geq v(\boldsymbol{y}_k)$ [11]. This relation addresses the feasibility of the relaxed minimization in Constraint 11.6.

11.2.3 Steps of GBD Algorithm

Following the description of the primal and dual problems, the algorithmic steps for GBD are as follows:

Step 0. Set the iteration counter $k = 1$. Set the lower and upper bounds to $-\infty$ and $+\infty$, respectively. Initialize the set of feasible and infeasible iterations to $F = \varnothing$ and $IF = \varnothing$, respectively. Initialize $\boldsymbol{y}_k$.

Step 1. Solve the primal problem [$\mathrm{P}^k_{\mathrm{GBD}}$] at $\boldsymbol{y} = \boldsymbol{y}_k$.

(a) If the primal is feasible,
- Update the feasibility set $F = F \cup \{k\}$.
- Update the upper bound $\mathrm{UB} = \min(\mathrm{UB}, f(\boldsymbol{x}_k, \boldsymbol{y}_k))$ where $\boldsymbol{x}_k$ is the optimal solution to [$\mathrm{P}^k_{\mathrm{GBD}}$].
- Use the optimal Lagrange multipliers $\boldsymbol{\lambda}_k$ and $\boldsymbol{\mu}_k$ to construct the Lagrange function $L_k(\boldsymbol{x}, \boldsymbol{y}, \boldsymbol{\lambda}_k, \boldsymbol{\mu}_k)$ using Equation 11.1.

(b) If the primal is infeasible,
- Update the infeasibility set $IF = IF \cup \{k\}$.
- Solve the feasibility problem [$\mathrm{PF}^k_{\mathrm{GBD}}$] to obtain $\boldsymbol{x}_k$ and Lagrange multipliers $\boldsymbol{\lambda}'_k$ and $\boldsymbol{\mu}'_k$.
- Construct the Lagrange function $L^{\mathrm{infeas}}_k(\boldsymbol{x}, \boldsymbol{y}, \boldsymbol{\lambda}'_k, \boldsymbol{\mu}'_k)$ using Equation 11.2.

Step 2. Solve the relaxed master problem [$\mathrm{RM}^k_{\mathrm{GBD}}$] to obtain $\boldsymbol{y}_{k+1}$ and u. Set the lower bound, $\mathrm{LB} = u$.

Step 3. If $\mathrm{UB} - \mathrm{LB} \leq \varepsilon$, where ε is a user-specified convergence tolerance, or if $\mathrm{LB} > \mathrm{UB}$ (i.e., crossover) then stop. Otherwise, update the iteration counter $k \leftarrow k + 1$ and return to Step 1.

It is worth noting that the relaxed master problem provides a valid lower bound on the solution of [$\mathrm{MINLP_{GBD}}$] only if the *convexity assumption* holds on set X and

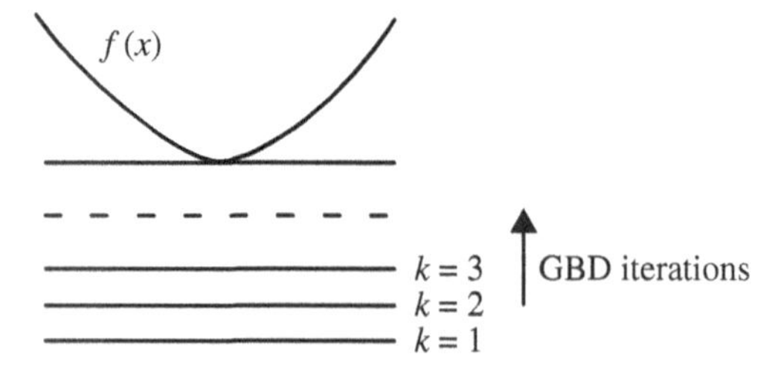

FIGURE 11.2 Benders approximation generates lower bounding hyperplanes independent of x that iteratively get closer to the optimal solution.

on functions f and $\boldsymbol{g}$, and the *linearity assumption* holds on $\boldsymbol{h}$ for a fixed value of $\boldsymbol{y}$. This means that while the GBD algorithm can be tried for any MINLP problem, convergence to a global optimal solution is guaranteed only if the stated assumptions hold. Given that the value of $\boldsymbol{x}$ is fixed in the master problem, the GBD algorithm generates lower bounding hyperplanes in X that iteratively get closer to the optimal solution (see Fig. 11.2). Because these hyperplanes have a fixed slope (equal to zero), the rate at which they converge to the optimum is generally slow. To remedy this, the outer approximation (OA) method was introduced that allows for a linearized contribution of the continuous variables $\boldsymbol{x}$ in the master problem.

11.3 OUTER APPROXIMATION

The OA algorithm was introduced by Duran and Grossmann [12] to solve the following class of MINLP problems:

$$\begin{aligned}
&\underset{\boldsymbol{x},\boldsymbol{y}}{\text{minimize}}\ f(\boldsymbol{x})+\boldsymbol{c}^{\mathrm{T}}\boldsymbol{y} \quad [\text{MINLP}_{\text{OA}}]\\
&\text{subject to}\\
&\boldsymbol{g}(\boldsymbol{x})+\boldsymbol{B}\boldsymbol{y}\leq \boldsymbol{0}\\
&\boldsymbol{x}\in X\\
&\boldsymbol{y}\in Y=\{0,1\}^{P}
\end{aligned}$$

where X is a nonempty, compact and convex subset of $\mathbb{R}^N$, $f(\boldsymbol{x})$ and $\boldsymbol{g}(\boldsymbol{x})=[g_1(\boldsymbol{x}),g_2(\boldsymbol{x}),\ldots,g_M(\boldsymbol{x})]^{\mathrm{T}}$ are continuously differentiable and convex in X, $\boldsymbol{c}$ is a $P\times 1$ vector, and $\boldsymbol{B}$ is a $M\times P$ matrix. Note that [MINLP$_{\text{OA}}$] belongs to the subclass of [MINLP$_{\text{GBD}}$] problems that are separable in $\boldsymbol{x}$ and $\boldsymbol{y}$ and are linear in $\boldsymbol{y}$. This is typically satisfied in metabolic network analysis applications, where binary variables that control the presence or absence of flux through reactions can be separated from nonlinear continuous terms, related to reaction fluxes and/or metabolite concentrations and expressed as linear terms using binary variable reformulation techniques (see Chapter 4). The original OA method does not allow for equality constraints. Therefore, all equality constraints must be eliminated before using the OA method [12, 14].

The basic idea of the OA method is similar to GBD as it iterates between the primal and master subproblems in order to generate converging upper and lower bounds on the optimal solution. The primal problem is the same as that in GBD, however, the master subproblem is an MILP based on an outer approximation (i.e., linearization) of the nonlinear objective function and constraints around the solution point of the primal problem.

11.3.1 The Primal Problem

The primal problem at iteration k is derived by fixing the binary variables at a particular 0–1 combination $\boldsymbol{y}_k$:

$$\begin{aligned}
&\underset{\boldsymbol{x}}{\text{minimize}}\ f(\boldsymbol{x})+\boldsymbol{c}^{\mathrm{T}}\boldsymbol{y}_k \quad \left[\mathrm{P}_{\mathrm{OA}}^{k}\right]\\
&\text{subject to}\\
&\boldsymbol{g}(\boldsymbol{x})+\boldsymbol{B}\boldsymbol{y}_k \le \boldsymbol{0}\\
&\boldsymbol{x}\in X
\end{aligned}$$

Problem $[\mathrm{P}_{\mathrm{OA}}^{k}]$ may be feasible or infeasible for $\boldsymbol{y} = \boldsymbol{y}_k$.

(i) *Feasible primal problem*: The upper bound is updated to be the lowest encountered so far, $\mathrm{UB} = \min(\mathrm{UB}, f(\boldsymbol{x}_k)+\boldsymbol{c}^{\mathrm{T}}\boldsymbol{y}_k)$ where $\boldsymbol{x}_k$ is the optimal solution to $[\mathrm{P}_{\mathrm{OA}}^{k}]$. Functions f and $\boldsymbol{g}$ are linearized around $\boldsymbol{x}_k$.

(ii) *Infeasible primal problem*: In this case, the following feasibility problem is formulated by introducing slack variables and relaxing all nonlinear constraints:

$$\begin{aligned}
&\text{minimize}\sum_{j=1}^{M} a_j \quad \left[\mathrm{PF}_{\mathrm{OA}}^{k}\right]\\
&\text{subject to}\\
&g_j(\boldsymbol{x})+B_j y_k \le a_j, \quad j=1,2,\ldots,M\\
&a_j \ge 0, \quad j=1,2,\ldots,M\\
&\boldsymbol{x}\in X
\end{aligned}$$

This problem is solved to determine a new $\boldsymbol{x}_k$ that is the closest possible to a feasible solution (see Fig. 11.1). Function $\boldsymbol{g}$ (but not the objective function) is next linearized around this solution point.

11.3.2 The Master Problem

Problem $[\mathrm{MINLP}_{\mathrm{OA}}]$ can be written as follows by separating the minimization operator over variables $\boldsymbol{x}$ and $\boldsymbol{y}$:

$$\begin{aligned}
&\underset{\boldsymbol{y}}{\text{minimize}}\ \underset{\boldsymbol{x}}{\text{minimize}}\ f(\boldsymbol{x})+\boldsymbol{c}^{\mathrm{T}}\boldsymbol{y} \quad \left[\mathrm{MINLP}_{\mathrm{OA}}\right]\\
&\text{subject to}\\
&\boldsymbol{g}(\boldsymbol{x})+\boldsymbol{B}\boldsymbol{y} \le 0\\
&\boldsymbol{x}\in X\\
&\boldsymbol{y}\in Y=\{0,1\}^{P}
\end{aligned}$$

Note that $\boldsymbol{c}^{\mathrm{T}}\boldsymbol{y}$ can be taken out of the inner minimization problem over $\boldsymbol{x}$. Let $v(\boldsymbol{y})$ be the solution to the inner problem:

$$
\begin{aligned}
&v(\mathbf{y}) = \boldsymbol{c}^{\mathrm{T}}\boldsymbol{y} + \underset{\boldsymbol{x}}{\text{minimize}}\ f(\boldsymbol{x}) \\
&\text{subject to} \\
&\boldsymbol{g}(\boldsymbol{x}) + \boldsymbol{B}\boldsymbol{y} \leq \boldsymbol{0} \\
&\boldsymbol{x} \in X
\end{aligned}
$$

Function $v(\boldsymbol{y})$ is equivalent to the optimal solution of $[\text{MINLP}_{\text{OA}}]$ for a given $\boldsymbol{y}$. Upon defining

$$
V = \{\boldsymbol{y} \mid \boldsymbol{y} \in \boldsymbol{Y}, \boldsymbol{g}(\boldsymbol{x}) + \boldsymbol{B}\boldsymbol{y} \leq \boldsymbol{0} \quad \text{for some}\ \boldsymbol{x} \in \boldsymbol{X}\}
$$

problem $[\text{MINLP}_{\text{OA}}]$ can be recast as follows:

$$
\begin{aligned}
&\underset{\boldsymbol{y}}{\text{minimize}}\ v(\boldsymbol{y}) \quad [\text{MINLP}_{\text{OA}}] \\
&\text{subject to} \\
&\boldsymbol{y} \in V
\end{aligned}
$$

As in GBD, both $v(\boldsymbol{y})$ and $\boldsymbol{V}$ are known only implicitly. However, in contrast to GBD, $v(\boldsymbol{y})$ is not expressed as a set of explicit constraints in OA. Instead both the nonlinear objective function and constraints are successively linearized thus converting $[\text{MINLP}_{\text{OA}}]$ to an MILP problem. Geometrically, the objective function and feasible region of $v(\boldsymbol{y})$ are expressed as the intersection of an infinite number of supporting half-spaces. These half-spaces correspond to the linearization of $f(\boldsymbol{x})$ and $\boldsymbol{g}(\boldsymbol{x})$ around all $\boldsymbol{x} \in X$. The convexity assumption on $f(\boldsymbol{x})$ and $\boldsymbol{g}(\boldsymbol{x})$ renders these linearizations valid supports (i.e., lower bounds) for these functions:

$$
f(\boldsymbol{x}) \geq f(\boldsymbol{x}_k) + \nabla^{\mathrm{T}} f(\boldsymbol{x}_k)(\boldsymbol{x} - \boldsymbol{x}_k), \quad \forall \boldsymbol{x}_k \in \boldsymbol{X} \tag{11.10}
$$

$$
\boldsymbol{g}(\boldsymbol{x}) \geq \boldsymbol{g}(\boldsymbol{x}_k) + \nabla^{T} \boldsymbol{g}(\boldsymbol{x}_k)(\boldsymbol{x} - \boldsymbol{x}_k), \quad \forall \boldsymbol{x}_k \in \boldsymbol{X} \tag{11.11}
$$

where ∇f is the gradient vector of f (an $N \times 1$ vector) and $\nabla \boldsymbol{g}$ is the Jacobian matrix of $\boldsymbol{g}$ (an $M \times N$ matrix). Under the convexity assumption on $f(\boldsymbol{x})$ and $\boldsymbol{g}(\boldsymbol{x})$, it can be shown that [12, 14]:

$$
v(\boldsymbol{y}) = \begin{bmatrix}
\underset{\boldsymbol{x}}{\text{minimize}}\ \boldsymbol{c}^{\mathrm{T}}\boldsymbol{y} + f(\boldsymbol{x}_k) + \nabla^{\mathrm{T}} f(\boldsymbol{x}_k)(\boldsymbol{x} - \boldsymbol{x}_k) \\
\text{subject to} \\
\boldsymbol{g}(\boldsymbol{x}_k) + \nabla^{\mathrm{T}} \boldsymbol{g}(\boldsymbol{x}_k)(\boldsymbol{x} - \boldsymbol{x}_k) + \boldsymbol{B}\boldsymbol{y} \leq \boldsymbol{0} \\
\boldsymbol{x} \in X
\end{bmatrix}, \quad \forall \boldsymbol{x}_k \in \boldsymbol{X} \tag{11.12}
$$

Problem [MINLP_{OA}] can thus be expressed as follows:

$$\begin{aligned}
&\underset{y}{\text{minimize}}\ \underset{x}{\text{minimize}}\ \boldsymbol{c}^{\mathrm{T}}\boldsymbol{y}+f(\boldsymbol{x}_k)+\nabla^{\mathrm{T}} f(\boldsymbol{x}_k)(\boldsymbol{x}-\boldsymbol{x}_k)\\
&\text{subject to}\\
&\boldsymbol{g}(\boldsymbol{x}_k)+\nabla^{\mathrm{T}}\boldsymbol{g}(\boldsymbol{x}_k)(\boldsymbol{x}-\boldsymbol{x}_k)+\boldsymbol{B}\boldsymbol{y}\le 0,\quad \forall \boldsymbol{x}_k\in \boldsymbol{X}\\
&\boldsymbol{x}\in X\\
&\boldsymbol{y}\in V
\end{aligned}$$

By combining the two minimizations and introducing a free variable u we obtain the following:

$$\begin{aligned}
&\underset{x,y,u}{\text{minimize}}\ \boldsymbol{c}^{\mathrm{T}}\boldsymbol{y}+u \qquad [\mathrm{M}_{\mathrm{OA}}]\\
&\text{subject to}\\
&u\ge f(\boldsymbol{x}_k)+\nabla^{\mathrm{T}} f(\boldsymbol{x}_k)(\boldsymbol{x}-\boldsymbol{x}_k),\quad \forall \boldsymbol{x}_k\in \boldsymbol{X} \qquad (11.13)\\
&0\ge \boldsymbol{g}(\boldsymbol{x}_k)+\nabla^{\mathrm{T}}\boldsymbol{g}(\boldsymbol{x}_k)(\boldsymbol{x}-\boldsymbol{x}_k)+\boldsymbol{B}\boldsymbol{y},\quad \forall \boldsymbol{x}_k\in \boldsymbol{X} \qquad (11.14)\\
&\boldsymbol{x}\in X\\
&\boldsymbol{y}\in V
\end{aligned}$$

This is the master problem (an MILP) for the OA method. Note that the minimum of u is achieved for the lowest valued term $f(\boldsymbol{x}_k)+\nabla^{\mathrm{T}} f(\boldsymbol{x}_k)(\boldsymbol{x}-\boldsymbol{x}_k)$ over x. The only remaining task is to approximate set V as a set of explicit constraints. Using again linearization, set V can be outer-approximated as follows:

$$V=\left\{\boldsymbol{y}\mid \boldsymbol{y}\in \boldsymbol{Y},\boldsymbol{g}(\boldsymbol{x}_k)+\nabla^{\mathrm{T}}\boldsymbol{g}(\boldsymbol{x}_k)(\boldsymbol{x}-\boldsymbol{x}_k)+\boldsymbol{B}\boldsymbol{y}\le 0 \ \text{for some}\ \boldsymbol{x}\in X\right\}$$

which is already encoded by Constraint 11.14 for any feasible primal problem. Note that the master problem has an infinite number of constraints as Constraints 11.13 and 11.14 must be satisfied for *all* $\boldsymbol{x}_k\in \boldsymbol{X}$. Duran and Grossmann [12] proposed that it is sufficient to impose these two constraints only for the optimal solutions of the primal problem for all $\boldsymbol{y}\in V$. Nevertheless, this still requires to exhaustively enumerate all feasible combinations of 0–1 variables $\boldsymbol{y}\in V$. A relaxation of this condition is achieved by imposing Constraints 11.13 and 11.14 at each iteration k only for the solution points of the primal problems from the previous $k-1$ iterations (*relaxed master*). In addition, we need to ensure that binary variable combinations rendering the primal problem infeasible are also infeasible in the master problem. To this end, the linearization of the nonlinear constraints around the solution of [$\text{PF}_{\text{OA}}^{k}$] must be included in the master problem as additional constraints. The relaxed master problem at iteration k is thus formulated as follows:

$$\begin{aligned}
&\underset{x,y,u}{\text{minimize}}\ \boldsymbol{c}^{\mathrm{T}}\boldsymbol{y}+u \qquad \left[\mathrm{RM}_{\mathrm{OA}}^{k}\right]\\
&\text{subject to}\\
&u\ge f(\boldsymbol{x}_p)+\nabla^{\mathrm{T}} f(\boldsymbol{x}_p)(\boldsymbol{x}-\boldsymbol{x}_p),\quad \forall p\in F \qquad (11.15)
\end{aligned}$$

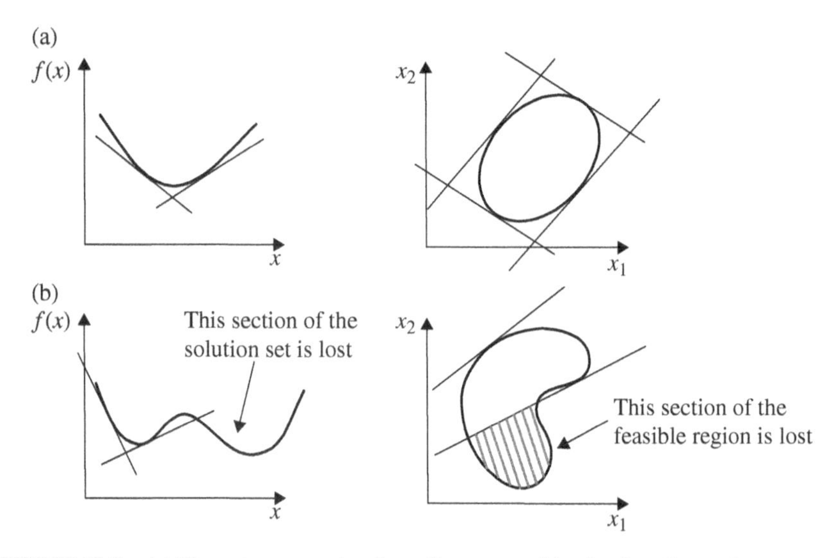

FIGURE 11.3 (a) The outer approximation of a convex objective function and convex constraints provides a valid support for the objective function and feasible region. (b) The outer approximation of a nonconvex objective function or nonconvex constraint set can result in a loss in the solution set or elimination of part of the feasible region, respectively.

$$\boldsymbol{g}(\boldsymbol{x}_p)+\nabla^{\mathrm{T}}\boldsymbol{g}(\boldsymbol{x}_p)(\boldsymbol{x}-\boldsymbol{x}_p)+\boldsymbol{B}\boldsymbol{y}\leq\boldsymbol{0},\quad \forall p\in F \tag{11.16}$$

$$\boldsymbol{g}(\boldsymbol{x}_q)+\nabla^{T}\boldsymbol{g}(\boldsymbol{x}_q)(\boldsymbol{x}-\boldsymbol{x}_q)+\boldsymbol{B}\boldsymbol{y}\leq\boldsymbol{0},\quad \forall q\in IF \tag{11.17}$$

$$\boldsymbol{x}\in X$$

$$\boldsymbol{y}\in Y$$

Here, F and IF denote the set of iterations with a primal problem that is feasible or infeasible, respectively. Furthermore, $\boldsymbol{x}_p$ and $\boldsymbol{x}_q$ are the optimal solutions to problems $[\mathrm{P}_{OA}^{k}]$ and $[\mathrm{PF}_{OA}^{k}]$, respectively. Observe that $[\mathrm{RM}_{\mathrm{OA}}^{k}]$ is an MILP problem over variables $\boldsymbol{x}$, $\boldsymbol{y}$ and u. Integer cuts can be added to the relaxed master problem to exclude identified integer solutions in the previous iterations. The objective function of $[\mathrm{MINLP}_{\mathrm{OA}}]$ is lower bounded in the relaxed master problem as its feasible region (inequality constraints) are inscribed using the polytope formed by the intersection of tangent hyperplanes, (i.e., the feasible region of $[\mathrm{MINLP}_{\mathrm{OA}}]$ is overestimated in $[\mathrm{RM}_{\mathrm{OA}}^{k}]$). Therefore, the solution to $[\mathrm{RM}_{\mathrm{OA}}^{k}]$ provides a valid lower bound on the optimal solution of $[\mathrm{MINLP}_{\mathrm{OA}}]$. The primal and relaxed master problems are iteratively solved until a user-specified termination criterion is satisfied.

Note that the linearized objective function and constraints are valid underestimators only under the convexity assumption for f and $\boldsymbol{g}$. Using these outer approximations of a nonconvex objective function or constraints can lead to the elimination of part of the feasible region and as a result the loss of confidence for global optimality (see Fig. 11.3).

Duran and Grossmann [12] argued that the addition of integer cuts excluding the previous integer solutions to the master problem alone is sufficient to replace $\boldsymbol{y} \in V$ with $\boldsymbol{y} \in Y$ (i.e., Constraint 11.17 was not included). However, it was shown later through a counterexample that this is not always true and we indeed need to include information from infeasible primal problems into the master problem (Constraint 11.17) [14, 15]. Given that the relaxed master problem of OA is solved with $\boldsymbol{x}$ as a variable, it provides a tighter lower bound compared to the relaxed master problem of GBD, which is solved for fixed values of $\boldsymbol{x}$.

11.3.3 Steps of the OA Algorithm

Considering the description of the primal and dual problems, the algorithmic steps for OA are as follows:

Step 0. Set the iteration counter $k = 1$. Set the initial lower and upper bounds to $-\infty$ and $+\infty$, respectively. Initialize the set of feasible and infeasible iterations to $F = \varnothing$ and $IF = \varnothing$, respectively. Initialize $\boldsymbol{y}_k \in Y$.

Step 1. Solve the primal problem $[\mathrm{P}_{\mathrm{OA}}^k]$ at $\boldsymbol{y} = \boldsymbol{y}_k$.
a. If the primal is feasible,

- Update the feasibility set $F = F \cup \{k\}$.
- Update the upper bound $\mathrm{UB} = \min(\mathrm{UB}, f(\boldsymbol{x}_k) + \boldsymbol{c}^{\mathrm{T}} \boldsymbol{y}_k)$ where $\boldsymbol{x}_k$ is the optimal solution to $[\mathrm{P}_{\mathrm{OA}}^k]$.
- Linearize $f(\boldsymbol{x})$ and $\boldsymbol{g}(\boldsymbol{x})$ around $\boldsymbol{x}_k$.

b. If the primal is infeasible,

- Update the infeasibility set $IF = IF \cup \{k\}$.
- Solve the feasibility problem $[\mathrm{PF}_{\mathrm{OA}}^k]$ to obtain $\boldsymbol{x}_k$.
- Linearize $\boldsymbol{g}(\boldsymbol{x})$ around $\boldsymbol{x}_k$.

Step 2. Solve the relaxed master problem $[\mathrm{RM}_{\mathrm{OA}}^k]$ to obtain $\boldsymbol{y}_{k+1}$ and u_k. Set the lower bound to $\mathrm{LB} = \boldsymbol{c}^T \boldsymbol{y}_{k+1} + u_k$.

Step 3. If $\mathrm{UB} - \mathrm{LB} \leq \varepsilon$, where ε is a user-specified convergence tolerance, or if $\mathrm{LB} > \mathrm{UB}$ (i.e., crossover), then stop. Otherwise, update the iteration counter $k \leftarrow k + 1$ and return to Step 1.

Note that if the relaxed master problem is infeasible, then $[\mathrm{MINLP}_{\mathrm{OA}}]$ is also infeasible, assuming that the assumptions made on set X and functions f and $\boldsymbol{g}$ are satisfied. If any of these conditions is not satisfied (i.e., for a nonconvex problem), the problem is generally resolved multiple times with a different initial $\boldsymbol{y}_k$.

11.4 OUTER APPROXIMATION WITH EQUALITY RELAXATION

The outer approximation algorithm was extended by Kocis and Grossmann [13] to handle nonlinear equality constraints. This algorithm, which was called *outer approximation with equality relaxation* (OA/ER) was designed to solve the following class of MINLP problems:

$$\begin{aligned}&\underset{\boldsymbol{x},\boldsymbol{y}}{\text{minimize}}\ c^{\mathrm{T}}\boldsymbol{y}+f(\boldsymbol{x}) \quad [\text{MINLP}_{\text{OA/ER}}]\\ &\text{subject to}\\ &\boldsymbol{g}(\boldsymbol{x})+\boldsymbol{B}\boldsymbol{y}\le\boldsymbol{0}\\ &\boldsymbol{h}(\boldsymbol{x})+\boldsymbol{A}\boldsymbol{y}=\boldsymbol{0}\\ &\boldsymbol{x}\in X\\ &\boldsymbol{y}\in Y=\{0,1\}^{P}\end{aligned}$$

where

- $\boldsymbol{B}$ and $\boldsymbol{A}$ are $M\times P$ and $L\times P$ matrices
- X is a nonempty, compact and convex subset of $\mathbb{R}^N$
- f is convex in X
- $\boldsymbol{g}(\boldsymbol{x})=[g_1(\boldsymbol{x}),g_2(\boldsymbol{x}),\ldots,g_M(\boldsymbol{x})]^{\mathrm{T}}$ such that
  - ∘ $\boldsymbol{g}_j(\boldsymbol{x})$ are convex, $\forall j\in\{j|g_j(\boldsymbol{x})<0\}$
  - ∘ $\boldsymbol{g}_j(\boldsymbol{x})$ are quasi-convex, $\forall j\in\{j|g_j(\boldsymbol{x})=0\}$
- $\boldsymbol{T}\,\boldsymbol{h}(\boldsymbol{x})$ are quasi-convex functions where $\boldsymbol{h}(\boldsymbol{x})=[h_1(\boldsymbol{x}),h_2(\boldsymbol{x}),\ldots,h_L(\boldsymbol{x})]^{\mathrm{T}}$ and $\boldsymbol{T}$ is a $L\times L$ diagonal matrix whose diagonal elements are defined as follows:

$$t_{ll}=\begin{cases}1 & \text{if }\lambda_l>0\\ 0 & \text{if }\lambda_l=0\\ -1 & \text{if }\lambda_l<0\end{cases}\quad l=1,2,\ldots,L$$

  Here, λ_l is the Lagrange multiplier associated with $h_l(\boldsymbol{x})$. Furthermore, functions f, $\boldsymbol{g}$ and $\boldsymbol{h}$ are assumed to be continuously differentiable. Note that the familiar case where $g_j(\boldsymbol{x})$ is convex for $j=1,2,\ldots,M$ and $h_l(\boldsymbol{x})$ is linear for $l=1,2,\ldots,L$ is a special case of the general conditions imposed on $g_j(\boldsymbol{x})$ and $t_{ll}h_l(\boldsymbol{x})$.

The main idea of the OA/ER algorithm is to relax equality constraints into inequalities and subsequently apply the OA algorithm. The key decision at each iteration is to surmise the active direction of the inequality (encoded within matrix $\boldsymbol{T}$). To this end, each equality constraint is written as two opposing inequalities:

$$\begin{cases}h_l(\boldsymbol{x})\le 0\\ h_l(\boldsymbol{x})\ge 0\end{cases}\quad l=1,2,\ldots,L$$

If the Lagrange multiplier associated with an equality constraint is positive then the equality can be relaxed as $h_l(\boldsymbol{x})\le 0$ with no effect on the optimal solution point or value. Similarly, if the Lagrange multiplier is negative then the same optimal solution will be obtained if the equality is relaxed as $h_l(\boldsymbol{x}) \geq 0$. Therefore, the sign of the Lagrange multiplier for an equality constraint at the optimal solution determines the permitted direction of equality relaxation with no effect on the optimal solution. The OA/ER algorithm attempts to choose the "correct" constraint directionality of the nonlinear equality constraints in order to relax them to inequalities. Here, we skip the description of the primal problem, as it is the same as that in the OA algorithm, with the only difference being the formulation of the infeasible primal problem, which is as follows:

$$\text{minimize} \sum_{j=1}^{M} a_j + \sum_{l=1}^{L} \left(b_l^+ + b_l^-\right) \quad \left[\text{PF}_{\text{OA/ER}}^k\right]$$

$$\begin{aligned}
&\text{subject to} \\
&h_l\left(\boldsymbol{x}, \boldsymbol{y}_k\right) + A y_k = b_l^+ - b_l^-, && l = 1,2,\ldots,L \\
&g_j\left(\boldsymbol{x}, \boldsymbol{y}_k\right) + B y_k \leq a_j, && j = 1,2,\ldots,M \\
&a_j, b_l^+, b_l^- \geq 0, && j = 1,2,\ldots,M, \quad l = 1,2,\ldots,L \\
&\boldsymbol{x} \in X
\end{aligned}$$

11.4.1 The Master Problem

The key idea for the derivation of the master problem is the following theoretical result from Kocis and Grossmann [13, 14]. If the conditions stated on X, f, $\boldsymbol{g}$, and $\boldsymbol{h}$ hold, then problem $[\text{MINLP}_{\text{OA/ER}}]$ for a fixed $\boldsymbol{y} = \boldsymbol{y}_k$ is equivalent to the following problem:

$$\begin{aligned}
&\underset{x}{\text{minimize}}\ \boldsymbol{c}^{\text{T}} \boldsymbol{y}_k + f(\boldsymbol{x}) \\
&\text{subject to} \\
&\boldsymbol{g}(\boldsymbol{x}) + \boldsymbol{B}\boldsymbol{y} \leq 0 \\
&\boldsymbol{T}_k\left[\boldsymbol{h}(\boldsymbol{x}) + \boldsymbol{A}\boldsymbol{y}\right] \leq 0 \\
&\boldsymbol{x} \in X \\
&\boldsymbol{y} \in Y = \{0,1\}^P
\end{aligned}$$

The relaxed master problem at each iteration k is similar to that in OA with the only difference being that an outer approximation of $\boldsymbol{T}_k \boldsymbol{h}(\boldsymbol{x}) \leq 0$ should be included in the set of constraints:

$$\underset{x,y,u}{\text{minimize}}\ \boldsymbol{c}^{\text{T}} \boldsymbol{y} + u \qquad \left[\text{RM}_{\text{OR/ER}}^k\right]$$

subject to

$$\boldsymbol{u} \geq f\left(\boldsymbol{x}_p\right) + \nabla^{\text{T}} f\left(\boldsymbol{x}_p\right)\left(\boldsymbol{x} - \boldsymbol{x}_p\right), \qquad \forall p \in F \tag{11.18}$$

$$\boldsymbol{g}\left(\boldsymbol{x}_p\right) + \nabla^{\text{T}} \boldsymbol{g}\left(\boldsymbol{x}_p\right)\left(\boldsymbol{x} - \boldsymbol{x}_p\right) + \boldsymbol{B}\boldsymbol{y} \leq 0, \qquad \forall p \in F \tag{11.19}$$

$$T_p\left[\boldsymbol{h}\left(\boldsymbol{x}_p\right) + \nabla^{\text{T}} \boldsymbol{h}\left(\boldsymbol{x}_p\right)\left(\boldsymbol{x} - \boldsymbol{x}_p\right) + \boldsymbol{A}\boldsymbol{y}\right]^{\text{T}} \leq 0, \quad \forall p \in F \tag{11.20}$$

$$\boldsymbol{g}(\boldsymbol{x}_q)+\nabla^{\mathrm{T}}\boldsymbol{g}(\boldsymbol{x}_q)(\boldsymbol{x}-\boldsymbol{x}_q)+\boldsymbol{B}\boldsymbol{y}\le 0, \qquad \forall q\in IF \tag{11.21}$$

$$T_q\left[\boldsymbol{h}(\boldsymbol{x}_q)+\nabla^{\mathrm{T}}\boldsymbol{h}(\boldsymbol{x}_q)(\boldsymbol{x}-\boldsymbol{x}_q)+\boldsymbol{A}\boldsymbol{y}\right]^{\mathrm{T}}\le 0, \quad \forall q\in IF \tag{11.22}$$

$$\boldsymbol{x}\in X$$

$$\boldsymbol{y}\in Y$$

The steps of the OA/ER algorithm mirror the ones for OA with the additional provision of having to store the optimal Lagrange multipliers associated with $\boldsymbol{h}(\boldsymbol{x})$ and construct the diagonal matrix $\boldsymbol{T}_k$ at each iteration k for the relaxed master problem. As in OA, integer cuts are added to the relaxed master problem to exclude the previously identified integer solutions. The OA/ER algorithm is accessible through the DICOPT solver in GAMS, wherein a number of different stopping criteria are available. Using option $stop = 1$ causes termination only when the gap between the upper and lower bounds is less than epsilon. Option $stop = 2$ causes termination whenever the value of the feasible primal worsens at an iteration and is reserved for long running problems when rigorous convergence is time prohibitive to reach. Option $stop = 3$ implies termination whenever $stop = 1$ or 2 criterion is met. In contrast, option $stop = 0$ causes termination only when the specified iteration limit is reached irrespective of the lower and upper bounds. This option is used for nonconvex problems where the relaxed master problem is not guaranteed to provide a valid lower bound. For complex problems, it is worthwhile to rely on a customized implementation of the OA/ER method to explore various reformulation techniques for solving the MILP problems associated with the relaxed master more efficiently (see Chapter 4).

In the next section, we present an MINLP application in metabolic networks extending the OptKnock algorithm for cases where some of the metabolic reaction fluxes are calculated with kinetic expressions.

11.5 KINETIC OPTKNOCK

The OptKnock procedure described in Chapter 8 relies ultimately on an MILP description to suggest targeted knockouts leading to the overproduction of a metabolite of interest using a stoichiometric description of metabolism. In this section, we describe an extension of the OptKnock, termed *k-OptKnock* (kinetic OptKnock), which uses an MINLP optimization formulation to identify both knockouts and enzymatic manipulation strategies while integrating kinetic expressions (whenever available) within stoichiometric models of metabolism. Kinetic expressions can remedy some of the shortcomings of genome-scale stoichiometric models by capturing the effect of concentration and enzyme levels on reaction throughput and regulation. As described in Chapter 3, kinetic models yield a system of ordinary differential equations (ODEs) that describe the time evolution of metabolite concentrations, enzyme activities and reaction fluxes. The k-OptKnock procedure extends the OptKnock algorithm [16] by integrating all available mechanistic details afforded by kinetic models within a constraint-based optimization framework (see Section 10.2). It uses a bilevel MINLP framework to identify genetic intervention strategies consistent with restrictions imposed by maximum enzyme activities, bounds on metabolite concentrations and kinetic expressions. k-OptKnock can be thought of as a special case of the k-OptForce procedure [2] which captures up- and downregulations in addition to knockouts for the stoichiometric part of the model.

11.5.1 k-OptKnock Formulation

Similarly to the hybrid stoichiometric and kinetic modeling approach described in Section 10.2, the set of reactions in the network J is divided into two subsets: reactions with kinetic information J^{kin} and those having only stoichiometric information J^{stoic}. The flux of reactions in J^{stoic} is constrained only by stoichiometric steady-state mass balances and reaction directionality restrictions as in FBA, whereas the flux of reactions in J^{kin} are determined by the corresponding enzyme activity, metabolite concentrations, and kinetic parameter values through kinetic rate expressions. While reactions in J^{stoic} are candidates for knockout only, reactions in J^{kin} can be continuously modulated (i.e, up-/downregulation) by manipulating their maximum enzymatic reaction rate $v_j^{\max}$. Two separate sets of binary variables are defined to capture interventions in the stoichiometric and kinetic parts of the model.

$$y_j^{\text{stoic}} = \begin{cases} 0 & \text{if reaction } j \text{ is eliminated} \\ 1 & \text{otherwise} \end{cases} \quad \forall j \in J^{\text{stoic}}$$

$$y_j^{\text{kin}} = \begin{cases} 0 & \text{if } V^{max} \text{ of reaction } j \text{ is allowed to vary} \\ 1 & \text{otherwise} \end{cases} \quad \forall j \in J^{\text{kin}}$$

k-OptKnock can then be formulated using the following bilevel MINLP:

$$\underset{y_j^{\text{kin}}, y_j^{\text{stoic}}, C_i, u_j}{\text{maximize}} \quad z = v_{\text{product}} \left[\text{k} - \text{OptKnock}\right]$$

subject to

$$u_j = u_j\left(v_j^{\max}, C_i, P_j\right), \quad \forall j \in J^{\text{kin}} \tag{11.23}$$

$$v_j^{\max,\text{ref}} y_j^{\text{kin}} \le v_j^{\max} \le v_j^{\max,\text{ref}}\left(y_j^{\text{kin}} + f\left(1 - y_j^{\text{kin}}\right)\right), \quad \forall j \in J^{\text{kin}} \tag{11.24}$$

$$C_i^{\text{LB}} \le C_i \le C_i^{\text{UB}}, \quad \forall i \in I^{\text{kin}} \tag{11.25}$$

$$\sum_{j \in J^{\text{kin}}} (1 - y_j^{\text{kin}}) + \sum_{j \in J^{\text{stoic}}} (1 - y_j^{\text{stoic}}) \le K \tag{11.26}$$

$$\left[\begin{array}{l} \underset{v_j}{\text{maximize}} \quad v_{\text{biomass}} \\ \text{subject to} \\ \sum_{j \in J} S_{ij} v_j = 0, \quad \forall i \in I \qquad (11.27) \\ LB_j y_j^{\text{stoic}} \le v_j \le UB_j y_j^{\text{stoic}}, \quad \forall j \in J^{\text{stoic}} \qquad (11.28) \\ v_j = u_j, \quad \forall j \in J^{\text{kin}} \qquad (11.29) \end{array}\right]$$

$$y_j \in \{0,1\}, v_j \in \mathbb{R}, \quad \forall j \in J$$

where v_{product} is the exchange flux of the target product, u_j is the rate of a reaction $j \in J^{\text{kin}}$ determined by its kinetic rate expression, C_i is the concentration of a metabolite i in the kinetic part of the model, C_i^{LB} and C_i^{UB} denote lower and upper bounds on the concentration of a metabolite i, P_j is the set of kinetic parameters related to a reaction $j \in J^{\text{kin}}$ and f denotes the maximum allowable departure (specified by the user) from the maximum enzymatic reaction rate in the reference (wild-type) strain $\left(v_j^{\text{max,ref}}\right)$.

The outer-level objective function maximizes the metabolic flux toward the desired metabolite consistent with reaction kinetics and stoichiometry. Constraint 11.23 sets the fluxes using the kinetic expressions for reactions in J^{kin}. Constraint 11.24 imposes enzymatic manipulations in the kinetic part of the network using binary variables y_j^{kin}. In particular, if $y_j^{\text{kin}} = 0$ then v_j^{max} can be modulated (continuously) between zero (indicating the removal of enzyme activity) and $fv_j^{\text{max,ref}}$ where f denotes the degree of enzyme amplification. If $y_j^{\text{kin}} = 1$ then v_j^{max} is kept constant at its reference (wild-type) value $v_j^{\text{max,ref}}$. Constraint 11.25 imposes lower and upper bounds on the concentrations for metabolites $i \in I^{\text{kin}}$ where I^{kin} is the set of metabolites in the kinetic part of the model. Constraint 11.26 restricts the total number of manipulations in the kinetic and stoichiometric parts of the network to be no more than a user-specified value K.

Constraint 11.27 describes the steady-state mass balance for all metabolites in the network. Constraint 11.28 imposes reaction removals for reactions in J^{stoic} consistent with the flux distribution in J^{kin}. Constraint 11.29 fixes the flux of reactions in J^{kin} at u_j determined by the outer problem. The constraint on the minimum required level of the biomass formation has been subsumed within Constraint set 11.28 with $y_{\text{biomass}}^{\text{stoic}}$ fixed at one. The same holds for the uptake limits of the limiting substrates in the growth medium (e.g., glucose and oxygen) as well as for the non-growth associated maintenance ATP. As in the original OptKnock procedure, the inner problem maximizes the biomass flux subject to the knockouts and enzyme level manipulations for reactions in J^{kin} imposed by the outer problem. Therefore, the inner problem is parametric in y_j^{stoic} and u_j.

11.5.2 Solution Procedure for k-OptKnock

The inner maximization problem in k-OptKnock can be recast as a set of explicit constraints exactly in the same manner as in OptKnock by imposing the linear programming (LP) strong duality theorem for the inner problem and accumulating the constraints of the primal and dual problems (see Chapter 8). Note that the dual problem is an LP as it is parametric in u_j and y_j^{stoic}. This transformation converts [k−OptKnock] to a single-level MINLP. There are two sources of nonlinearities in the resulting problem. The first arises due to the presence of the kinetic expressions, which are typically nonlinear functions of the metabolite concentrations (Constraint 11.23). The second source of nonlinearity is due to the dualization of the inner problem leading to the presence of bilinear terms, which are the product of u_j and the dual variables. Note that in contrast to the product of a continuous variable and binary variables, which can be linearized in a straightforward manner (see Chapter 4), the product of two continuous variables cannot be exactly linearized. The bilinear terms due to the dualization can be avoided by imposing the KKT optimality conditions for the inner problem instead of using the LP strong duality theorem (see Chapters 2 and 8). This, however, comes at the expense of introducing a new set of binary variables to enforce the complementary slackness (CS) condition. In both cases, the resulting single-level MINLP can be solved using any of the techniques described in this chapter, however, these methods will give only a local optimum as here we have a nonconvex MINLP. A strategy for finding a (near) global solution is to solve the MINLP using multiple initial points and choose the one with the best objective function value. Alternatively, one can use global MINLP solvers such as BARON [17] and ANTIGONE [18], which are both accessible through GAMS. Similar to OptKnock, k-OptKnock is successively solved for an increasing number of interventions (i.e., by increasing the value of K in Constraint 11.26) until the target yield is met (see also Chapter 8).

Example 11.1

Use k-OptKnock to identify enzymatic modulations and reaction removals leading to the overproduction of acetate in *Escherichia coli* using the *i*AF1260 stoichiometric model [19] and a kinetic model of *E. coli* central metabolism presented in [20] under aerobic conditions in a minimal medium with glucose as the carbon source. Identify as many interventions as needed to achieve a nonzero acetate production yield. Use the following input parameters:

- Minimum required biomass formation in the network: 10% of the theoretical maximum $\left(v_{\text{biomass}} \geq 0.1 v_{\text{biomass}}^{\text{WT}}\right)$.
- Maximum allowable upregulation of $v_j^{\max}$, $f = 10$ for acetate kinase (ACK) which is the terminal reaction in the acetate production pathway and $f = 2$ for the rest of reactions in the kinetic part of the model.
- Bounds on reactions in the stoichiometric part of the model: See Table 3.2.
- Bounds on metabolite concentrations: Allow departure from the concentrations in the reference (wild-type) strain by at most twofold (i.e., $0.5\, C_i^{\text{ref}} \leq C_i \leq 2\, C_i^{\text{ref}}$).

Generate the trade-off plot for the wild-type and identified mutant strain. Is acetate production coupled with biomass formation in the mutant strain?

Solution: k-OptKnock suggests five interventions for a nonzero production of acetate, all of which are present in the part of the metabolic network with available kinetic information. The identified manipulations include upregulation of phosphogluconate dehydrogenase (GND) and phosphoenolpyruvate carboxylase (PPC) enzymes by two-fold of their reference activity along with a tenfold upregulation of acetate kinase (ACK). In addition, k-OptKnock suggests an almost twofold downregulation of phosphofructo-kinase (PFK) and removal of pyruvate kinase (PYK) activity. Downregulation of PFK reduces the flux through phosphoglycerate kinase (PGK), which in combination with the removal of PYK, reduces the overall ATP yield of EMP glycolysis. This reduced ATP production efficiency renders the acetate production pathway the major source of ATP generation (ACK produces one mole of ATP) thereby coupling acetate and biomass production. These interventions enable an acetate production flux of $15.36 \dfrac{\text{mmol}}{\text{gDW} \cdot \text{h}}$ for $10 \dfrac{\text{mmol}}{\text{gDW} \cdot \text{h}}$ fed glucose corresponding to 60.03% of the theoretical maximum.

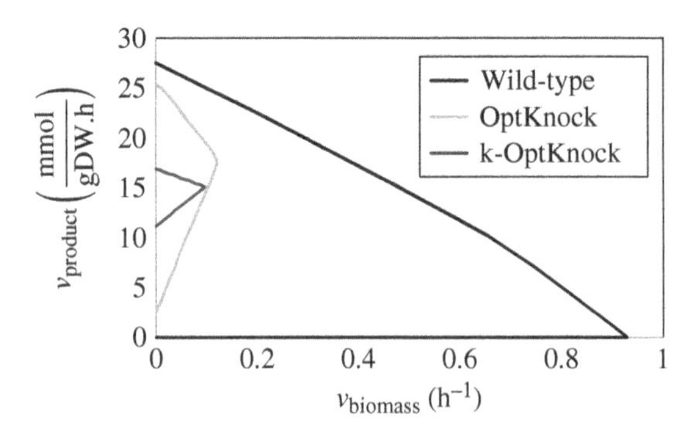

FIGURE 11.4 Trade-off plot for the wild-type and the mutant strain designed by OptKnock (see Fig. 8.4b) and k-OptKnock (see Example 11.1) for the overproduction of acetate in *E. coli*.

The trade-off plot for the mutant strain (see Fig. 11.4) shows that acetate production is fixed at maximum biomass (0.096 h^{-1}) for the designed mutant, implying that acetate production and biomass formation are fully coupled. It is worth noting that the incorporation of kinetic constraints reduces the variability in acetate production reached for a lower level of biomass yield (0.096 h^{-1}) compared to that achieved by OptKnock (0.12 h^{-1}) for the double knockout mutant (ΔATPS, ΔENO) (see Example 8.1). A GAMS implementation of this example is available on the book's website. □

More complex strain design case studies can be explored using MINLP by following the reformulation techniques and concepts introduced in this example.

EXERCISES

11.1 Solve the following MINLP using the OA/ER algorithm accessible through the DICOPT solver in GAMS (this problem was adapted from Ref. 13):

$$\text{minimize } z = 4.5y_1 + y_2 + 2.5y_3 + 1.9x_1 + 8x_5 + x_6 + 1.3x_7 - 11.5x_8$$

subject to

$$x_6 - \ln(1.5 + x_2) = 0$$
$$x_7 - 1.5\ln(1 + 2x_3) = 0$$
$$x_8 - 0.8x_4 = 0$$
$$-x_4 + x_5 + x_6 + x_7 = 0$$
$$x_1 - x_2 - x_3 = 0$$
$$x_4 - 6y_1 \le 0$$
$$x_2 - 6y_2 \le 0$$
$$x_3 - 6y_3 \le 0$$
$$x_6 \le 6$$
$$x_8 \le 1$$
$$y_1, y_2, y_3 \in \{0,1\}$$
$$x_1, x_2, x_3, x_4, x_5, x_6, x_7, x_8 \ge 0$$

11.2 Consider Example 11.1. Repeat the study after replacing the maximization of biomass formation as the inner problem with the MOMA criterion [1]. Use as an approximation of the wild-type steady-state metabolic fluxes those obtained by solving the system of ODEs or the corresponding system of equations for the kinetic part of the network. Next, fix the flux of these reactions in an FBA problem to approximate the wild-type steady-state fluxes for the stoichiometric part of the network. Apply the KKT conditions to express the optimality relations for the inner problem. What are the differences in the results when compared with those obtained in Example 11.1?

REFERENCES

1. Segrè D, Vitkup D, Church GM: Analysis of optimality in natural and perturbed metabolic networks. *Proc Natl Acad Sci U S A* 2002, **99**(23):15112–15117.
2. Chowdhury A, Zomorrodi AR, Maranas CD: k-OptForce: Integrating kinetics with flux balance analysis for strain design. *PLoS Comput Biol* 2014, **10**(2):e1003487.
3. Nikolaev EV: The elucidation of metabolic pathways and their improvements using stable optimization of large-scale kinetic models of cellular systems. *Metab Eng* 2010, **12**(1):26–38.
4. Pozo C, Guillen-Gosalbez G, Sorribas A, Jimenez L: A Spatial Branch-and-Bound Framework for the Global Optimization of Kinetic Models of Metabolic Networks. *Indust Eng Chem Res* 2011, **50**(9):5225–5238.
5. Pozo C, Miro A, Guillen-Gosalbez G, Sorribas A, Alves R, Jimenez L: Gobal optimization of hybrid kinetic/FBA models via outer-approximation. *Comput Chem Eng* 2015, **72**:325–333.
6. Guillén-Gosálbez G, Miró A, Alves R, Sorribas A, Jiménez L: Identification of regulatory structure and kinetic parameters of biochemical networks via mixed-integer dynamic optimization. *BMC Syst Biol* 2013, **7**:113.
7. Kirkpatrick S, Gelatt CD, Jr., Vecchi MP: Optimization by simulated annealing. *Science* 1983, **220**(4598):671–680.
8. Goldberg DE: *Genetic algorithms in search, optimization, and machine learning*. Boston, M.A.: Addison-Wesley Professional; 1999.
9. Dorigo M, Stützle T: *Ant colony optimization*. Cambridge, M.A.: MIT Press; 2004: 1 online resource (xi, 305 p).
10. Kennedy J, Eberhart R: IEEE: Particle swarm optimization. *IEEE International Conference on Neural Networks Proceedings,* Vols 1–6, 1995: 1942–1948.
11. Geoffrion A: Generalized Benders Decomposition. *J Optiz Theory App* 1972, **19**(4):237–260.
12. Duran MA, Grossmann IE: An outer-approximation algorithm for a class of mixed-integer nonlinear programs. *Math Program* 1986, **36**(3):307–339.
13. Kocis GR, Grossmann IE: Relaxation strategy for the structural optimization of process flow sheets. *Indust Eng Chem Res* 1987, **26**(9):1869–1880.
14. Floudas CA: *Nonlinear and mixed-integer optimization : fundamentals and applications*. New York: Oxford University Press; 1995.
15. Fletcher R, Leyffer S: Solving mixed integer nonlinear programs by outer approximation. *Math Program* 1994, **66**(3):327–349.

16. Burgard AP, Pharkya P, Maranas CD: Optknock: a bilevel programming framework for identifying gene knockout strategies for microbial strain optimization. *Biotechnol Bioeng* 2003, **84**(6):647–657.
17. Sahinidis NV: BARON: A general purpose global optimization software package. *J Global Optim* 1996, **8**(2):201–205.
18. Misener R, Floudas CA: ANTIGONE: Algorithms for coNTinuous/integer global optimization of nonlinear equations. *J Global Optim* 2014, **59**(2–3):503–526.
19. Feist AM, Henry CS, Reed JL, Krummenacker M, Joyce AR, Karp PD, Broadbelt LJ, Hatzimanikatis V, Palsson B: A genome-scale metabolic reconstruction for *Escherichia coli* K-12 MG1655 that accounts for 1260 ORFs and thermodynamic information. *Mol Syst Biol* 2007, **3**:121.
20. Khodayari A, Zomorrodi AR, Liao JC, Maranas CD: A kinetic model of *Escherichia coli* core metabolism satisfying multiple sets of mutant flux data. *Metab Eng* 2014, **25C**:50–62.

APPENDIX A

CODING OPTIMIZATION MODELS IN GAMS

General Algebraic Modeling System (GAMS, www.gams.com) is a modeling interface for optimization. It provides a convenient way to

- build an optimization model
- interface with a variety of different optimization solvers
- create a customized output
- check the validity of the obtained solution

GAMS provides a simplified interface to input an optimization problem in a similar way as the problem is defined. GAMS is accessed by the user through a *keyword input file* with the gms extension. The optimization model is supplied in the form of *indexed algebraic equations* that describe the objective function and constraints of the model without having to define matrices and/or vectors of coefficients or functions for non-linear terms. The advantage of indexed algebraic equations is that a constraint such as

$$\sum_{i=1}^{50} a_{ij} x_j \leq b_j, \quad j = 1,2,\ldots,100$$

can be encoded in GAMS with a single statement

```
SUM(i, a(i,j) * x(i)) =L= b(j)
```

Optimization Methods in Metabolic Networks, First Edition. Costas D. Maranas and Ali R. Zomorrodi.

where, i and j are declared as sets with specified elements. This description avoids creating multiple declarations for 100 separate constraints each containing 50 terms. GAMS then compiles the model and interfaces automatically with the solver of one's choice. Finally, the execution results (global optimality, local optimality or infeasibility) are reported to the user through an output file.

A.1 RUNNING GAMS

Create the input file using your favorite text editor and save it as a flat *text* file with the extension .gms (i.e., *filename.gms*). Issue the command "gams filename. gms" to compile and execute the optimization model. If there are any syntax errors in the input file, GAMS will provide appropriate and extensive error messages in the output file. The output file has the same filename as the input file with the extension lst (i.e. *filename.lst*) and can be read using any text editor. After correcting all errors in the input gms file and re-running the model the solution results will be included in the *filename.lst* file. Customized output files can be created using the PUT statement described later in this chapter. Note that multiple tries may be needed before one generates an error-free code. Do not get discouraged by the large number of errors seen at first. Most of them are due to propagation of the original mistakes.

A.2 THE INPUT FILE

GAMS input files are generally organized as follows:

- Options
- Sets
- Parameters
- Variables
- Equations
- Model and output files
- Solve

It is good programming practice to comment within the input file in order to inform others as to the purpose of the model and specific tasks within the optimization task. In GAMS, commenting can be achieved in two ways:

1. The character "*" in the first column of the input file implies that the rest of the line is a comment to be ignored by GAMS. Be sure that a line, which has been text wrapped, does not begin with an "*" indicating a comment instead of multiplication.

2. The $ONTEXT/$OFFTEXT commands can also be used to comment out all lines between the initiation and termination of these commands. Every $ONTEXT must have a matching $OFFTEXT and both of the commands must be on their own line.

Another good programming practice is to use capital letters for GAMS keywords (as done throughout this appendix) and lower case letters for user-defined variables, equations and data. However, GAMS does not distinguish between lower case and capital letters in the input file.

A.2.1 Options

Options and specifications can be defined to change the default values assigned by GAMS pertaining to the tolerance, iteration number, space allotted, etc. These statements are specified at the top of the input file under a section declared as "OPTIONS". Each one of the options is separated by either a space or, more typically, placed in the next line. A semicolon is used to denote the end of all newly specified options. The following is a list of frequent entries:

- OPTCA: The absolute optimality tolerance. The default value is zero, but a more conservative value of 10^{-6} is often used (OPTCA = 0.000001).
- ITERLIM: The maximum number of iterations. The default is 2000000000, but it can be redefined to a lower value (e.g., ITERLIM = 100000).
- LP/NLP/MIP/MINLP: The type of optimization problem (LP = linear programming, NLP = nonlinear programming, MIP = mixed-integer linear programming, MINLP = mixed-integer nonlinear programming) followed by the solver to be used. For example LP = cplex.

A.2.2 Sets

In GAMS, sets typically are ordered indices over which constraints and/or sums can be written. A set S that contains elements i, j and k (i.e., $S=\{i, j, k\}$) is written in GAMS notation as follows:

```
SET  S  example set
/i, j, k/;
```

A set R of integers can be simply defined in GAMS using the asterisk as follows:

```
SET  R  example set of integers
/10 * 75/;
```

Defining a set in this manner will assign a new element for every integer between 10 and 75 increasing by 1. Finally, sets can be defined from any external text file imported into GAMS. The text file which must follow proper GAMS syntax is imported using the "$INCLUDE" command:

```
$INCLUDE [path/]filename.txt
```

A.2.3 Parameters

The parameters section in GAMS is used to specify parameters with known values defined over the previously defined sets. Three different data types can be used to enter data in this section: scalar, parameter and table statements. Generally for larger data sets, the parameter statement is used to enter data in a list format. The scalar statement is used to define single data entries and the table statement is used for two-dimensional data. Parameters that depend on two different sets, for example *S(i,j)*, can be entered using the following format:

```
PARAMETER S(i,j)
i1.j1  value
i2.j1  value/;
```

A.2.4 Variables

Variables are defined similarly to sets and parameters. Each variable must be defined before use on a separate line with a semicolon at the end indicating the conclusion of all variable declarations as shown:

```
VARIABLES
x(i,j)  Declaration of variable x over indices i and j
y    Declaration of variable y
```

POSITIVE, NEGATIVE, BINARY and INTEGER are the allowed variable types with the FREE variable type being the default. Variables can be assigned lower, upper, fixed or level values.

- The lower .LO and upper bounds .UP are useful for providing problem-specific variable bounds without defining separate constraints.
- Fixing variables at a defined value can be accomplished with the .FX suffix. For example,

  ```
  x.FX('5') = 0;
  ```

  fixes x_s to be equal to zero. This can be useful when debugging a model.
- The current value or level .L of a variable can be defined using the .L suffix. This is used to set an initial guess for a variable. The variable level stores the value of the variable upon execution of the optimization model.

- The marginal value or dual value of a variable is denoted using the .M suffix. Similarly to the level value, the marginal value is updated when the model is solved. The marginal value of a variable corresponds to the sensitivity of the optimized objective function value to the variable value (see Chapter 2).

A.2.5 Equations

The equations are algebraic relationships used to generate the constraints within the model. The equation names must first be declared before the definition is provided. Similar to the way the variables are defined, the equations are declared as follows:

```
EQUATIONS
eqn_name1  description of eqn1
eqn_name2(i)  description of eqn2 defined over index i;
```

When defining equations, the equation name is followed by '..' and subsequently mathematical description for the equation closing with a semicolon. The GAMS keywords =E=, =L= and =G= should be used instead of =, ≤ and ≥, respectively, to define an equation. The dollar operator ("$") provides a convenient way to exclude elements of a set from a summation or an equation declaration. In a logical statement, keywords "EQ", "NE", "LT", "GT", "LE" and "GE" are used to represent logical relationships "=," "≠," "<," ">," "≤," and "≥," respectively.

Consider the following objective function and constraints:

$$f = \sum_i x_i$$

$$x_i \leq 5, \quad i = 4,5,6,7$$

They can be written in GAMS as follows:

```
EQUATIONS
obj  the objective statement of the model
eq(i) the inequality constraints of the problem ;
obj..    f =E= SUM(i, x(i));
eq(i)$((ORD(i) GE 4) AND (ORD(i) LE 7)).. x(i) =L= 5;
```

Note that the following functions and statements are commonly used to define equations in GAMS:

- The alias statement ALIAS(i,i_star); is used to access the original set by multiple running indices. This is equivalent to using i' in math representation when both i and i' assume values from the same set.

- The ORD function ORD(set element) returns the placement order of an element in a set.
- Similarly, the CARD function CARD(set name) returns the total number of elements in a set.
- The summation notation SUM(index, equation to sum over); sums over all variables in a set. More than one indices can be summed over at once as shown in the following example: $\sum_i \sum_j x_{ij}$ can be written as SUM((i,j), x(i,j)).
- In the power function POWER(x,y) exponent y must be an integer.
- The "**" operator (x**y) instead is used only when y is real and positive. Correct use of parenthesis is essential when using the "**" operator. The operation –x**y will be treated as –(x**y).

A.2.6 Model and Output Files

The model must be defined prior to solving the problem. To define the model, the MODEL statement is used followed by a collection of all equation names defined previously. Alternatively, the ALL statement can be used to employ all equations previously defined in the model, however, it is a good programming practice to explicitly list all equations present in a model to avoid unwanted inclusion of equations.

```
MODEL model_name /eqn_name1, eqn_name2 /;
```

To create and direct to an output file of your choice, the FILE and PUT commands can be used:

```
FILE res /[path/]output_filename.txt /;
PUT res
```

The PUT command followed by a value or command is used to place the results within the output file. A "/" represents a line break in the output file and text can be written by enclosing it with " marks. Different characteristics of the variable or set element, here represented by x, can be placed in the output file using the following common variable extensions:

- x.TL defines the title or name of the variable or set element
- x.L defines the level or value of the variable or set element
- x.M defines the marginal or dual value of the variable or set element

A.2.7 Solve

The solve statement in GAMS is formatted in an intuitive manner. The solve statement makes use of the SOLVE, USING and MAXIMIZING/MINIMIZING commands.

```
SOLVE model_name USING solve_command MINIMIZING
objective_variable;
```

The appropriate solver must be chosen for the optimization problem being solved. The most common solvers are displayed in the following text. For other solvers and a full list of all model classifications, you need to refer to the GAMS manual. Depending on the problem, iterative statements such as the LOOP, IF-ELSEIF-ELSE and WHILE statements may be used to either encompass the SOLVE statement or used after the SOLVE statement to create the desired output file. All iterative statements have a similar syntax with a comma following the condition and a semicolon after each statement. An example IF–ELSEIF–ELSE statement is shown below:

```
IF(condition,
 statements;
ELSEIF condition,
   statements;
ELSE
   statements;);
```

A.2.8 Model Completion and Status

Once the model is solved, it is important to check the termination condition. The solver status and model status can be checked by printing the numerical value associated with the model_name.SOLVESTAT and model_name.MODELSTAT variables. Common values and interpretation are shown in Tables A.1 and A.2.

A.3 FREQUENTLY ENCOUNTERED PROBLEMS IN GAMS

- Do not use tabs in the input file. They introduce errors in the parsing of the input file
- During debugging make sure that you make your corrections in the *filename.gms* and not in the *filename.lst* file

TABLE A.1 Interpretation of the Frequently Occurring Solver Status Values

SOLVESTAT Number	Meaning
1	Normal completion
2	Maximum number of iterations reached (adjust ITERLIM)
3	Maximum allowable time reached (adjust RESLIM)
5	An evaluation error occurred (e.g., division by zero)

See the GAMS user's guide for a complete list.

TABLE A.2 Interpretation of the Frequently Occurring Model Status Values

MODELSTAT Number	Meaning
1	Optimal solution is achieved (for LP and MILP problems).
2	Local optimal solution is achieved (for NLP and MINLP problems).
3	Solution is unbounded. (This is reliable for a linear problem but not always accurate in nonlinear problems.)
4	The model is provably infeasible.
8	An integer solution is found (This indicates that a feasible solution for a problem with integer variables has been found, but it may not be the optimal.)
19	The model is infeasible, and no solution can be provided. (This commonly occurs when the problem is locally infeasible.)

See the GAMS user's guide for a complete list.

- Do not forget to *declare* variables and equations
- Do not forget semicolons ";" at the end of each statement
- Nonlinear models typically have multiple minima/maxima. Do not forget to initialize your variables and provide lower and upper bounds. For example, x.L(i) = 5.23; x.LO(i) = 2.0; x.UP(i) = 5.0;

For successful entry, compilation, debugging and execution of optimization problems with GAMS familiarity with the material covered in the User's Manual is necessary. In particular, Chapter 2 of Rosenthal's GAMS manual is a very good primer to get you started with GAMS.

A.4 EXAMPLE PROBLEM

The concepts introduced in the previous section are demonstrated with a simple example in this section. Consider a number of tasks (e.g., computer programs) $i = 1,\ldots,4$ that must be assigned to a set of available computers $j = 1,\ldots 4$. The computing costs c_{ij} in Gb of assigning program i to computer j and the amount of time t_{ij} required for each program i to be completed on a computer j are given in Tables A.3 and A.4, respectively.

Assuming that each program can be split between multiple computers, variable x_{ij} declares the fraction of each program i assigned to computer j. Each computer can run for 24 hours a day. We would like to ensure that each task is fully completed over a 1-day period by assigning it to multiple computers. The problem of minimizing the total assignment cost can then be formulated as the following LP problem:

TABLE A.3 Computing Costs c_{ij} in Gb of Assigning Program i to Computer j

	Computer			
Program	1	2	3	4
1	5	0.2	2	1.6
2	2	0.4	1.8	2
3	4.1	3	2.4	1.6
4	5.6	1	1.9	1.5

TABLE A.4 Amount of Time (t_{ij}) Required for Each Program i to be Completed on a Computer j

	Computer			
Program	1	2	3	4
1	2	30	4	12
2	25	10	21	6
3	4	2	13	26
4	11	8	25	15

$$\text{minimize } z = \sum_{i=1}^{4}\sum_{j=1}^{4} c_{ij} x_j$$

subject to

$$\sum_{j=1}^{4} x_{ij} = 1, \quad i = 1,2,3,4$$

$$\sum_{j=1}^{4} t_{ij} x_{ij} \le 24, \quad i = 1,2,3,4$$

$$0 \le x_{ij} \le 1, \quad i = 1,2,3,4 \quad \& \quad j = 1,2,3,4$$

Note that the first constraint ensures that each task is fully assigned, while the second ensures that each task i is fully completed over a 24-hour period. The third constraint ensures that the fractional assignments are between zero and one. This optimization model can be written in the GAMS environment as follows:

```
*--- begin input file---
$ONTEXT
This code will be used to solve the optimization code to
minimize the computational cost associated with solving
each program between the 4 computers provided.
$OFFTEXT
```

```
OPTIONS

*A maximum of 1000 iterations is set

ITERLIM = 1000

*cplex is chosen as the LP solver
LP = cplex;

SETS i programs /1*4/
     j computers /1*4/;

TABLE c(i,j) costs of assigning program i to computer j
       1         2        3        4
 1     5         0.2      2        1.6
 2     2         0.4      1.8      2
 3     4.1       3        2.4      1.6
 4     5.6       1        1.9      1.5;

PARAMETER
t(i,j) time for executing program i in computer j
/
1.1   2
1.2   30
1.3   4
1.4   12
2.1   25
2.2   10
2.3   21
2.4   6
3.1   4
3.2   2
3.3   13
3.4   26
4.1   11
4.2   8
4.3   25
4.4   15
/;

VARIABLES x(i,j), z;

*Set lower and upper bounds for x. (i.e. 0<=x<=1)
x.LO(i,j) = 0;
x.UP(i,j) = 1;
```

```
EQUATIONS
obj  the objective function of the problem
const1(j)  each program i must be fully assigned to the
set of computers j
const2(i)  time constraint of each computer having one
day to operate
;

obj.. z=E=SUM((i,j),c(i,j)*x(i,j));
const1(j).. SUM(i,x(i,j))=E=1;
const2(i).. SUM(j,t(i,j)*x(i,j))=L=24;

MODEL jobs
      /obj
      const1
      const2/;

*Declare that when output is directed to res1, it will
*be written in the file job_out.txt

FILE res1 /jobs_out.txt/

SOLVE jobs USING LP MINIMIZING z;

*Identify where to send the output of the program

PUT res1;

PUT "Objective function value=",z.l; PUT /;

*Create a table of x_ij values
PUT "x_ij values are: ";
PUT /;
PUT "          ";
LOOP(i, PUT i.tl); PUT /;
LOOP(i, LOOP(j, PUT x.l(i,j); ); PUT /; );

*Create a table of values corresponding to the fractional
assignments
PUT "T_ij*x_ij values are: ";
PUT /;
PUT "          ";
LOOP(i, PUT i.tl); PUT /;
LOOP(i, PUT const2.l(i);); PUT /;
```

```
* Confirm that all processes are fully complete
PUT "Sum(x_ij) values are: ";
PUT /;
PUT "          ";
LOOP(j, PUT j.tl); PUT /;
LOOP(j, PUT const1.l(j);); PUT /;
*---end input file---
```

The cost and time parameters can both be entered in the table form or the list form as shown above. The file above (jobs.gms) can be run using the command "gams jobs.gms" GAMS creates file jobs.lst that contains standard output and file jobs_out.txt that contains the requested formatted output using the PUT command. The output file for this code will return the following:

```
*---Begin output file ---
Objective =          5.97
x_ij values are:
        1           2           3           4
     0.00        0.78        0.16        0.00
     0.96        0.00        0.00        0.00
     0.04        0.00        0.00        0.92
     0.00        0.22        0.84        0.08

T_ij*x_ij values are:
        1           2           3           4
    24.00       24.00       24.00       24.00

Sum(x_ij) values are:
        1           2           3           4
     1.00        1.00        1.00        1.00
*--- End output file ---
```

INDEX

Note: Page numbers in *italics* refer to Figures; those in **bold** to Tables.

Optimization Methods in Metabolic Networks, First Edition. Costas D. Maranas and Ali R. Zomorrodi.